Philosophy of Molecular Medicine

Philosophy of Molecular Medicine: Foundational Issues in Theory and Practice aims at a systematic investigation of a number of foundational issues in the field of molecular medicine. The volume is organized around four broad modules focusing, respectively, on the following key aspects: What are the nature, scope, and limits of molecular medicine? How does it provide explanations? How does it represent and model phenomena of interest? How does it infer new knowledge from data and experiments? The essays collected here, authored by prominent scientists and philosophers of science, focus on a handful of mainstream topics in the philosophical literature, such as *causation, explanation, modeling,* and *scientific inference.* These previously unpublished contributions shed new light on these traditional topics by integrating them with problems, methods, and results from three prominent areas of contemporary biomedical science: *basic research, translational and clinical research, and clinical practice.*

Giovanni Boniolo (degrees in Physics and in Philosophy) is Professor of Philosophy of Science and Medical Humanities at the Università di Ferrara, and Anna Boyksen Fellow at the Institute for Advanced Study, Technische Universität München. He has published 13 books (plus 12 books edited) and about 200 research articles.

Marco J. Nathan is Assistant Professor of Philosophy at the University of Denver. His work has been published in various philosophical and scientific venues, including *Noûs, Philosophy of Science, British Journal for Philosophy of Science, Nephrology Dialysis and Transplantation, Biology and Philosophy,* and *Synthese.*

Philosophy of Molecular Medicine

Foundational Issues in Research
and Practice

Edited by Giovanni Boniolo and Marco J. Nathan

LONDON AND NEW YORK

First published 2017 by Routledge

2 Park Square, Milton Park, Abingdon, Oxon, OX14 4RN
605 Third Avenue, New York, NY 10017

*Routledge is an imprint of the Taylor & Francis Group, an
informa business*

First issued in paperback 2020

Library of Congress Cataloging in Publication Data
Names: Boniolo, Giovanni, editor. | Nathan, Marco J., editor.
Title: Philosophy of molecular medicine : foundational issues in
research and practice / edited by Giovanni Boniolo and Marco
J. Nathan.
Description: New York : Routledge, 2016.
Identifiers: LCCN 2016015413 | ISBN 9781138940673
(hardback) | ISBN 9781315674162 (e-book)
Subjects: LCSH: Molecular biology–Philosophy. | Pathology,
Molecular.
Classification: LCC QH506 .P483 2016 | DDC 572.8–dc23
LC record available at https://lccn.loc.gov/2016015413

ISBN: 978-1-138-94067-3 (hbk)
ISBN: 978-0-367-73669-9 (pbk)

Typeset in Times New Roman
by Deanta Global Publishing Services, Chennai, India

Contents

Contributors

Tudor Baetu is a lecturer in Philosophy of Science at the University of Bristol. He holds degrees in biology and philosophy, and his research focuses on issues in philosophy of medicine and biology.

Federico Boem is Adjunct Professor of Logic and Critical Thinking at the University of Milan and researcher at the European Institute of Oncology in Milan. His research focuses on philosophy of molecular and computational biology.

Giovanni Boniolo is Professor of Philosophy of Science and Medical Humanities at the University of Ferrara and A. Boyksen Fellow at the Institute for Advanced Study, Technische Universität München. His research focuses on ethical and philosophical foundations of biomedical research and clinical practice.

Marie Darrason is completing her medical residency in Paris, while holding an honorary postdoctoral fellowship at the IHPST (Institut d'Histoire et de Philosophie des Sciences, Paris). Her research focuses on the philosophical foundations of human medical genetics.

Melinda Bonnie Fagan is Associate Professor of Philosophy at the University of Utah, where she holds the Sterling McMurrin Chair. Her research focuses on experimental and modeling practices in biology (particularly stem cell and systems biology), explanation, and social epistemology.

Mariacarla Gadebusch Bondio is Director of the Institute for History and Ethics of Medicine at Technische Universität München, chair of the University Hospital Clinical Ethics Committee, and member of the Munich Center for Technology in Society. Her research lies at the intersection of medicine and philosophy.

Pierre-Luc Germain currently holds a postdoctoral position in computational biology in the laboratory of stem cell epigenetics of the European Institute of Oncology (IEO), where he pursues research both in molecular biology and in the philosophy of science.

Maël Lemoine is Associate Professor of Philosophy of Biomedical Science at the University of Tours and associate researcher at the IHPST in Paris. His research focuses on the philosophy of medicine and psychiatry, mostly on problems involving the modeling of pathological processes.

Katherine E. Liu is a PhD candidate in Ecology, Evolution, and Behavior at the University of Minnesota. Her research focuses on the concepts and methods used in cancer translational research and experimental evolution.

Alan C. Love is Associate Professor of Philosophy at the University of Minnesota and director of the Minnesota Center for Philosophy of Science. His research concentrates on concepts, methods, and reasoning in developmental and evolutionary biology with a special focus on interdisciplinary explanation.

Marco J. Nathan is Assistant Professor of Philosophy at the University of Denver. His research focuses on the philosophy of science, with particular emphasis on topics in molecular biology, neuroscience, cognitive psychology, and economics.

Anya Plutynski is Associate Professor of Philosophy at Washington University in St. Louis. Her research is in history and philosophy of science, primarily the biomedical sciences.

Emanuele Ratti holds a postdoctoral position in History and Philosophy of Biology at the University of Notre Dame's Center for Theology, Science and Human Flourishing. He is currently working on the context of discovery of big data biology (with special focus on biomedicine) and on the topic of virtues in science.

Federica Russo is Assistant Professor in Philosophy of Science at the University of Amsterdam. Her interests span the social and biomedical sciences, with special attention to issues related to causality, modeling, and technology.

Francesco Spöring (PhD 2014, Swiss Federal Institute of Technology) is a researcher at the Institute for History and Ethics of Medicine at Technische Universität München. His current research focuses on the history of psychopharmacology.

David Teira is Associate Professor of Philosophy at the Spanish Open University (UNED). He currently leads a research project on the correction of subjective biases in social and medical experiments. For further details, please visit: http://www.uned.es/personal/dteira/

Michael Travisano is Professor of Ecology, Evolution, and Behavior at the University of Minnesota and a member of the BioTechnology Institute. His research uses methods from microbial experimental ecology and evolution to answers longstanding questions about how complex biological entities develop and evolve.

Paolo Vineis is Professor of Environmental Epidemiology at Imperial College, London, and head of the Unit of Molecular and Genetic Epidemiology at the HuGeF Foundation, Turin.

Introduction

Giovanni Boniolo and Marco J. Nathan

1 Welcome to the New Biomedical Frontier!

Over the last few decades, momentous theoretical and technological advancements in the biomedical sciences have enabled an increasingly pervasive and systematic exploration of organisms—including human beings—at the molecular level. This deeper *understanding* has been accompanied by a more accurate and effective *manipulation* of biological systems, which, in turn, underlies a more efficacious and individualized approach to patients. The attempt to study conditions and treat diseases from a molecular perspective is often referred to as "molecular medicine." This more precise and "individualizable" molecular approach is widely considered to lie at the frontier of clinical and medical practice (Collins and Varmus 2015). While, for the time being, much of this cutting-edge research still only carries a promise of future therapeutic payoffs, we already have examples of concrete clinical results as witnessed by molecular approaches in cancer therapies (www.cancerresearchuk.org/health-professional/cancer-statistics/survival).

Contemporary science often relegates philosophical reflection to an ancillary role. To borrow an expression from Liu, Love, and Travisano's chapter, in a world where technological innovation, experimental results, and practical implications are rapidly becoming a necessity, conceptual reflection is widely deemed an expensive "luxury" that few of us can afford. While a general diagnosis of how this anti-intellectualist—and, in the long term, self-defeating—attitude became established lies beyond our present concerns, we agree with Liu and colleagues that discounting the significance of conceptual reflection is myopic. Philosophy has an important role to play in scientific research and practice, especially at new and exciting frontiers such as molecular medicine. The development of the health sciences requires both a strong analytic stance and empirically informed theoretical reflection to provide solid foundations for the ongoing research. Conceptual analysis is not a mere luxury but can lead to more deliberate and well-informed diagnosis, prognosis, and therapy for present and future patients. The path for philosophers of molecular medicine is by no means easy or straightforward. Conceptual clarity and rigorous methodological analysis requires one to achieve a high standard of scientific literacy, blended with an informed historical background and knowledge of contemporary philosophy. This is the ambitious aim that we should strive to achieve.

Philosophical reflection on biology has grown exponentially, at least since the 1970s, when the life sciences were established as a legitimate target for philosophical inquiry. Moreover, there is a longstanding tradition in the philosophy of medicine, which mainly focuses on the conceptual analysis of *health* and *disease*, in addition to various ethical, epistemological, and methodological aspects. Yet, the philosophy of contemporary *biomedicine* is surprisingly underdeveloped, especially considering the amount of interest and resources that this area of the health sciences is currently drawing. To be sure, the philosophy of molecular biology is growing and, arguably, thriving as a field. Furthermore, the fascinating history underlying the "molecularization of biology" (Allen 1975; Kay 1993; Morange 1998; de Chadarevian 2002) and the related and somewhat parallel "molecularization of medicine" (de Chadarevian and Kamminga 1998) have been the subject of various monographs and collections. However, systematic philosophical reflections on the metaphysical and epistemological foundations of contemporary biomedical practice have been scant. The present volume constitutes one of the first systematic attempts to bring together various aspects of contemporary philosophical research with the enormous mass and richness of cutting-edge results in molecular medicine.

The essays collected here focus on a handful of topics that have been extensively discussed in the philosophy of science literature, such as *causation*, *explanation, modeling*, and *scientific inference*. These previously unpublished contributions shed new light on these traditional topics by integrating them with problems, methods, and results from three prominent areas of contemporary biomedical science: *basic research, translational and clinical research*, and *clinical practice*. Before delving into a more detailed overview of these essays, let us briefly address the rise of molecular medicine and how it eventually became the object of philosophical reflection.

2 The Rise of Molecular Medicine

Establishing the precise starting point of any discipline is a daunting, and perhaps futile, task. Paraphrasing Popper (1945), history only acquires meaning through interpretative work. Yet, historians typically provide conflicting interpretations of the key events that mark the foundation of new cultural *milieux* and that, over time, become crystalized into academic fields. Consequently, a univocal narrative is often hard to find. The relatively short—albeit rich—history of molecular medicine is no exception.

An important step in the foundation of molecular medicine was certainly the introduction of the term *molecular biology* by Warren Weaver, a mathematician famous for his pioneering work on automatic translation, in a report written in 1938 for the Rockefeller Foundation (Weaver 1970). The term, however, was not adopted widely until the 1950 Harvey Lecture, delivered by William T. Astbury, one of the founding fathers of x-ray molecular diffraction studies, focusing particularly on DNA. Here, Astbury presents his approach to "molecular biology" and later defends it from the competing label "ultrastructural biology," which was preferred by influential biologists such as C.H. Waddington (1961). Molecular biology, Astbury maintains, "is concerned particularly with the *forms* of biological molecules and with the evolution, exploitation and ramification of these forms in the ascent to higher levels of organisation" (Astbury 1961, p. 1124, italics in original).

Important as they are, one should not overestimate the significance of the introduction of new terms or concepts, or their official recognition. Studies of the molecular foundations of life predate the coinage of the expression *molecular biology*. Similarly, molecular medicine was steadily developing by the time John C. Kendrew—a biochemist and crystallographer who was awarded the 1962 Nobel Prize for chemistry—founded the *Journal of Molecular Biology* in 1959. Nonetheless, the prophetic visions of Weaver and Astbury, just like Kendrew's pioneering editorial strategy, are unmistakable signs of the development and consolidation of a cultural process that traces its roots decades— indeed, centuries—earlier.

A depiction, no matter how sketchy, of the historical context in which contemporary biomedicine arose would be incomplete without tracing its roots to early molecular studies in physics. The concept of *molecule* has a long and hallowed history, going back (at least) to Robert Boyle (1661), who argued that matter is composed of clusters of particles and that chemical change results from the rearrangement of the clusters. One hundred fifty years later, Amedeo Avogadro (1811) concluded that "The smallest particles of gases are not necessarily simple atoms, but are made up of a certain number of these atoms united by attraction to form a single molecule." Only a few decades had passed before James Clark Maxwell (1873), following a hallowed tradition that goes back to the Presocratics, explicitly defined an atom as "a body which cannot be cut in two" and a molecule as "the smallest possible portion of a particular substance." One could also mention the birth of physiological chemistry and biochemistry due, to a large extent, to the pioneering work of Justus von Liebig and Eduard Buchner, respectively. In short, contemporary biomedicine was founded on the pillars of physical studies of molecules and molecularization, which had taken huge strides since Boyle's early experiments, culminating with the awarding of the 1929 Nobel Prize to Jean Baptiste Perrin for "his work on the discontinuous structure of matter."

It is important to note that, until the first few decades of the twentieth century, the study of matter fell under the umbrella of physics. Molecules were entities that physicists originally posited to study the deep structure of matter and its principles of aggregation. Over time, these same entities gradually became the object of interest of biophysicists and biochemists. Molecules effectively became the fundamental units for understanding the constitution of organisms and their development, and the modes of the heritability of traits. Here, the history of physics and the history of biology overlap, in a sense realizing the vision of August Weismann, who, toward the end of the nineteenth century, conjectured that heredity could be based on the transmission of nuclear material. From this perspective, one should not forget the work of the early geneticists at the turn of the twentieth century: Hugo de Vries's contribution to the redis-covery of Mendel's work on the transmission of "factors"; William Bateson's introduction of the term *genetics* to designate the newborn science of heredity; Wilhelm Johansen's coinage of the expressions *gene, genotype,* and *phenotype,* which are now part of mainstream biology; and Archibald Garrod's work on the heritability of alkaptonuria.

While the foundations of the science envisioned by Weismann were solidly in place early in the century, the development and maturity of molecular biol-ogy, as we currently understand it, had to wait a few more years. Two important milestones are worth emphasizing. The first was reached in 1943, when Oswald T. Avery, Colin M. MacLeod, and Maclyn McCarty provided definitive experi-mental proof that DNA was the vehicle of genetic "information." The second occurred when Linus Pauling and colleagues (1949) uncovered the molecu-lar underpinnings of sickle-cell anemia and dubbed it "a molecular disease." Soon thereafter, in 1953, James D. Watson, Francis Crick, Maurice Wilkins, and Rosalind Franklin developed the double helix as a model for DNA, leading Watson and Crick to famously "suggest a structure for the salt of deoxyribose nucleic acid (D.N.A.) [that] has novel features which are of considerable bio-logical interest" (Watson and Crick 1953, p. 237).

Before, betwixt, and subsequently lie myriad other significant discoveries that paved the way and allowed us to begin to understand the fundamental con-stitution of biological organisms and what is transmitted from one generation to the next (Kay 1993; Morange 1998; de Chadarevian and Kamminga 1998; de Chadarevian 2002). Incidentally, this also provided the key to understanding how some diseases could be caused by a "faulty" replication or transmission of genetic material or could be due to some "mistakes" in such material. This important realization constitutes the birth of contemporary medicine—the medi-cine of the present and, many would agree, of the future. This book focuses on the philosophical foundations of this revolution in biological and biomedical practice.

3 Philosophical Reflections on Science, Biology, and Medicine

Historically, philosophy and science have been strongly connected; indeed, their paths substantially overlapped for millennia. Ancient and Medieval thinkers would not distinguish between philosophy and science. It was only during the scientific revolution, which marks the "official" birth of contemporary science, that the wedge between scientific and philosophical research began to appear. Yet, one of the late central figures in this momentous historical chapter, Isaac Newton, still defined his physical theories as part of "natural philosophy." A clear distinction between science and philosophy was not drawn before the nineteenth century, when the two areas started following independent paths. After Kant, few philosophers were trained in the bourgeoning sciences, such as mathematics and physics. Similarly, the increasing specialization of empirical knowledge and the crystallization of the experimental "scientific method" caused scientists to lose interest in philosophical speculation. Science was finally established as a self-standing discipline, independent of its intellectual parent.

Philosophy of science, as we understand it today, was born as a reaction to the separation of science and philosophy. As noted, by the end of the nineteenth century, philosophy was far removed from scientific research. Spurred by ongoing dissatisfaction with then-popular intellectual developments, a new philosophical movement was born in post-World-War-I *Mitteleuropa*. This intellectual endeavor grew out of new seminal results in logic, mathematics, and physics and was inspired, more or less explicitly, by the philosophical reflections of Europe's leading scientists: Mach, Botzmann, Poincaré, Duhem, Peano, Einstein, Campbell, and many others. This *nouvelle vague* of philosophy was named *logical positivism* (and later *logical empiricism*) and revolved around the so-called "Vienna Circle."

When the logical empiricists reoriented the direction of philosophy of science in the 1920s and 1930s, they generally focused their attention on the most developed sciences of their time: mathematics and physics. These fields set both the agenda and the tone of the discussion for the years to come. Nonphysical sciences—biology, chemistry, and the social sciences—were often assumed to be less general and to employ different criteria of rigor. With some notable exceptions (e.g., Neurath), logical positivists assumed, more or less explicitly, the special sciences to be less "scientific" than their physical counterparts.

The once-received view that logical positivism completely neglected philosophical analyses of biology has recently been subject to criticism (Byron 2007). In addition, philosophical debates such as reductionism and holism were at the core of the work of prominent biologists such as Haldane and Hogben. Yet, despite the exponential growth of biological thought in the twentieth century (Mayr 1982), with a few notable exceptions (Beckner 1959; Grene 1959; Gouge 1967), mainstream philosophy of science—which, in the meanwhile, due to a massive escape from Nazism, had relocated from Europe to the United States—remained generally focused on physics (Braithwaite 1953; Nagel 1961; Pap 1962). The marginalization of the life sciences becomes even more evident in the case of molecular biology. Even milestones such as the discovery of the DNA double-helix and the first successes in the "molecularization" of biology left philosophers relatively unmoved.

The situation began to change in the 1960s and 1970s thanks the work of philosophers such as Hull, Wimsatt, Schaffner, and Ruse. Since the 1970s, philosophy of biology has had a continuous and increasing presence in the philosophy of science. However, in the late 1970s and early 1980s, philosophy of biology became almost exclusively concerned with evolutionary theory (with notable exceptions, of course, such as some works by Kitcher, Keller, and Rosenberg). While initially the almost exclusive focus on evolution might have been a useful guidance for philosophers entering this newborn field, in time, it became a hindrance. As a result, philosophy of biology was ignoring much of what was going on in contemporary biology: molecular and developmental biology, immunology, ecology, theories of mind and behavior, and so on. In addition, it partly caused among molecular biologists a lack of concern for philosophical critiques of their enterprise. This became evident in the case of the Human Genome Project, where philosophers of science played virtually no part, unlike historians, bioethicists, and social scientists. Since the 1990s, however, philosophical writing on biology has extended its scope to cover many areas within biology beyond evolutionary theory, covering ecology, molecular and developmental biology, and experimentation. Molecular medicine is likely to become one of the next exciting frontiers of philosophical research in biology.

4 An Overview of the Essays

This volume is organized around four broad modules focusing, respectively, on the following key aspects of molecular medicine: What are its *nature, origins,* and *scope*? How does it provide *explanations*? How does it *represent* and *model* phenomena of interest? How does it *infer* new knowledge from data and experiments?

4.1 Molecular Medicine: Nature, Origins, and Scope

The first module collects three chapters centering on the nature, origins, and scope of molecular medicine. These essays raise questions such as how should molecular medicine be characterized? When did it begin, and how is it related to its methodological predecessors? Is molecular medicine best conceptualized as a "revolutionary" novelty or as a collaborative transition from related medical approaches? These issues have clear historical interest but, in addition, they are relevant—indeed crucial—for anyone who wants to have a solid grasp on the current state of medical sciences.

In his chapter "Molecular Medicine: The Clinical Method Enters the Lab," Giovanni Boniolo begins by asking how this approach should be characterized in order to understand whether it is truly a novelty in the historical development of medicine. Boniolo's essay focuses on methodological issues: how molecular medicine has developed in time and how it has affected other ways of understanding and doing medicine. He argues that if molecular medicine is intended, somewhat simplistically, as the study of diseases and potential therapies at the molecular level, then it cannot be considered a "novelty" at all, especially from a methodological point of view. However, closely examining one of the fields in which the molecular approach has been most successful—that is, molecular oncology—suggests a different and more promising perspective. Boniolo discusses tumor heterogeneity and the difference between cancer cell lines and primary tumor cultures. He concludes that the field of tumor heterogeneity reveals a real methodological novelty characterized by the "fusion" of the clinical method of traditional medicine with the experimental method of contemporary science.

As mentioned at the outset, a more individualized, molecular approach is widely assumed to be the frontier and future of clinical and medical practice. However, it would be a mistake to consider personalized medicine as a recent endeavor. Sure, individualized medicine, as we currently understand it, is founded on molecular medicine. Yet, personalized medicine has a long and hallowed tradition that traces its roots deep into the history of medicine, going back to Hippocrates. In a collaborative piece aptly entitled "Personalized Medicine: Historical Roots of a Medical Model," Mariacarla Gadebusch Bondio and Francesco Spöring discuss the development of this widespread—and constantly growing—medical model, from the *constitutional medicine* of the 1920s to the pioneering work of Vogel, Motulsky, and Lalow in the birth of *pharmacogenetics*. A historical approach, the authors argue, highlights forgotten aspects in the consideration of challenging approaches to quantifying qualities and features of individuals. The value of this work, however, is not "merely" historical. Conceptualizing the methodological roots and experimental foundations of personalized medicine provides us with tools for understanding important aspects of contemporary medicine, such as psychodynamics, culture-specific aspects of illness, and the limits of genetic reductionism.

This first module on the nature, origins, and scope of molecular medicine concludes with a discussion by Marie Darrason—"From the Concept of Disease to the Geneticization of Diseases"—that focuses on an important paradox affecting contemporary medical genetics. As late as the 1960s, the concept of *genetic disease* was taken to be synonymous with the concept of a *monogenic disease*, a pathological condition depending on a Mendelian mutation in a single gene. Since then, two notable shifts have occurred. On the one hand, the concept of genetic disease has extended to the point that virtually any disease can now be dubbed *genetic* (a process known as the *geneticization of diseases*). On the other hand, the model of simple monogenic diseases has disintegrated, breaking down the once intuitive distinction between monogenic and polygenic diseases and, consequently, leaving us without a plausible definition of *genetic disease*. Darrason's essay traces the history and development of the geneticization of diseases, leading to what she calls the "paradoxes of contemporary medical genetics." Next, she examines three extant philosophical solutions, arguing that each of them leaves us wanting. Finally, she offers a novel way to dispel the paradox. This discussion effectively shows how the progress of molecular medical sciences does not only depend on experimental and technological advancements but also requires a clear conceptual framework for categorizing and discussing these findings—in a word, it needs *philosophical* work.

4.2   Explanation in Molecular Medicine

The second group of essays collected in this volume focuses on *explanation*. While explanation is one of the most widely discussed and popular topics in the philosophy of science, explanatory practices in molecular medicine have not yet received the attention they deserve.

In Chapter Four, Maël Lemoine focuses on explanation in the burgeoning field of molecular psychiatry. Specifically, Lemoine's discussion is sparked by an intriguing observation and an ensuing question: Psychiatry has not (yet) been revolutionized by genomics; why is this the case? His diagnosis focuses on genome-wide association studies (GWASs), one of the leading trends in molecular psychiatry. GWASs have so far been successful at various tasks, including confirming the role of a gene in the onset of a disease, discovering new candidate genes, suggesting new molecular pathways to diseases or to new candidate biomarkers or treatments, and challenging the received nosology. However, none of these achievements has so far been reached in psychiatric research on schizophrenia, bipolar disorders, and major depressive disorders, the three most investigated mental disorders in molecular psychiatry. Lemoine identifies four main causes for this situation. (1) First, mental disorders are not and cannot just be translated into reliably identifiable phenotypes. (2) Second, their heritability is particularly difficult to assess. (3) Third, the size of groups needed to reach statistical power and thus detect effects may be difficult, if not impossible to get. (4) Fourth and finally, the entrenchment of genomic, etiological, pathophysiological, and diagnostic complexity—in a condition the mechanisms of which have not yet been investigated with sufficient results to limit possibilities—might well hinder any further progress with GWASs. Lemoine is skeptical about (1) being associated with "phenotypic complexity" and rather blames it on "descriptive complexity," that is, an epistemological problem rather than an ontological one. Furthermore, he argues that (3) should be blamed not only on inheritance complexity (2), but also on mechanistic complexity (4). In other terms, genomics requires that enough of the molecular details of a condition are known before providing more. Whereas identifying mechanisms is the ultimate goal, the first step in the molecularization of mental disorders has to be statistical and probabilistic models.

Despite massive investments and substantial strides in cancer research, the development of successful treatments that reduce mortality is still lagging behind. In their chapter, "How Cancer Spreads: Reconceptualizing a Disease," Katherine Liu, Alan Love, and Michael Travisano suggest that conceptualizing cancer as an infectious disease that we give ourselves might shed new light on properties of cancerous cells that are associated with the metastatic dimensions of cancer and are responsible for the majority of cancer-related deaths. In their chapter, the authors draw an analogy between cancer and cystic fibrosis in order to illustrate the value of modeling diseases as genetic or infectious in character, especially for identifying effective treatments. They note that while conceptual reflection is often deemed a luxury, especially in matters of life and death, "philosophical" reflection on modeling practices might spur important clinical applications. Shifting attention away from *how cancer grows*, that is, cellular proliferation and tumor growth, to *how cancer spreads*—cellular motility and metastasis—has the potential to advance both our theoretical understanding and clinical treatments of cancer.

In Chapter Six, "Evolutionary Perspectives on Molecular Medicine: Cancer from an Evolutionary Perspective," Anya Plutynski investigates the ways that cancer progression is like and unlike evolution in other contexts. In her chapter, she endorses a multilevel perspective on cancer, investigating the levels at which selection may be acting, the unit or target of selection, the relative roles of selection and drift, and the idea that cancer progression may be a byproduct of selection at other levels of organization. The aim is to integrate data and theory from molecular biology and *in situ* studies of cancer progression as well as dynamical models of cancer that represent progression as a multistage process.

4.3 Representation and Modeling in Molecular Medicine

The third module focuses on representation and modeling. Once again, this is a topic that has been widely discussed in philosophy of science. However, experimental, technological, and conceptual advancements coming from molecular medicine provide a plethora of new concepts that need to be integrated with mainstream philosophical work.

In Chapter Seven, "Toward a Notion of Intervention in Big-Data Biology and Molecular Medicine," Federico Boem and Emanuele Ratti argue that contemporary molecular biology has changed the nature of *manipulation* and *intervention*. While molecular biology and molecular studies in medicine have been traditionally associated with experimentation in the form of material manipulation and intervention, "big science" biological projects focus on the practice of data mining of databases. Material manipulation still has a role in the practice of data mining; yet, that role is merely confirmatory in nature. In the course of their chapter, Boem and Ratti defend two principal claims. First, data mining in biological databases can achieve goals similar to traditional interventionist and experimental practices but without engaging in material manipulation. Second, such a practice is indeed a form of intervention, but it must be distinguished from traditional philosophical accounts.

In Chapter Eight, "Pathways to the Clinic: Cancer Stem Cells and Challenges for Translational Research," Melinda Bonnie Fagan examines cancer stem cells (CSCs): a small subpopulation of self-renewing cells within a tumor or blood-borne cancer, posited to be responsible for maintaining and growing the malignancy. The CSC model of cancer has profound clinical implications and has been a focus of research and controversy for over a decade. But while experimental work on CSCs is ongoing, results are equivocal, and clinical translation has been lacking. Fagan examines conceptual and evidential challenges blocking clinical translation of CSCs and proposes a way forward. Alongside *methodological pluralism* advocated by CSC researchers, she proposes *conceptual pluralism*: distinguishing between two CSC concepts with different substantive content, suited to their respective purposes and criteria for success. Successful clinical translation, Fagan argues, requires empirical validation of the clinical CSC concept, while existing experimental support for the other, basic, CSC model indicates how such evidence could be obtained.

This section on representation and modeling concludes with a chapter by Marco J. Nathan entitled "Counterfactual Reasoning in Molecular Medicine." This essay focuses on an important class of scientific inferences, *counterfactual explanation*, and how they function in the context of molecular medicine. The first part of this contribution suggests that subjunctive conditionals play a key role in the diagnosis and prognosis of molecular diseases and other pathological conditions. Diagnoses, Nathan claims, are inferences to the best explanation; as such, they require considering not only what *is* the case but also what *could* be the case. Similarly, prognoses, which the author characterizes as "inferences to the best prediction," have an essential modal component. In the second part of the essay, Nathan argues that traditional semantic analyses of subjunctive conditionals are guilty of conflating two scientific goals—*truth* and *explanation*—that should be kept distinct. He sketches a "placeholder" analysis, according to which counterfactuals stand in for predictions and explanations that a speaker commits to producing, in principle or in practice. Finally, he concludes by arguing that this account, which requires some substantial departures from traditional philosophical analyses, captures the role of counterfactual reasoning in molecular medicine and, more generally, in science.

4.4 Inference in Molecular Medicine

The last module focuses on the kind of knowledge that molecular inference provides and, specifically, the advantages and related challenges of the various kinds of inferences that are required in order to make molecular data relevant and applicable to clinical practice.

Biomedical research is built upon inferences transposing knowledge across systems—be it across species, between an experimental system and another, or from controlled clinical studies to specific health-care applications. While similarities and differences between these forms of extrapolation have long been discussed, they have seldom been given a systematic philosophical treatment or considered in the specific context of molecular medicine. In their chapter, "Forms of Extrapolation in Molecular Medicine," Pierre-Luc Germain and Tudor Baetu characterize the general form of extrapolation, its implementations, and the challenges they face. The authors show that different forms of extrapolation are often intermingled. In addition, they clarify and problematize the validity of these inferences in the face of contemporary developments in clinical and biomedical research. Finally, they argue that, barring idealized cases, all forms of extrapolation are based on the assessment of relevant similarities.

Should conventional randomized clinical trials provide the standard of safety and efficacy when testing targeted treatments for cancer? Should we make amendments to our current regulatory standard, stick to it, or dispense with it? In his chapter, "Testing Oncological Treatments in the Era of Personalized Medicine," David Teira maintains that, under certain circumstances, smaller phase II trials provide good enough grounds to grant regulatory approval for targeted therapies. His argument hinges on the size of trial population, showing how this size is important not only for scientific considerations but also for ethical and political reasons. The current system was designed to provide massive consumer protection at a point when our understanding of the biology of cancer was still relatively poor and statistical tests gave the only solid evidence about treatment effects. With targeted therapies, risks are hedged in a way that allows patients (if well informed) to make decisions for themselves instead of deferring to pharmaceutical regulators.

In the final essay collected in this volume, "Opportunities and Challenges of Molecular Epidemiology," philosopher Federica Russo and scientist Paolo Vineis present some challenges and opportunities for *molecular epidemiology*, the study of distribution and variation in exposure and disease in populations at the molecular level. They begin by discussing how molecular epidemiology emerged from traditional epidemiology. Next, they introduce "exposome research"—the molecular roots of exposure to environmental factors. In the final part of the essay, they focus on problems and opportunities for molecular epidemiology. Specifically, they maintain that data interpretation and the use of technology poses a challenge to the conceptualization of productive causality and to the design of public health policies.

5 Concluding Remarks

The essays collected here barely scratch the surface of a rich and fast-growing field. While comprehensiveness was never an attainable goal that we had in mind, we hope that readers will be inspired by the present discussion to contribute to this exciting research project that lies at the intersection of cutting-edge philosophical reflection and the new frontier of biomedical research.

Acknowledgments

We would like to express our gratitude to Routledge's editorial staff for their help in developing this volume. We are also very grateful to the University of Denver for their financial support.

References

Allen, G.E. (1975) *Life Science in the Twentieth Century*. Cambridge: Cambridge University Press.

Astbury, W.T. (1961) "Molecular biology or ultrastructural biology?" *Nature* 190, p. 1124.

Avogadro, A. (1811) "Essai d'une manière de déterminer les masses relatives des molécules élémentaires des corps," *Journal de Physique* 73, pp. 58–76.

Beckner, M. (1959) *The Biological Way of Thought*. New York: Columbia University Press.

Boyle, R. (1661) *The Sceptical Chymist*. London: Cadwell.

Braithwaite, R.B. (1953) *Scientific Explanation*. Cambridge: Cambridge University Press.

Byron, J.M. (2007) "Whence philosophy of biology?" *British Journal for Philosophy of Science* 58, 409–22.

Collins, F.S. and H. Varmus (2015) "A new initiative on precision medicine," *New England Journal of Medicine* 372, pp. 793–95.

de Chadarevian, S. (2002) *Designs for Life. Molecular Biology After World War II*. Cambridge University Press.

de Chadarevian, S. and H. Kamminga (Eds.) (1998) *Molecularizing Biology and Medicine. New Practices and Alliances, 1910s–1970s*. London: Harwood Academic Publishers.

Gouge, T.A. (1967) *The Ascent of Life*. Toronto: University of Toronto Press.

Grene, M. (1959) "Two evolutionary theories, I–II," *British Journal for the Philosophy of Science* 9, 110–27, pp. 185–93.

Kay, L.E. (1993) *The Molecular Vision of Life: Caltech, the Rockefeller Foundation, and the Rise of the New Biology*. New York: Oxford University Press.

Maxwell, J.C. (2003 [1873]) "Molecules." In W. Niven (Ed.), *The Scientific Papers of James Clerk Maxwell* 2, pp. 361–78. New York: Dover.

Mayr, E. (1982) *The Growth of Biological Thought*. Cambridge: Harvard University Press.

Morange, M. (1998) *A History of Molecular Biology*. Cambridge: Harvard University Press.

Nagel, E. (1961) *The Structure of Science*. New York: Harcourt Brace.

Pap, A. (1962) *An Introduction to the Philosophy of Science*. New York: Free Press.

Pauling, L., H.A. Itano, S.J. Singer, and I.C. Wells (1949) "Sickle cell anemia, a molecular disease," *Science* 110, pp. 543–48.

Popper, K.R. (1945) *The Open Society and Its Enemies*. London: Routledge.

Waddington, C.H. (1961) "Molecular biology or ultrastructural biology?" *Nature* 190, p. 184.

Watson, J.D. and F.H. Crick (1953) "A structure for deoxyribose nucleic acid," *Nature* 171, pp. 737–738.

Weaver, W. (1970) "Molecular biology: Origin of the term," *Science* 170, pp. 591–92.

Part I
Nature, Origins, and Scope

1 Molecular Medicine

The Clinical Method Enters the Lab

Giovanni Boniolo

Abstract

In this chapter, I propose that the novelty characterizing molecular medicine consists in a "fusion" of the clinical method and of the experimental method. In order to show this, I proceed, first, historical-critically and, then, by emphasizing tumor heterogeneity and the difference between cancer cell lines and primary tumor cultures. The historical-critical analysis will allow seeing whether molecular medicine is truly a novelty, since nothing is a cultural novelty except in the light of history of culture. The emphasis on the molecular approach to tumor heterogeneity will permit to consider what I will argue to be the very novelty, that is, the investigational method here adopted, which is, as said, a sort of fusion between the standard clinical method and the standard experimental method. This is the reason why I claim that "the clinical method enters the lab."

1 Introduction

Over the last five to six decades, biomedical knowledge has taken an enormous leap forward, thanks to discoveries in the field of molecular biology and to amazing biotechnological innovations. On the one hand, it is almost a platitude to assert that we are facing a new era in medicine. On the other hand, there is not much philosophical analysis on whether we are spectators of a truly new manner of practicing medicine, that is, on whether *molecular medicine* is a *bona fide* novelty. This means that there is a pressing necessity to inquire into its philosophical foundations. I wish to address this question through a historically driven approach, since it is impossible to grasp the novelties of a human product without considering the historical background in which it has been incubated and from which it departs: *nothing is a cultural novelty except in the light of history of culture.*[1] On the other hand, history, as we know, is the only benchmark to understanding whether a sociocultural event—in particular a scientific event—truly represents a new perspective.

Although there are dozens of alternative, sometimes intersecting definitions and interpretations of what molecular medicine is, there is no shared consensus.

Certainly, it could be totally acceptable to say that the core problem of molecular medicine consists in understanding the molecular bases of diseases and, thus, in how to prevent and possibly cure them, as it is also claimed in the home page of *Molecular Medicine*, one of the journals of the field:

> *Molecular Medicine* strives to understand normal body functioning and disease pathogenesis at the molecular level, which may allow researchers and physician-scientists to use that knowledge in the design of specific molecular tools for disease diagnosis, treatment, prognosis, and prevention.
> (http://molmed.org/home; accessed October 2015)

Nobody, I think, doubts this. Yet, this definition does not help us in understanding whether the molecular approach constitutes an innovative way of considering and practicing medicine. Claiming that now we know more about diseases since we know more from the molecular perspective is not sufficient, as I will show, to assess the effective step forward that is claimed. If I am correct, we should look in a different direction. In particular, I propose to consider the *methodology*. But in what sense? Is molecular medicine new because it is characterized by an entirely novel method? I do not think so. As I will argue, molecular medicine, especially whenever it has to cope with tumor heterogeneity, is new because it is characterized by a sort of fusion of the classical observational method of clinical medicine with the classical experimental method of research medicine. More precisely, I propose that contemporary molecular medicine is a new way of doing medicine on the basis of the fact that the method adopted at the patient's bedside enters the lab, where it is fused with the usual experimental procedure canonized over the centuries and typical of any good experimental science. In order to justify my claim, I will resort to what could be thought of as a real turning point in contemporary medical studies: the discoveries in the field of tumor heterogeneity and the differences between usual cancer cell lines and primary tumor cultures.

2 Let's Go Molecular

Let us begin to inquire why explaining the pathogenesis resorting to the molecular level is not sufficient to claim that molecular medicine is something new.

On this point, we could follow K.R. Popper and state that "we are not students of some subject matter but students of problems. And problems may cut right across the borders of any subject matter or discipline" (Popper 1969, Chapter 2). We could agree on the idea that disciplines are historical and social constructions produced mainly for academic scopes. Unfortunately, if we slavishly followed him on his focus on problems, we would have an easy way out that nevertheless, might lead us to a wrong conclusion about what molecular medicine is. We should simply say that it consists in studying the molecular bases of diseases and, therefore, indicates how to prevent and possibly cure them. Unfortunately, this is not sufficient to justify the claim that we have a big novelty. To better address the point, we must come back to some historical events, as we always should do whenever we wish to understand what is truly new and what is only a different dress for an old matter.

Many identify the birth of molecular medicine with an extremely influential paper that L. Pauling cowrote in 1949: "Sickle Cell Anemia: A Molecular Disease." It is said that it is the beginning of molecular medicine since, for the first time, both the title and the content deal with a "molecular disease." But is this really the founding event?

Going back about a decade, we find that, in 1936, R. Peters spoke of "biochemical lesions" when, working with pigeons as animal models, he discovered the effects of thiamine deficiency on the nervous system (see also Peters 1969). Are we entitled to claim that this was the very dawn of molecular medicine, since thiamine is a molecule? Or should we affirm that Peters was doing biochemical medicine?

Taking another step back, in 1935, we find a book by H. Schade on *Molekularpathologie*, wherein he claimed that diseases do not affect organs or tissues but molecules, and that their causal explanation should be found at this level. Does this mean that the birth of molecular medicine dates back to work?

Actually, a serious look at the history of medicine reveals many other moments that would increase our perplexity about the idea that molecular medicine is new because it offers a molecular explanation of pathogenesis. To mention another example, it is worth recalling what is possibly the first Mendelian analysis of diseases affecting lineages, that is, the 1902 work by A. Garrod on the causes of alkaptonuria. Beyond the fact that Garrod was one of the first who worked on medical genetics, what is interesting for us is that, in his paper, he explicitly wrote about an "alternative course of metabolism" and about "chemical abnormalities." Should we say, therefore, that he founded metabolic medicine or chemical medicine? Or should we affirm that he was at the very birth of molecular medicine, just because it was a "matter" of molecules in particular enzymes?

Summing up, from the history of medicine, we know that the "going down" from tissues and organs to molecules and passing through cells, as R. Virchow taught in 1858, was recognized long before Pauling. Therefore, if molecular medicine wants to be a novelty, it should be something more than the study of diseases and their treatments at the molecular level. Otherwise, it would not be something different from what Garrod, Peters, and Pauling did, or from what Schade described in his 1935 textbook.

3 On the Method in Medicine

Since the focus on "problems," and therefore on a molecular-based explanatory approach, does not help us in understanding the real novelty of contemporary molecular medicine, let us see if the focus on "method" offers better arguments. Actually, pursuing this avenue means immediately distinguishing between the clinical method, usually adopted at the patient's bedside; and the experimental method of biomedicine, which is characteristic of what occurs at the lab benches. Moreover, we should understand if the experimental method of molecular medicine is really different from the experimental method of other sciences, such as physics.

3.1 The Clinical Method Between Disease Diagnosis and Disease Discovery

Science has its method. Medicine, in particular clinical activity, has its method. Let us focus on this. Skipping its old story—begun with Hippocrates: the first *klinikòs* of Western history—the idea that clinical method concerns how "to acquire, analyse and report clinical data derived from the patient" is plausibly sharable (Morgan and Engel 1969). According to this interpretation, it is the set of procedural steps for a correct interpretation and integration of the subjective symptoms reported by a patient during the anamnesis and of his objective bodily signs (and now of his lab results) in order to conjecture a diagnosis and a plausible prognosis, and to predispose and implement a successful therapy.

Actually, there is a second reading of "clinical method." This time it does not deal with diagnosis but with disease discovery. Indeed, most of the discoveries in the fields of pathogenesis and nosology have started from the observation of an individual patient suffering from a still unknown disease. That is, beside the *clinician*, more interested in attending to and possibly curing an individual patient, there is the figure (sometimes overlapping the former) of the *pathologist*, more interested in identifying and describing a new disease and its natural history. In both cases, the starting point is given by the patient's body, with its own pathological signs and with its own physiological parameters. From the observation of this individual "object," on the one hand, the clinician tentatively infers his disease; on the other hand, the pathologist individuates the new disease which all individuals possessing more or less the same pathological signs and the same physiological parameters could be claimed to be affected by.

3.2 The Methodenstreit and its Relevance for Medicine

To fully appreciate, from a conceptual perspective, the epistemological structure of both aspects of the clinical method, let us indulge in a controversy that occurred at the end of the nineteenth century. It was called the *Methodenstreit* (i.e., the debate on method) and it revolved around the possibility of a science of individual events, as individual diseased patients are. The focus of discussion was the distinction between *Naturwissenschaften* (natural sciences) and *Geisteswissenschaften* (humanities). The former were thought to be characterized by a *nomothetic* approach, based on objective generalizations (or natural laws); and the latter by an *idiographic* approach, based on the subjective comprehension of unique or contingent events.[2] This debate continued up to the 1960s, even if in a slightly different form, with the dispute (called *Positivismusstreit*) between the supporters of a unique method, guided by K.R. Popper; and the supporters of a typicality of the humanities, led by T.L.W. Adorno (Adorno et al. 1969).

Concerning the unique method, Popper argued that any individual event (an individual car crash, an individual volcano eruption, or an individual episode covered by social sciences or medical sciences, etc.) can be nomologically–deductively explained as any other scientific event. That is, we should first indicate the right general laws (the universals), and then we should display the individual conditions of that event so that we can go deductively from the general laws to its explanation. In this way, we move top-down, starting from the universal (the law) and arriving at the particular (the individual event) *via* introducing all the individual features necessary to subsume that individual into that universal.

Moving the other way round, we can move bottom-up by "taking away" one by one the characterizing features of that individual to arrive at the universal, that is, thanks to a process of continuous abstraction from what pertains to the individual—a method that the medievals called *abstrahere ab aliquo*.

This "methodological back and forth" between general laws (the universal) and the individual event (the particular) also underlies the dynamics between *general pathology*, dealing with the universal (symptomatic, anatomic, pathophysiological, aetiological) features of a certain disease; and *clinical pathology*, centered on an individual patient with an individual implementation of a certain disease. To have a paradigmatic example of this way of conceiving the methodological correlation between the universal and the individual in medicine, I wish to recall the position held at the beginning of the twentieth century by the Italian pathologist and methodologist of medicine G. Viola.[3] In 1932, Viola wrote *La costituzione individuale* (*The individual bodily constitution*), wherein he proposed that clinics are where natural sciences, dealing with universals, meet individuals; and in particular where general pathology, thought of as a natural science, meets individual patients with individual diseases. Indeed, Viola said that, on the one hand, there is the general knowledge offered by the pathologists concerning a given disease and, on the other hand, there is the individual patient with his own disease. Yet, how should one go from the abstract and universal to the concrete and individual? Starting from this problem, Viola proposed that the clinician can move top-down, from the universal knowledge constructed by the pathologists to the individual diseased patient, through a process of progressive individualization realized *via* the continuous addition of particularizing conditions (exactly as Popper proposed 30 years later). Actually, Viola indicated the possibility of even moving the other way round: the clinician can move also bottom-up. That is, he can move from the individual diseased patient to the universal disease by a continuous *abstrahere ab aliquo* process. This is the double orientation of the core of what he called the "individual approach to clinic," based on the "study of the generalisations concerning the individual anatomo-pathological variants and of their classification, which then should be linked with the study of the correlations between the variability of the pathological frameworks and the variation of the different individuals" (p. 16).

Not only are Viola's methodological remarks extremely useful for understanding how a diseased patient can be subsumed under the universal knowledge established by the general pathologists, but also for grasping how to construct the latter. In his effort to subsume what is affecting that individual patient under a universal knowledge, the pathologist could realize that such a universal knowledge does not yet exist. In this case, he finds himself in the context of discovery, in particular concerning a possible new disease. Therefore, he could propose a new one that, once individualized, can be applied to that original diseased patient.

Summing up, according to this perspective on clinical method, the universal knowledge concerning pathologies has to be particularized to the individual patient in order to explain his pathologies. Vice versa, the diseased individual patient could be the starting point for the discovery of a universal knowledge concerning pathologies.

3.3 The Experimental Method in Medicine

Roughly in the same period in which we had the debate on the relationship between universal and individual knowledge, we had also a real methodological acquisition in the field of medicine. In 1865, the French physician and physiologist C. Bernard, famous for his studies on glucose metabolism and diabetes mellitus, published *Introduction à l'étude de la médecine expérimentale*. First of all, in that work he showed that he had learned the methodological objection that Virchow raised against G.B. Morgagni in 1847: *morphology cannot put into light the causal relationships; this can be done only by physiology*. That is, only experimental knowledge of the correct (and wrong) processes occurring in the human body can reveal the causes of the diseases and, possibly, their course. Differently from observation (concerning patients' external signs or cadavers' internal signs), experimentation allows the identification, by directly intervening and circumscribing, of the causal processes, starting from the triggering event(s) and ending with the disease. In short, with Bernard, the experimental method "officially" entered into medicine and gave rise to the creation of a duality between clinical medicine, based on the observation of individual patients, and experimental medicine, grounded on experiments.

This was a turning point in the history of medicine. We know that the experimental method entered science at the beginning of the Middle Ages, and it had its first systematic expression in the seventeenth century, paradigmatically with G. Galilei, who emphasized the necessity of removing all the possible perturbations (the scientist "*must deduce the material hindrances*"; Galilei, 1638, pp. 208–209; my italics) in order to realize an experimental set up isolated as much as possible from all the uninteresting factors possibly disturbing the analysis of the phenomenon to be studied. Moreover, only in this manner is it possible to create an objective experimental situation that can be reconstructed by any other researcher to control the results obtained.

Galilei and Bernard had a clear awareness of the epistemological structure and of the methodological and epistemological load of experiments. In particular, they were deeply aware that a good experiment—that is, an experiment that provides intersubjectively controllable results—has to be realized with an experimental set up that isolates as much as possible the phenomenon to be studied by eliminating as much as possible all the possible perturbing effects (i.e., the causes of the systematic errors).[4] In a sense, they explicitly codified that an experiment (i) involves human manipulations; (ii) has to be performed in a totally controlled environment where the construction of the experimental set up is central in order to avoid possible external noninteresting perturbations (the Galilean "material hindrances"); and (iii) has to provide results that are intersubjectively controllable.

Bernard's book on the methodology of experimental medicine was published in an extremely rich cultural environment in which science, in particular medicine, had a strong and proud conceptual ally: the philosophy of science. Bernard's work can be conceived as a watershed for medical studies and practice. Before its publication, there was clinical pathology, which had an observational approach and was addressed to patients' diseases, plus a non-precisely defined theoretical pathology aimed at finding the causes of the diseases. After its publication, medicine also had, as physics had had for a long time, a precise procedure based both on experiments and on the statistical indication of the validity of the outcomes. Furthermore, it was precisely in that period that it was recognized that an experimental result has to be accompanied by a nonambiguous statistical indication of its validity.[5] Moreover, Bernard's book also marked the birth of two ways of doing medicine: that at the patient's bedside and that at the lab bench, very often without any contact with patients and hospitals.

4 On the Experimental Method in Medicine and in Physics

If we pay attention to the accurate philosophical analysis of Bernard's studies on diabetes and on other pathologies in his *Introduction à l'étude de la médecine expérimentale*, we realize that he was putting into practice exactly Galilei's experimental method. The first step is to realize a suitable experimental set up, such that the phenomenon under investigation is isolated from external perturbations by reducing them as much as possible, and, in any case, by absorbing them within the error interval related to the experimental outcomes. What is measured during the experiment should not be considered valid just for the individual case under investigation, but as a prototypical result valid for any similar phenomenon or process. Moreover, the result must be intersubjectively controllable.

There is another interesting consideration offered by Bernard in the second part of his 1895 work, that is, that the experimental method is unique and that there is no difference between experimental medicine and experimental physics. To illustrate this point, let us compare two different experimental situations: one from contemporary biomedicine, which is the focus of the present volume, and one from physics.

4.1 A Classical Biomedical Experimental Set Up: Cell Lines

Let us consider the experiments performed in molecular biology, specifically the experimental set up adopted in the majority of the studies concerning metabolism, aging, drug efficacy and toxic effects, mutagenesis, carcinogenesis, and so on. That is, let us focus on *cell lines*.

A cell line (here, I deal with *cancer cell lines*) can be conceived as an experimental set up designed to isolate the biological phenomenon under consideration (a molecular or cellular mechanism, process, or pathway) from external perturbations. Actually, in order to achieve this ideal methodological situation, we have to work "a little bit." First of all, we have to remove a cell population from a tissue or from an organ, either directly through an explant or through, for example, techniques of enzymatic digestion. This population, at this stage called *primary culture*, is put into a favorable artificial environment (i.e., into suitably prepared vessel: a Petri dish, a flask, a roller bottle).

Note that, a primary culture is not yet usable for many reasons. First of all, it is *nonhomogeneous* and *nonclonal*. It is nonhomogeneous since it consists of a mixture of all the epithelial components of the organ, of the fibroblasts, of the endothelial cells of the vessels, of the hematopoietic cells that might be present in the organs, especially lymphocytes (in the case of cancer, there is almost invariably an infiltrate), and macrophages. It is nonclonal since it lacks the capacity to be cloned. Thus, it has to be transformed into a homogeneous and clonal culture.[6]

There is another serious drawback regarding primary cultures: they do not last forever.[7] Note that if the population died, we would lose the possibility of having an experimental component with standard characteristics utilizable again. Fortunately, it has been discovered how to immortalize a cell population.[8] Of course an immortalized cell *in vitro* has very little to do with the original cell that lived in a tissue belonging to a human being. However, after all of these manipulations, we have a *cell line*, that is, something that is virtually immortal, homogenous, and clonal.

From what just said, it should be intuitive that the establishment of a cell line is really the establishment of an extremely artificial experimental set up, which—and this has to be taken into consideration—is a very remote proxy (model) for the tumor from which it originated. But we are not at the end of the story.

When we work with cell lines, we have to maintain them alive and avoid contamination. Concerning maintenance, we should put them in a vessel in which there is a substrate (the medium) containing nutrients (amino acids, carbohydrates, vitamins, minerals), growth factors, hormones, and gases (O_2, CO_2). Moreover, we have to regulate the physicochemical environment (for example, there should be the right pH, the right osmotic pressure, and the right temperature[9]).

Concerning contamination, the cell lines, like any other component of an experimental set up, have to be shielded from external perturbations that could systematically alter the outcomes. This means that we should protect them from adulteration due to both chemical contaminants (impurities in the media, water, endotoxins, plasticizers, detergents, etc.) and biological contaminants (bacteria, molds, yeasts, viruses, etc.). Moreover, we should also avoid cross contamination between two or more different cell lines. This means that the environment in which they are cultivated and used for experiments has to be as aseptic as possible.[10] Experimenters must pay attention and wash their hands and use clean gloves. Of course, the hood under which cells are manipulated has to be in an area that is restricted to cell cultures and to cell culture personnel, and protected from air currents coming from doors and windows.

Summing up, not only is the cell line an extremely artificial and controlled set up; in addition, the environment in which it lies (and lives) is extremely artificial and controlled, as is the case (or should be the case) with all experimental set ups and experimental environments (be they physical, chemical, biological, etc.). To conclude, let me mention that cell lines can now be produced in any lab or bought from specialized companies.[11]

4.2 A Classical Physical Experimental Set Up: The Torsion Balance

As a physical case study, let us consider the remake of the experiment performed (or recounted) by Galilei. In particular, I focus on the 1922 attempt by R. von Eötvös, D. Pekár, and E. Fekete (EPF) to establish whether there is a numerical difference between inertial and gravitational mass (Boniolo 1992).

Let us concentrate on the construction of the experimental set up and on the devices required to eliminate possible external perturbations. First of all, EPF decided to use a torsion balance (Figure 1.1). One body was attached directly to the end of one beam; the other was suspended from a wire tied at the other extreme. Possibly (though this is somewhat controversial), in this way, they wanted to take into account possible effects due to gravitational gradients (a balance like that is particularly sensitive to gravitational gradients).

They then predisposed the torsion balance to eliminate possible terrestrial magnetic effects (even if they did not say what they really did). As a third step, they closed the torsion balance into a box realized with three metallic shells, in order to eliminate (according to them!) possible radiation perturbations. They spread soot on the inner shell to exclude possible electrostatic effects between the box and the balance. The three shells and the soot were also meant to shield possible temperature variations. After the preparation of the set up, they performed the experiment many times, putting at the ends of the beams bodies of

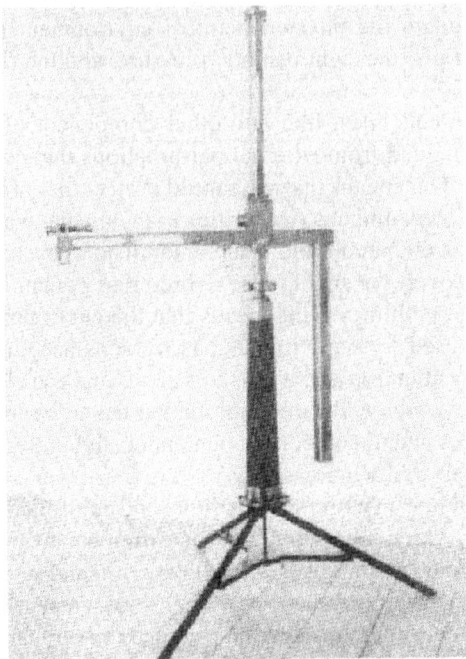

Figure 1.1 The torsion balance used by R. von Eötvös.

different nature (water, copper, asbestos, tallow, magnalium, snakewood, etc.), and they found that if there were a difference between the gravitational and the inertial mass, this would have been less than 10^{-9}.

What they achieved was not valid only for that particular torsion balance, for the particular bodies used, and for that particular experimental set up. It was a result valid in general: whatever body you consider and whatever experimental technique you use to check the difference between inertial and gravitational mass, you should find the same outcome (as it in fact occurred).

4.3 Comparing the Experimental Methods

EPF paid great attention to the experimental set up in order to isolate the phenomenon to be studied from possible perturbative effects. They also paid great attention to collecting the experimental data and to their statistical elaboration. A similar degree of attention is also given to the preparation of a cell line. Actually, from a methodological point of view, there seems to be no difference between EPF's torsion balance and the cell lines. Both of them are experimental set ups realized to avoid any possible external noninteresting perturbation (the Galilean "material hindrances"). Only in this way is it possible to perform a repeatable experiment and to find reobtainable results.

The same great methodological attention on the construction of the experimental set up is found in Bernard's experiments with dogs and other animal models, and in Pauling's experiments with human blood cells taken from sickle cell–anemic individuals. All of them realized an experimental set up and an experimental procedure such that the results obtained were intersubjectively controllable. The Galilean method was instantiated by EPF and similarly it was instantiated by Bernard and by Pauling, as it has been (and still is) instantiated millions of times in the course of the history of science and in contemporary empirical research, in particular in contemporary biomedical research whenever cell lines are used.

In short, there is no novelty from the perspective of experimental method between EPF, Bernard, Pauling, or other contemporary molecular scientists, even if the former are physicists and the latter are biologists, even if some were working around a century ago and others in the present day. On the other hand, (i) to construct an experimental set up isolating the phenomenon to be studied from external perturbations, (ii) to perform an experiment that could be repeated, and (iii) to find results that are reobtainable are requirements that any scientist has to satisfy if he wants his work to be accepted by the international research community (Boniolo and Vaccari 2012).

5 The Novelty Characterizing Molecular Medicine

The situation just described does not apply to the case of the clinical method, especially when its disease-discovery aspect is considered. A single patient cannot be considered in any way an isolated experimental set up, even if he or she is located in an environment that is as controlled as possible: the hospital. Every individual patient has his own genome, his own way in which this is expressed due to his lifestyle, the environment in which he has lived and is living, and so on. Every individual patient has his own way in which the given disease is instantiated in space (its body) and in time (the course) and his own way in which it is existentially and socially perceived. We could claim that the patient changes a little every day, due to the cure he is receiving or simply by virtue of his status as a "living" being. Thus, as described by Viola, the clinician who is looking for a diagnosis has to tentatively individuate that universal under which that individual patient could be subsumed. Vice versa, the general pathologist, if he is looking for a possible new disease affecting that individual patient, has to take away (*abstrahere ab aliquo*) what is typical of that patient and to arrive at a conclusion valid for all those who have the (more or less) same set of pathological signs.

But what is occurring in molecular medicine, and what could be considered as a real conceptual novelty? We have seen that considering the usual molecular medicine through historical lenses means abandoning the superficial idea that its novelty lies in the molecular level of its analysis. If the molecular level were really the relevant feature, the birth of its history should be dated back to Peters's paper on thiamine deficiency, or to Schade's book on *Molekularpathologie*, or to Garrod's work on alkaptonuria, or even before. Even from a methodological point of view, there is nothing particularly new. Molecular medicine proceeds through a typical Galilean experimental method, which was introduced in medicine by Bernard. This means that a certain separation between experimental (molecular) medicine and clinical medicine still exists. The former was always interested in finding (at molecular level) causes of and therapies for diseases, while the latter was always interested in the individual disease of an individual patient. And the two levels methodologically interact in the way highlighted by Viola (and, 30 years later, by Popper).

So far, there is nothing new from the perspective of the molecular level of inquiry. Similarly, there is nothing new from the methodological perspective, concerning both the characteristics of the experimental method of molecular medicine and the methodological relations between the level of general knowledge pertaining to it and the level of the individual instantiation pertaining to clinical practice. Thus, should we really conclude that contemporary molecular medicine does not constitute any novelty? Probably until 10 years ago, we should have answered positively. But now there is a set of discoveries that should change our mind. And these have to do with *tumor heterogeneity*.

5.1 Tumor Heterogeneity and Primary Tumor Cultures

The human cancer cell lines discussed above have had an important role in preclinical studies. Thanks to this artificial experimental set up, we have been able to design and realize experiments both *in vitro* and *in vivo* that, in the course of the last decades, have allowed us to obtain intersubjectively controllable results that have expanded the understanding of a lot of aspects to do with cancer initiation, cancer progression, and cancer metastasis. As a consequence, new targets for therapies have been individuated and new drugs have been produced, thanks to a more correct comprehension of drug sensitivity and resistance. Unfortunately, after the initial encouraging successes that have spurred many to foresee a victory in what was called "the war against cancer," we are now aware that the matter is not at all as simple and as we thought (on cancer, see also Plutynski; Fagan; Liu, Love and Travisano, this volume). On the one hand, there is the sad awareness of drug-resistant tumors and tumor relapses. On the other hand, there are the discoveries of new features of cancer due to technologies such as high-throughput screenings. All of this has led to the conclusion that the usual experimental set up based on artificial and very far from real-life cancer cell lines is no longer sufficient. It should be superseded by something different, something less artificial and closer to real cancers. This has been individuated in the so-called *primary tumor cell cultures*. This move, as a consequence, has begot a totally new methodological scenario that has truly heralded the age of the personalized approach to molecular medicine. Let us try to understand this point.

First of all, since the advent of cancer nosology, an organ-based classification system has been used with good success in clinical practice and research. Thus, for a long time we have spoken about "breast cancer," "lung cancer," "prostate cancer," "liver cancer," and so on. However, with the coming of molecular technologies, both genetic (for example, KRAS mutation for colon and lung cancer; EGFR mutation for lung cancer; BRAF mutation for colon cancer and melanoma; see Febbo et al. 2011) and epigenetic (for example, DNA methylation alterations; see Baylin and Jones 2011), biomarkers have begun playing a major role both in the diagnostic phase and in the drug discovery process that could lead to new therapies (see Boem, Pavelka and Boniolo 2015). Moreover, the introduction of high-throughput gene expression profiling has led to the discovery of the genetic and epigenetic heterogeneity among tumors and thus that not only are no two tumors identical, but not even two samples from the same patient's cancer are identical. Thus, step by step, we have come to understand that cancer is not a single disease but that any cancer is a "different disease" and that in the same cancer actually we have "many cancers," each of which has its own histopathological and biological features.

Let us dwell on this important point a bit longer. We know that the major features of cancer are uncontrolled cellular growth, proliferation, invasion, and metastasis. In order to acquire these capabilities, preneoplastic and neoplastic cells must undergo genetic (somatic) and epigenetic changes. It should be noted that preneoplastic or neoplastic cells do not live in isolation but constantly interact with their microenvironment and thus with their human host (and thus with his nonneoplastic cells such as inflammatory and immune cells, vascular endothelial cells, and fibroblast and other mesenchymal cells). It is this interaction that provides oxygen, nutrients, growth factors, and other chemical mediators necessary for growth, proliferation, and invasion. Of course the tumor microenvironment is heavily influenced by exogenous factors such as the dietary regime of the human host, the environment in which he lives, and the lifestyle he adopts. Needless to say, this individuality has substantial repercussions for the typicality of the tumor cells. In a slogan, just as every individual human being is unique (see Boniolo and Testa 2012; Boniolo 2013), every tumor is unique, or, rather, every tumor cell population in an individual tumor is unique. Such uniqueness has two facets: the *intertumor heterogeneity* (the variability occurring between tumors arising in the same organ but in different individuals) and the *intratumor heterogeneity* (the variability occurring in the same tumor of the same individual).

Very briefly, tumor heterogeneity synthesizes the fact that the same individual tumor is composed of several different cell populations having different tumorigenic potential and different markers.[12] This means that different cells of the same tumor can show distinct morphological and phenotypic profiles, including cellular morphology, gene expression, metabolism, motility, proliferation, and metastatic potential.

Certainly, despite tumor heterogeneity, the histopathology and genetic expressions of tumors often remain relatively stable during progression and, even if each tumorigenesis pathway is unique, there are similarities in some pathways. But, as is intuitive, tumor heterogeneity poses serious challenges in personalized cancer prevention and care, beyond explaining the failure of the usual drug therapy based on single targets.

If we really want to take into due consideration tumor heterogeneity, especially intratumor heterogeneity, the usual cancer cell lines are no longer reliable for the study of cancer *in vitro*. As we have seen, they are composed of homogeneous cells and virtually each line is a standardized population. Actually they have been cultured precisely to be homogeneous and standardizable. When they were proposed, the knowledge of cancer was at a certain stage; now it is at a stage in which tumor heterogeneity is at the core of the attention of both researchers and clinical oncologists, and their artificiality and "simplicity" is no longer satisfying. They are too detached from what is really happening in a real human tumor. Now we have realized that we should be dealing with the complexity and with the individuality of any single cancer. This situation has forced, and is forcing, an increasing number of researchers to switch from cancer cell lines to the *primary tumor cells*, that is, cell cultures directly taken from patients' cancers and maintained in the original state of heterogeneity.

To satisfy this aim, new methods for acquiring, preparing, growing, and maintaining these primary cell cultures have been developed (Mitra et al. 2013). Of course the same attention that was paid for the isolation of the cancer cell lines from contamination is now paid for the primary tumor cells, which become the new controllable experimental set up for oncological studies.

Nevertheless, there is substantial discontinuity with the usual cancer lines. Because of the heterogeneity of its cell subpopulations, the primary tumor cell culture is no more stable and standardizable than it was previously. Actually, due to the interactions among the different cell populations and with the environment in which it is maintained, it changes continuously. Moreover, it is an individual culture, which cannot be considered as a good prototypical instantiation of all the cancer lines. It is like a patient's body, or rather it is a "part of a patient's *body in a dish*." It is located in a controlled and controllable environment—the dish in the lab—exactly as the patient is located in a controlled and controllable environment—the hospital. It reacts in an individual manner depending on the particular genetic and epigenetic characteristics of the different subpopulations of cancer cells it is composed of, exactly as the patient reacts in an individual manner depending on his or her particular genetic and epigenetic characteristics.

5.2 Tumor Heterogeneity and Methodological Novelty

A clinician in a ward, observing the heterogeneous and continuously changing body of a patient, and the researcher in a lab, observing the heterogeneous and continuously changing specimen of a patient's tumor, meet similar methodological problems. If they want to arrive at a universal knowledge starting from those individuals, they both have to take away what is particular; both have to work by *abstrahere ab aliquo*.

This means that what Viola understood concerning the bottom-up side of the relationship between the universal level of general pathology and the individual level of the instantiation of the disease works also inside a lab whenever tumor heterogeneity is concerned. By considering this, it is not so implausible to conclude that with tumor heterogeneity, the discovery side of the clinical method described by Viola enters the lab. But here it fuses with the classical Galileian experimental method characterized by a controlled and controllable experimental environment and by controlled and controllable experimental techniques.

All of this could be seen from a different angle that allows another epistemological question to be emphasized. Since the Aristotelian discussion of the architectonic of knowledge, it has been accepted almost as a platitude that there is no science of the individual. This idea has been almost always accepted along the entire history of Western epistemology. For example, it was at the basis of the interpretation of the Porphyrian tree, and it was used by the supporters of the differences between nomothetic and idiographic knowledge. But could it be maintained without any doubt? Maybe not, especially in the case we are discussing. Whenever you study the primary cancer cells of a given patient, you are also studying tumor heterogeneity, that is, something at the universal level. But you are also studying the particular disease of that particular patient, that is, you are studying the individual. And since this study is performed according to the classical Galilean method, it is done in a scientific manner. It means that you are doing science of the individual! Put in a different way, within the field of tumor heterogeneity we have the possibility of doing science of the individual, since the tumor cancer cell population actually is an individual (patient) *in vitro*.

6 Concluding Remarks

A physician at the patient's bedside should continue to be a *klinikòs* who observes signs on the patient's body. Probably now he should be something more: a *klinikòs* trained in molecular biology. A molecular scientist in his lab should continue using the Galilean method. But now, in the field of molecular medicine, something new has happened with the discoveries concerning tumor heterogeneity. Certainly all efforts are being to find out the molecular bases (also considering the systemic level) of neoplastic diseases and, possibly, to identify molecular ways of caring for or curing patients suffering from them. Yet this is not classic molecular biology, since it has to study real diseased heterogeneous cell populations coming from an individual patient's body. Of course it is not even classic clinical medicine, since it uses experimental techniques adopted in a molecular biology lab. It is something in between the two; something that, in order to achieve results, has to put together the two methods that, respectively, characterize classical molecular medicine and classical clinical practice. That is, and this is the methodological novelty, with the discovery of tumor heterogeneity and the advent of the primary tumor cells in molecular medicine, there is a fusion between Viola's clinical method and Galileo's experimental method. Rather, there is fusion of the two characteristics of the former:

(i) Its being based on the observation of an individual heterogeneous "object" varying in time (in clinical medicine, the patient's body; in molecular medicine, the patient's primary tumor cells)
(ii) Its proceeding through *abstrahere ab aliquo* (in clinical medicine, the patient's individual features have to be removed away to reach the universal; in molecular medicine, the patient's primary tumor cells individual features should be taken away to achieve the universal)

with the characteristic of the latter:

(i) There are human manipulations.
(ii) It is performed in a totally controlled environment.
(iii) It offers intersubjectively controllable results.

Viola's clinical method in the field of discovery	*Galilei's experimental method*
• Based on the observation of an individual heterogeneous "object" varying in time (in clinical medicine, the patient's body; in molecular medicine, the patient's primary tumor cells) • Proceeding through *abstrahere ab aliquo* (in clinical medicine, the patient's individual features have to be removed away to reach the universal; in molecular medicine, the patient's primary tumor cells individual features should be taken away to achieve the universal)	• Based on human manipulations • Performed in a totally controlled environment • Giving intersubjectively controllable results

As a result, we have a sort of hybrid method totally unthinkable for a scientist of Bernard's age but also unthinkable for molecular scientists like Pauling. It is something different, something new, something that allows us to claim that contemporary molecular medicine coping with tumor heterogeneity is really a new way of doing medicine.

Acknowledgments

I wish to thank M. Lemoine, M. Mameli, M. Nathan, and D. Teira for their comments on preliminary versions of this chapter.

Notes

1 This is a rephrasing of the famous title of T. Dobzhansky (1973), which, in turn, is a reuse of a sentence of Dobzhansky (1964, 449).

2 Actually, the *Methodenstreit* is a sort of spin-off of the debate that there was in the field of economics between the Austrian School, guided by K. Menger, and the German Historical School, defended by G. von Schmoller. It was a controversy concerning whether any science (as opposed to only history) could be used to explain the dynamics of human actions. We might individuate its origin in 1893 with the publication of *Einleitung in die Geisteswissenschaften* by W. Dilthey, immediately followed, in 1894, by the introduction by another Kantian philosopher—W. Windelband—of the distinction between *nomothetic* and *idiographic*. In this way, two different approaches to knowledge were characterized: nomothetic (from the Greek *nomos*, that is, law) to indicate an approach typical of the natural sciences; and idiographic (from the Greek *idios*, that is, own or private) typical of the humanities (history, psychology, social, and political sciences, etc.). This idea concerning a deep methodological difference between the *Naturwissenschaften* (natural sciences) and the *Geisteswissenschaften* (humanities), supported also by H. Rickert (1896), arose against the position epitomized by J.S. Mill (1843), according to which there is only one method for all sciences.

3 Despite his fame in Italy, especially for his studies on the correlation between bodily traits and diseases, Viola was totally removed from the international debate on the philosophical foundations of medicine. Still, it is worth mentioning him for his seminal methodological insights.

4 Nevertheless, we have to wait until the late nineteenth/early twentieth century to find a definitive methodological and epistemological understanding as to how an experimental set up has to be realized and an experiment performed. In those years, a number of philosopher-scientists, like P. Duhem (1906), E. Mach (1905), N. Campbell (1920), H. Dingler (1928), and E. Schrödinger (1955), began discussing these questions in depth.

5 It is worth noting that Bernard had a clear idea of the importance of the statistical methods in research medicine (Part II, Ch. 2, Sec. 9). Note that it was precisely in that period that it was recognized that an experimental result has to be accompanied by a nonambiguous statistical indication of its validity. This awareness was reached thanks to the seminal work of F. Gauss (1809) and the innovative contributions by R. Dedekind (1860a, 1860b), C. Lüroth (1869, 1876), P. Pizzetti (1889), K. Pearson (1895), Student (1908), and R.A. Fisher (1912, 1922, 1925).

6 The nonhomogeneity can be resolved by a number of techniques that allow purification (or enrichment) of one component over the others. For practical purposes, the major interest is frequently in the isolation of the epithelial population, and this can be achieved with procedures that are rather standardized. The nonclonality might not be a problem in itself, but it becomes a real issue in cancer studies. There are several sources of nonclonality. The first is that, even if we limit ourselves only to the epithelial cells, there are several types of epithelial differentiated cells in an organ. For example, in the case of the breast, there are at least two components: luminal cells and myoepithelial cells. If one wants to study a single population, one has to devise strategies for further purification (e.g., FACS methodology based on immunophenotypical markers). Even with a purified population of this kind, by definition there is nonclonality. Actually, clonality cannot be achieved in a primary culture: all technologies that we use to obtain clonal populations necessitate several rounds of replication in culture, and this inevitably clashes with the fact that primary cultures have limited life span, as we will see. If clones are obtained, their number is never enough to allow extensive biochemical studies.

7 This fact was first established by L. Hayflick (1965), who observed that a population of normal human foetal cells divides *in vitro* (in a primary culture) between 40 and 60 times (the so-called *Hayflick's limit*). Note that there is an additional aspect linked to the limited life span of a primary culture: only the stem cell compartment can refuel the proliferating populations. Therefore, when we establish a primary culture (assuming, as has been said, that it is a pure epithelial population) we have essentially three types of cells: (i) terminally differentiated cells (which do not divide), (ii) progenitors (which divide actively but eventually differentiate), and (iii) stem cells (which divide rarely in an asymmetric fashion or in a more symmetric way in cancer). *De facto*, the proliferating population is represented by the progenitors. However, as has been recalled, these cells have a limited number of replications available and the repopulating unit (the stem cells) is present with scarce frequency. Moreover, these cells eventually lose replicative potential.

8 Various techniques are available. For example, the effect could be obtained by taking them from naturally occurring cancer by inducing suitable mutations, by introducing viral genes that modify the cell cycle, and so on.

9 For mammalian cells, the pH should be about 7.4. The temperature should be the same as that of the living tissue from which the cells have been taken, which, in the case of human tissue, is about 36°C.

10 For example, the ordinary and the experimental manipulations are performed under a hood that maintains a constant and unidirectional flow of HEPA (high-efficiency particulate absorption) filtered air. Furthermore, each item (pipettes, reagents and media containers, etc.) that can be used to manipulate the cells has to be sprayed with 70% ethanol.

11 As it happens for *HeLa* cells, obtained from a cervical cancer taken, in 1951, from Henrietta Lacks, an Afro-American patient who died of her cancer about eight months afterward; or for the *A549 cells*, created in 1972 from a cancerous lung tissue of a 58-year-old Caucasian male. Concerning a catalogue of cancer cell lines, see http://cancer.sanger.ac.uk/cancergenome/projects/cell_lines/. Regarding one of the first reviews on cancer cell lines, see Rockwell (1980).

12 For a first approach to tumor heterogeneity, see Schilsky 2010; Gillet et al. 2011; Visvader 2011; Gerlinger et al. 2012; Meacham & Morrison 2013; Marjanovic et al. 2013; Sommer 2014. For a philosophical analysis, see Boniolo, 2017.

References

Adorno, Th. et al. (1969) *Der Positivismustreit in der deutschen Soziologie*. Darmstadt: Luchtterland.

Baylin, S.B. and P.A. Jones (2011) "A decade of exploring the cancer epigenome—biological and translational implications," *Nat. Review Cancer* 11, pp. 726–734.

Bernard, C. (1865) *Introduction à l'étude de la médecine expérimentale*. Paris: J.B. Baillière et Fils.

Boem, F., Z. Pavelka, and G. Boniolo (2015) Stratification and biomedicine. How philosophy stems from medicine and biotechnology. In *The Future of Scientific Practice: 'Bio-Techno-Logos'*. Edited by M. Bertolaso. London: Pickering & Chatto, pp. 103–115.

Boniolo, G. (1992) "Theory and experiment. The case of Eötvös' experiments," *The British Journal for the Philosophy of Science* 43, pp. 459–486.

Boniolo, G. (2013) "Is an account of identity necessary for bioethics? What post-genomic biomedicine can teach us," *Studies in History and Philosophy of Biological and Biomedical Sciences* 44, pp. 401–411.

Boniolo, G. (2017) "Patchwork narratives for tumour heterogeneity." In H. Leitgeb, I. Niiniluoto, E. Sober, and P. Seppälä. *CLMPS 2015 Proceedings*, London: College Publications.

Boniolo, G. and T. Vaccari (2012) "Alarming shift away from sharing results," *Nature* 488, p. 157.

Boniolo, G. and G. Testa (2012) "The identity of living beings, epigenetics, and the modesty of philosophy," *Erkenntnis* 76, pp. 279–298.

Campbell, N. (1920) *Physics. The Elements.* Cambridge: Cambridge University Press.

Dedekind, R. (1860a) "Mathematische Mittheilungen III: Ueber die Elemente der Wahrscheinlichkeitsrechnung," *Vierteljahrsschrift der Naturforschenden Gesellschaft in Zürich* 5, pp. 66–75.

Dedekind, R. (1860b) "Mathematische Mittheilungen: IV Ueber die Bestimmung der Präcision einer Beobachtungmethode nach der Methode der kleinsten Quadrate," *Vierteljahrsschrift der Naturforschenden Gesellschaft in Zürich* 5, pp. 76–83.

Dilthey W. (1883) *Einleitung in die Geisteswissenschaften. Versuch einer Grundlegung für das Studium der Gesellschaft und Geschichte.* Leipzig: Verlag von B.G. Teubner.

Dingler, H. (1928) *Das Experiment. Sein Wesen und seine Geschichte.* München: Verlag Ernst Reinhardt.

Dobzhansky, T. (1964) "Biology, molecular and organismic," *American Zoologist* 4, pp. 443–452.

Dobzhansky, T. (1973) "Nothing in biology makes sense except in the light of evolution," *American Biology Teacher* 35, pp. 125–129.

Duhem, P. (1906) *La Théorie physique. Son objet, sa structure.* Paris: Chevalier & Rivière.

Febbo, P.G. et al. (2011) "NCCN Task Force report: Evaluating the clinical utility of tumor markers in oncology," *J Natl Compr Cancer Netw* 9:S1–S32, quiz S33.

Fisher, R.A. (1912) "On an absolute criterion for fitting frequency curves," *Messenger of Mathematics* 41, pp. 155–160.

Fisher, R.A. (1922) "On the mathematical foundations of theoretical statistics," *Philosophical Transactions of the Royal Society A* 222, pp. 309–368.

Fisher, R.A. (1925) *Statistical Methods for Research Workers.* Edinburgh: Oliver & Boyd.

Galilei, G. (1632) *Dialogue Concerning the Two Chief Word Systems—Ptolemaic and Copernican.* Berkeley: University of California Press 1968. (Translation by S. Drake of *Dialogo sopra i due massimi sistemi del mondo*, in G. Galilei, *Le opere.* Firenze: Giunti Barbera, Vol. VII, 1953).

Galilei, G. (1638) *Discorsi e dimostrazioni matematiche intorno a due nuove scienze attinenti la mecanica e i movimenti locali.* In G. Galilei, *Le opere.* Firenze: Giunti Barbera, Vol. VIII, 1953.

Garrod, A.E. (1902) The Incidence of Alkaptonuria: A Study in Chemical Individuality. *Lancet* II, pp. 1616–1620.

Gauss, F. (1809) *Theoria motus corporum coelestium in sectionibus conicis solem ambientium.* Hamburg: Perthes et Besser.

Gerlinger, M. et al. (2012) "Intratumor heterogeneity and branched evolution," *The New England Journal of Medicine* 366, pp. 883–892.

Gillet, J.P. et al. (2011) "Redefining the relevance of established cancer cell lines to the study of mechanisms of clinical anti-cancer drug resistance," *PNAS* 108:18708–18713.

Hayflick, L. (1965) "The limited in vitro lifetime of human diploid cell strains," *Experimental Cell Research* 37, pp. 614–636.

Lüroth, C. (1869) Bemerkung über die Bestimmung des wahrscheinlichen Fehlers. *Astronomische Nachrichten* 73, pp. 187–190.

Lüroth, C. (1876) "Vergleichung von zwei Werten des wahrscheinlichen Fehlers," *Astronomische Nachrichten* 87, pp. 209–220.

Mach, E. (1905) *Erkenntnis und Irrtum.* Leipzig: Johann Ambrosius Barth.

Marjanovic, N.D., R.A. Weinberg, and C.L. Chaffer (2013) "Cell plasticity and hetero-geneity in cancer," *Clinical Chemistry* 59, pp. 168–179.

Meacham, C.E. and S.J. Morrison (2013) "Tumour heterogeneity and cancer cell plastic-ity," *Nature* 501, pp. 328–337.

Mill, J.S. (2012) *A System of Logic, Ratiocinative and Inductive,* Cambridge: Cambridge University Press (First published 1843).

Mitra, A., L. Mishra, and S. Li (2013) "Technologies for deriving primary tumor cells for use in personalized cancer therapy," *Trends in Biotechnology* 6, pp. 347–354.

Morgan, W.L. and G.L. Engel (1969) *The Clinical Approach to the Patient.* Philadelphia: W.B. Saunders.

Pauling, L. et al. (1949) "Sickle cell anemia. A molecular disease," *Science* 110, pp. 543–548.

Pearson, K. (1895) "Contributions to the mathematical theory of evolution. II. Skew vari-ation in homogeneous material," *Philosophical Transactions of the Royal Society A* 186, pp. 343–414.

Peters, R.A. (1936) "The biochemical lesion in vitamin B1deficiency. Application of modern biochemical analysis in its diagnosis," *Lancet* 1, pp. 1161–1164.

Peters, R.A. (1969) "Biochemical lesion and its historical development," *British Medical Bulletin* 25, pp. 223–226.

Pizzetti, P. (1889) Sopra il calcolo dell'errore medio di un sistema di osservazioni. *Atti Reale Accademia dei Lincei* 5, pp. 740–744.

Popper, K.R. (1969) *Conjectures and Refutations.* London: Routledge and Kegan Paul.

Rickert, H. (1896) *Die Grenzen der naturwissenschaftlichen Begriffsbildung, Eine logis-che Einleitung in die historischen Wissenschaften.* Tübingen: Verlag von J.CB. Moh.

Rockwell, S. (1980) "In vivo-in vitro tumour cell lines: Characteristics and limitations as models for human cancer," *Br J Cancer Suppl* 4, pp. 118–122.

Schade, H. (1935) *Die Molekularpathologie der Entzündung, ihre Bedeutung für das Krankheitsverstehen und Krankheitsheilen. Eine Einführung für Studierende und Ärzte,* Dresden: Verlag von Theodor Steinkopff.

Schilsky, R.L. (2010) "Personalized medicine in oncology: The future is now," *Nat Rev Drug Discov* 9, pp. 363–366.

Schrödinger, E. (1955) "The philosophy of experiment," *Il Nuovo Cimento* 10, pp. 5–15.

Sommer, L. (2014) "Open questions: Development of tumor cell heterogeneity and its implications for cancer treatment," *BMC Biology* 12, p. 15.

Student (W.S. Gosset) (1908) "The probable error of a mean," *Biometrika* 6, pp. 1–25.

Viola G. (1932) *La costituzione dell'individuale.* Bologna: Cappelli.

Virchow R.C. (1858) *Die Cellularpathologie in ihrer Begründung auf physiologische und pathologische Gewebelehre.* Berlin: Hirschwald.

Virchow, R.L.K. (1847) "Über die Standpunkte in der Wissenschaftlichen Medicin," *Virchows Archiv* 1, pp. 3–19.

Visvader, J.E. (2011) "Cells of origin in cancer," *Nature* 469, pp. 314–322.

von Eötvös, R., D. Pekár, and E. Fekete (1922) "Beiträge zum Gesetz der Proportionalität von Trägheit and Gravität," *Annalen der Physik* 68, pp. 11–66.

Windelband, W. (1894) *Geschichte und Naturwissenschaft.* Strassburg: Strassburger Rektoratsrede.

2 Personalized Medicine

Historical Roots of a Medical Model

Mariacarla Gadebusch Bondio
and Francesco Spöring

Abstract

In order to visualize the key developments of personalized medicine, this chapter firstly discusses Theodor Brugsch's (1878–1963) *Personallehre*, which developed within the context of the constitutional medicine of the 1920s. Subsequently, a closer look will be taken at the pioneers of pharmacogenetics, such as Friedrich Vogel (1925–2006), Arno G. Motulsky (born 1923), and Werner Kalow (1917–2008). This history highlights partially forgotten aspects in the consideration of challenging approaches to quantifying qualities and characteristics of the individual. In a concluding section, these proto approaches are compared to contemporary personalized medicine. Brugsch and contemporaries are particularly instructive with regard to the consideration of psychodynamic and culture specific aspects of illness, which relativize simple concepts of health and highlight the limits of a genetic reductionism. Furthermore, they contain a self-critical potential that may dispel fallacies.

1 Introduction

The "personalization" or "individualization" of medicine is currently experiencing a boom. This is evidenced not only by recent publications from the sphere of politics,[1] but also by articles[2] listed under the keywords "personalized medicine" and "individualized medicine" on PubMed (cf. Figure 2.1), where these two "MeSH terms" are considered as synonyms referring to the same results.[3] Following the intensive discussions on the necessity of a terminological specification of these two terms held until 2012, but which now, however, are decreasing, the preference for the term *personalized medicine* (PM) has established itself at present (Michl 2015a, 47). The majority of studies on this subject listed on PubMed are drawn from the areas of genetics, molecular medicine, and oncology. From the outset, they refer to PM's guiding objective: "the right treatment for the right person at the right time" (Kroemer 2013, 12).

Looking back at the history of PM allows for a critical approach to its central tenets, but also the misgivings of its advocates. This history highlights problems

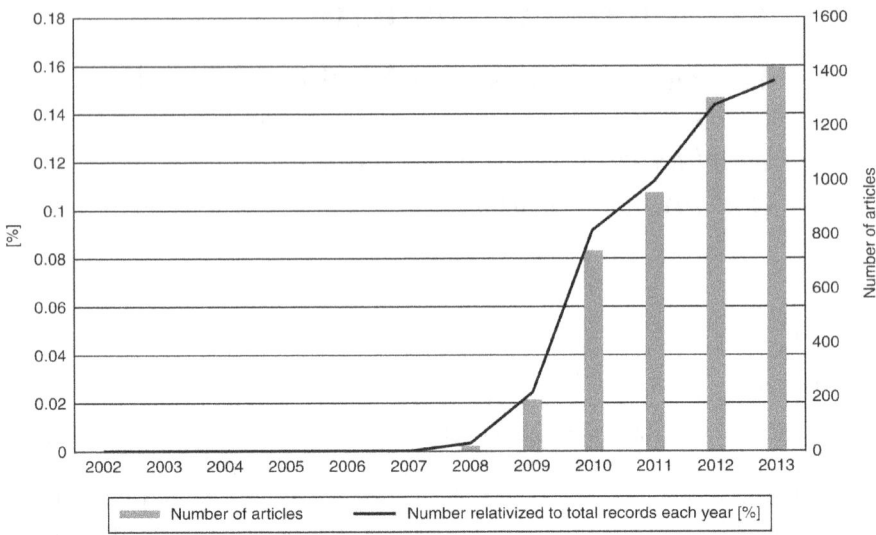

Figure 2.1 Number of articles registered on PubMed under the MeSH terms *individualized medicine* and *personalized medicine*.

in quantifying qualities and characteristics of the individual. In order to visualize the key developments in PM, the present discussion will deal with the two broad streams that led to its arrival in the late twentieth century.

The first of these was Theodor Brugsch's (1878–1963) *Personallehre*,[4] which was developed within the context of the constitutional medicine of the 1920s. This was an ambitious attempt to systematically understand a person's individuality through quantitative indicators and qualitative descriptive nomenclatures. The approach can be read as a pioneering version of contemporary PM. However, a few crucial differences stand out. In contrast to his contemporaries, such as Giacinto Viola (1932) from Italy (see Boniolo, this volume), Brugsch strove for a radically interdisciplinary approach.[5] So far, no project of comparable ambition in this time period is known to us. Equally important in the development of PM were the early pharmacogenetics of the 1950s as well as the work of the pioneers of preventive and predictive medicine. Here, we take a closer look at the criteria for differentiating individuals employed by trailblazers such as Roger Williams (1893–1988), Friedrich Vogel (1925–2006), Arno G. Motulsky (born 1923), Werner Kalow (1917–2008), and Emanuel Cheraskin (1916–2001). Most of them were German researchers who migrated from Germany to the U.S. to continue their research. In addition to this common background, they also depicted their work as a continuity of Hippocratic ideals. To what extent were the appeals to a primarily Hippocratic medical tradition convincing? Why does this impetus toward a personalization of medicine emerge at the beginning and then again in the middle of the twentieth century? Is today's PM to be seen more as a groundbreaking field of research, the current fulfillment of an old dream?

Following the example of earlier approaches to PM, our chapter aims to show how from the very beginning its advocates aspired to see themselves as perpetuators of a tradition that is no longer remembered today in this area of research. The reference to Hippocrates seems commonplace today; seen, however, in a *pars pro toto* sense, it stands for the revival of a concept of medicine that had its cradle in ancient Greek times, and which in the twentieth century is attempting to assert itself under completely different conditions. The multifaceted nature of the various specialized disciplines that have been particularly important for the rise of PM since the end of the 1970s and the start of the 1980s will, of necessity, have to remain in the background.

2   *Personallehre* in the 1920s and the Problem of Normal Values

The First World War, described by George F. Kennan as the "great seminal catastrophe" of the twentieth century (Kennan 1979, 3), affected societies on a global scale. It also had a dramatic effect on medicine, which had to deal with its consequences. Amongst the lessons learned from this catastrophe, that of the German internist Theodor Brugsch is particularly noteworthy. Looking back, Brugsch stated:

> The individual should be evaluated according to completely different criteria than those placed into our hands by the state of peace before 1914. Many suffering from heart defects, asthenics, neurasthenics and many men regarded till now as sick have proved themselves to be exceptionally physically fit, whilst those marked out as "normal" here and there have physically failed completely.
>
> (Brugsch 1926, 19)

The war had shaken the physiopathological values that, up until then, had been seen as normal, and had extended the spectrum of so-called "constitutional experiences," for example in view of the effects of climate and malnutrition on humans. During the war, several physicians working with Brugsch had already questioned why the genesis, progression, and treatment of a disease varied from person to person (Bauer 1917).[6] Unlike the proponents of laboratory-oriented bacteriology, internists such as Friedrich Kraus (1858–1936) and Brugsch endeavored to establish a project that they alternately referred to as *Biologie der Person* [personalized biology], *Personallehre*, and *Individualmedizin*.[7] This project aimed at remedying deficits in the prevention and treatment of diseases (Lindner 1999, 116) and called for a re-exploration of the individual.[8] In particular, it was the proponents of constitutional and individualized medicine who applied themselves to this task.

This attempt could be found most extensively in the work of Brugsch who, together with the German Jewish neurologist Friedrich H. Lewy (1885–1950), published the four-volume handbook *Biologie der Person* between 1926 and 1931 (Brugsch and Lewy 1926, 1929, 1930, 1931).[9] Their purpose here was to place the perceived singularity of people on an empirical basis and thus to critically consider notions such as "normality" and "average."[10] The collected articles came from various human scientific disciplines such as "general and specific genetics"; "collective theory of degree" (as part of medical statistics); "individual anatomy" and "individual pathology"; "constitutional and racial theory"; toxicology, pharmacology, and sport science (all in relation to constitution); criminology (and body type); thanatology; graphology; "medical characterology"; "psychophysical type research"; research into environmental influences; "ability, education and selection"; and seven cultural studies on topics such as "The Individual in India" and analogously in Japan, in Judaism, Christianity, Islam, and Catholicism.

For the proponents of so-called *Individualmedizin*, it was important to refer to the Hippocratic tradition. They attributed a central role to Hippocrates as the progenitor of a form of medicine that applied itself to the individual and his characteristics (Brugsch 1926, 5–19; Williams 1956, 1–2). Brugsch (1926) also described the traditions from antiquity to the twentieth century as the basis of the evolving constitutional science. Indeed, individualizing approaches can already be found in antiquity. For Greek physicians and their successors, there was a clear purpose in recognizing the right treatment for each individual, and those aspects that, according to age, gender, profession, climate, and mental and physical wellbeing, could help fight pathogenic causes. In view of the varying individuals and the countless factors and variables that influenced them, for physicians led by Hippocrates, the difficulty in developing a rational procedure to minimize mistakes was a key challenge (Hippocrates 2006a, 108–09).[11]

It is, however, worthwhile to recall the Hippocratic-Galenic reflections on individualizing considerations, as their views enable a comparison with contemporary trends. Hippocratic medicine saw health and sickness as a continuum in which in each case an individually appropriate, dynamic good condition can be determined (Hippocrates 2006a, Ch. 3). Galen also regarded a healthy condition as a dynamic balance that was always to be related to the specific situation of the living being, and not to be reduced to mathematical standard values.[12] Galen's understanding of variable health implies that health is not an absolute entity, but rather a relative one. The countless variations which, according to Galen, are to be found within the range of a healthy condition could even contain and tolerate minor pathologies ("healthy dyscrasias").[13] The Galenic method of diagnosis, prognosis, and, above all, of the treatment of diseases was based upon case descriptions. The starting point was sensory perception (*aísthesis*). Through the connection of *aísthesis* and *lógos* (rationality), Galen pursued a gradual perfecting of systematic "speculation." This meant a cautious approach to the individual situation of the sick person in order to be able to understand and treat him appropriately (Galen De sanitate tuenda K, VI, 2, Ch. 7). Admittedly, Galen's theory of signs contained a variety of diagnostic, prognostic, and anamnestic distinctive features of a general nature; however over and again he discussed the difficulty of applying these in individual clinical cases. For Galen, the task of a physician was to examine the situation of the individual patient meticulously. The broad catalogue of interpretable variables culminated in a highly individualized therapy—the foundation of the Hippocratic-Galenic approach.

On the one hand, the proponents of the *Personallehre* of the 1920s described ancient medicine with its purely qualitative and emphatically individualized approach as a model; on the other, they had to take a new context into account (Gadebusch Bondio 2015, 23–25). With the social permeation of statistical methods, numerical average values had also established themselves in medical research, particularly in the nineteenth century. In 1835, Adolphe Quetelet (1796–1874) demonstrated the applicability of statistics and the significance of "standard values" for physiology and medicine with his *homme moyen*. The medical significance of this arithmetical construct was reinforced by the biometricians Francis Galton (1822–1911) and Karl Pearson (1857–1936) at the close of the nineteenth century (Büttner 1997, 21). August Comte's (1798–1857) standards for the determination of mean values for the numerical definition of the biological normal (average type) and abnormal (deviant type) became the standard in almost all areas of medicine. At the same time, however, at the turn of the century, the theory of evolution and Mendel's laws had promoted work on biological variations. Standard values had to be reconsidered, and quantitative criteria for the determination of variants (variation statistics) had to be perfected. With the experience of the First World War, this task seemed to be even more urgent. Only eight years after the First World War, when Brugsch was reflecting on these processes in medical thinking, he came to see himself, together with Friedrich Kraus, as a proponent of the new school of thought:

> The approach in our work was not directed at meta-individual constitution typing in general, or constitution anomalies in particular, but was rather to attempt to capture the constitutional core of each person, and thus be able to give a judgement about their performance in every medical respect.
>
> (Brugsch 1926, 20)

At the center of Brugsch's project lay the unity, wholeness, and uniqueness of the individual. In order to assess an individual's performance, new data were required: instead of an enumeration of anatomical-morphological differences in characteristics, "an individual's overall norms of reaction" had to be discussed and understood. The individual characteristics should be recorded with the help of a set of "cinematic," "dynamic," "statistic," "genetic," and "personal" considerations that should be determined on the basis of a database of systematic examinations of a large range of people (Brugsch 1926, 22).

Overall, two forms of description dominated the four volumes of *Biologie der Person*: on the one hand, *quantitative* data collections provided information in correlations between differing anthropometric body measurements, which registered factors such as posture, distribution of subcutaneous fat, or hair shape of individuals. On the other hand, in the opus, several *qualitative* description systems differentiated themselves. Following the model of the well-known constitutional medicine physician Ernst Kretschmer (1888–1964), who made an extremely detailed record not only of physical, but also of morphological characteristics such as the presence of a moveable tenth rib or an elevated patella reflex, the proponents of psychiatry, psychology, and characterology in particular committed themselves to this qualitative description.[14] Examples are the articles of the existentialist psychiatrist Rudolf Allers (1883–1963), of the psychiatrist Karl Birnbaum (1878–1950), who dealt with "biopsychic personality development" (Allers 1926, 511–658), or the graphologist Elisabeth Flatow-Worms (dates unknown).[15] The latter concluded the character of a person from their handwriting and, like Lewy, dealt with motor expressions—we might even refer here to "kinetic markers." Whilst Flatow-Worms presupposed an intuitive ability for expertise in graphology, Lewy, as coeditor of the complete work, denigrated precisely intuitive evaluations as teleologically biased and therefore unscientific (Lewy 1931, 851). Instead, he attempted to typify photographically recorded human movement in specific situations on abstract mathematical descriptions (Flatow-Worms 1931, 699). In doing so, he was running the risk of typifying individuals on the basis of a one-off reaction without considering confounding variables such as time of day, diet, and alcohol intoxication into account. In contrast, the graphologist took numerous effects of mood, changes in emotion, and other variables into consideration (1931, 696). Furthermore, she had already emphasized the possibility of detecting and counteracting early defects through an analysis of children's handwriting. Thus, Flatow-Worms (1931, 733) described the linking of letters through what she referred to as "arcadian bonding" as a sign of "mental inhibitions," which, with timely intervention, could be treated, especially in childhood. Although at the start of her observations, she also addressed the possibility of cultural influences and oriented herself toward Freudian psychoanalysis, such acquired characteristics hardly came into play in her interpretations. In fact, her observations were characterized by a pronounced determinism, a static "self" furnished, through the graphology, with resolute evaluations such as "impotent," "weak-willed," or "psychopathic." Whereas the intercultural viewpoint she initially claimed in her article presented more a prospect than a program, there were seven cultural studies articles in the collection examining the challenge of individuality in different cultures.

This colorful package of observations, already indicated in these few examples, were intended by Brugsch to contribute toward the determination of the overall "reaction norm" of each individual (Brugsch 1926, 2). To this purpose, it was necessary to reinterpret the concept of the norm due to the fact that medical standard values, as numerically determined criteria, contradicted the principle of individualization. In Brugsch's opus, the Munich physician Ignaz Kaup (1870–1944) discussed the "significance of the concept of the norm in the theory of the personality" specifically (Kaup 1926, 191–225).[16] His point of departure was the "responsivity" of the individual as defined by the internist Louis Ruyter Redcliffe Grote (1886–1960):[17]

> Every person's standard is set according to the parameters of his biological performance. A person's responsivity should express the congruence of the physiological performance which actually exists and which is necessary for the individual, responsivity signifies individual, in other words relative health; it is formal good health without regard to the quality or quantity of morphology and function. The genuinely normal person will always be responsive, amongst the countless who are not normal, the majority are also responsive; they live in the best possible way for them.
>
> (Kaup 1926, 192)[18]

Grote had already introduced the concept of responsivity in 1918 and used it to identify a central problem of medicine (Grote 1918, 353–58): its subject, the individual who defied all attempts at a clinical approach toward typification, demanded an "individual approach" on the part of the physician (Grote 1961, 77–93).[19] However, this desired individualization could not be carried out without reference to the orientation-giving criteria derived from abstract typification. The tension within this coexistence, so difficult to harmonize, had to be borne by every physician as a scientist and doctor in everyday practice. Its object was the "specific human being" who was not to be "understood completely" by scientific means (Grote 1922). The fiction of statistical normality, of the *homme moyen*—as Grote emphasized in an article published on the norm concept in 1922 (Grote 1922, 370)—had its justification in science and not in medical practice. The Austrian constitutional physician Bernhard Aschner (1883–1960) demanded the relinquishment of numerical instruments even more firmly. He did this by advocating humoral pathology enriched by endocrinology and serology (Aschner 1928; Timmermann 2001, 717–739). His rejection of quantifying methods seemed to be supported by the impression that these measurements always lagged behind the holistic subjective perception of the physician acting intuitively.

The criticism of standard notions was a highly topical subject in German medicine in the 1920s.[20] Kaup also dealt primarily with two questions in his article: firstly, whether in view of this variability a norm might be determined; and secondly, which criteria for determining the norm could be exercised for this purpose.[21] To this end, he created an overwhelming number of tables, indices, percentages, variants and constants, "proportionality factors," and formulas. Although he was able to determine a mathematical correlation between heart, liver, kidneys, and pancreas in this process, he could not distinguish any significant correlation between internal organs and bodily habits. Kaup, who was searching for a degree of correlation for characteristic pairs, specified that it was related to a "functional-dynamic" state, to a "similarity of demand" rather than to a static one. As an aid for the physician, Kaup (1926, 225) proposed the "inner organ habitus equilibrium," which he identified as the "normal type." Thus the "normal type" continued to persist in the *Personallehre*; no longer, however, as a mathematical average, but as the degree of correlation of an individual with his environment characterized by relationality, responsivity, and dynamics. To promote a dynamic view of human beings, Theodor Brugsch introduced the expression "constitution variation on a conditional basis" (Brugsch 1921, 43), in which the temporal dimension of life processes was emphasized.[22] The notion of normality was maintained but was redefined on the basis of dynamic parameters and complex criteria for identification: not only relationality and responsivity but also the performance of an individual in their given environment and within a population group were essential determinants in this respect.

With its 4,000 pages, Brugsch's four volumes on the biology of a person presented a heterogeneous program—hardly any area of medicine was left out. With the publication of the last volume in 1931, however, the project, in which much time and effort had been invested, appeared to have reached an end. Despite the impetus and conviction that can be felt in each line of Brugsch's work, the "science of the individual" remained a topic never again taken up in the career of the internist. After the seizure of power by the National Socialists, Friedrich H. Lewy emigrated to the U.S. in 1934. Brugsch had to resign his academic position in 1935 (Brugsch 1986, 242).[23] He had a private practice in Berlin from 1935 to 1945.[24] After the war, his career at the Charité, in the GDR, experienced a second heyday.[25] He never again reactivated his old project of individual medicine. In 1958, four years before his death, Brugsch reported a standstill in "the theory of the personality," due, as he explained, to the success of experimental therapeutic medicine (1958, 932–935). This temporary standstill also correlated with the fact that the individual contributions had hardly any points of contact, and thus this major project of the *Personallehre* left an overall eclectic impression. From the wide range of approaches, hardly any practical therapeutic approaches could be distilled which would have satisfied the holistic demand that enabled physicians to make an holistic "personalized assessment," in particular "on questions of education of the people, advice, choice of profession, on the question of practical life aptitude" (Brugsch 1958, 23). There seemed to be a lack of technological support for establishing interrelationships between the extensive theses of the various areas. People like Aschner, who propagated a return to an *aísthesis* supplemented by a few modern testing procedures, pointed out the unease felt that the numerical descriptive corset would only insufficiently record the complexity of individuality.

Despite Brugsch's disenchantment, in the second half of the 1950s, new approaches to research were developed in the fields of pharmacology, biochemistry, and genetics that gave personalization in medicine fresh impetus.

3 The Rise of Emerging Pharmacogenetics in the 1950s

Friedrich Vogel, Arno G. Motulsky, and Werner Kalow are regarded as pioneers of pharmacogenetics and thus as progenitors of present-day PM. They looked particularly for molecular variations in connection with varying effects of pharmaceuticals (Motulsky 1957, 835–37). With his statement "everyone a deviate," as early as 1956 the American biochemist Roger Williams, who became known for his discovery of pantothenic acid, had pleaded for a biochemical detection of variability. In his book *Biochemical Individuality*, he postulated the so-called "genetotrophic approach," according to which genetic defects in individual metabolisms should be detected and compensated for by special nutrition. This approach can basically be understood as a form of dietetics which, in contrast to ancient models, was based upon biochemical descriptive forms of nourishment.[26] This involved a narrowing of perspective: if constitutional medicine strove toward a preferably holistic approach, then this biochemical analytics tended toward a focus on organic compounds. For human senses, these molecules were no longer accessible without additional technological aids but were determined with the help of gasometry, spectrophotometry, paper chromatography, or electrophoresis (Vogel 1959, 110–11).

These types of abstract methods were able to establish themselves as a result of their ability to explain clinical problems—for example, intolerability with medication such as the local anesthetic Procain (Novocain®) or the antimalarial drug Primaquin (Primaquine®). There was reason to suspect that such phenomena extended to conspicuously deviating enzyme activities (Kalow et al. 2001, 122–134).[27] In the 1950s, Kalow had investigated these occurrences and soon established that deviations in enzyme activity in certain populations, in particular in certain families, occurred frequently. Combined with this, questions relating to population genetics played a particular role (Jones 2013, 1–48). Pharmacogenetics opened the question of an optimized therapeutic efficiency, where formerly the detection of rare "abnormal enzyme structures" had been the primary consideration (Kalow et al. 2001, 2).

Whilst Vogel, Motulsky, and Kalow were primarily concerned with conspicuous extreme examples outside the standard deviation of a normal distribution, Williams (1956, 123) dealt increasingly with the variability within the "normal" range. The focus of his argument was aimed at the individual as variation carrier. This characterizes Williams's work and distinguishes it, as far as his reflection on medicine is concerned, from the work of the pioneers of pharmacogenetics. The scientific program of Friedrich Vogel or Arno Motulsky[28], with whom Vogel published the impressive work *Human Genetics* in 1979, focused on populations.[29] Early publications between the end of the 1950s and the beginning of the 1960s initially emphasized the significance of pharmacogenetics for families, patient groups, parts of a population, or races (Kalow 1926, 1). Williams's insistence on individual variability within the "normal" range was at first largely overlooked. Despite translations into Russian, Italian, and Polish, Williams's work was only reissued in 1998 when the return of PM to the leading scientific centers of Europe and the U.S. was announced. The initial weak response may be due to the fact that the Biochemical Institute of the University of Texas offered Williams no ideal access to medical schools, as Motulsky (2002, 687) observed with hindsight when he accused Williams of understanding little of genetics, despite his emphasis on it.

Furthermore, although Williams (1956, 15–17) took a personal dislike to purely mechanistic views, he reinforced a deterministic worldview. He believed, for example, that through experimentation on rats, he had proven that he could protect alcoholics from overconsumption with a special cocktail of vitamins (Williams, Berry, and Beerstecher 1949, 286). In an early study from 1949, he traced his observations on the drinking patterns of various rat communities back to a presumed variability of inherited metabolic patterns. In his experiments, rats on vitamin-supplemented diets consumed significantly less alcohol than rats on marginal diets. He concluded from these findings that the maintenance of a "wisdom of the body"[30] would require certain vitamins and micronutrients (Williams 1956, 162). Although aware of the limits of his unrefined experiments, Williams's explanations of human and animal behavior remained mechanistic (Williams et al. 1949, 288). Particularly noteworthy in this regard is his emphasis on causal similarities between humans and animals as well as his choice not to mention other possible explanations.[31] Nevertheless, Williams was the forerunner of a long-term trend: as well as his focus on "normal variability," his rejection of psychoanalysis in favor of biological explanations for behavior and his promise of efficient prevention are conspicuous.

In addition to the work of pioneers such as Williams, Motulsky, and Kalow, whose geographical focus was the U.S., the contribution of the German human geneticist Friedrich Otto Vogel to the field of recent pharmacogenetics also stands out. In his book *Moderne Probleme der Humangenetik*, published in 1959, he had programmatically introduced the term pharmacogenetics (Vogel 1959, 52–125).[32] In this work, Vogel, who was then working at the Free University of West Berlin, was able over the course of 125 pages to weave together population-genetic and biochemical methods. Vogel was a representative of German human genetics, which had to struggle to be reinstated into the scientific community following the ideological distortions and biopolitical escalations prior to the Second World War, and to reposition itself internationally:

> The expression "pharmacogenetics" of human beings [...] is so far more a program than a description of a field of work. Up until now, only approaches are available towards a recognition of hereditary variants in the reaction to medication and other externally introduced substances. For the most part,

this is probably due to the fact that these relationships have still hardly been examined; individual observations appear to indicate that in future the whole area should become increasingly important.

(Vogel 1959, 117)

Vogel's article came with a rich bibliography "limited" to 12 pages, representing a concentration of investigations and single studies that had emerged after the war in genetics and biochemistry. Most were in English. Vogel's primary interest was in issues of population genetics but also covered Roger Williams's individually centered "genetothropic principle" (Vogel 1959, 123). Retrospectively, however, his discussion of "phenogenetics" is particularly noteworthy. This term, which goes back to Valentin Haecker (1864–1927) and Eugen Fischer (1874–1967), was used by Vogel (1959, 113) to address the fact that new biochemical methods give information about diseases at a point at which the disease has not yet manifested itself. This was based not least upon the growing arsenal of methods for the visualization of biochemical differences between various individuals. Paper chromatography, electrophoresis, coagulation physiology, cell structures, and serogenetics belong to the most important of these methods. Together with physical methods such as the electroencephalograph (EEG), such instruments opened up the possibility of describing different enzyme activities (Vogel 1959, 107–13). By grasping this variability as a phenotypical expression of a particular genotype, Vogel was able to speculate about specific genotypes. Although he was familiar with the double helix structure discovered by Watson and Crick in 1953, as a model he treated it with caution. Due to the still modest amount of information on possible reference sequences, his comments concerning possibly altered base pairs remained consciously speculative (Vogel 1959, 109).

Nevertheless, these biochemical procedures already point toward a pronounced phenogenetic variability in humans. When Vogel published his *Lehrbuch der allgemeinen Humangenetik* in 1961, he was concerned, amongst other things, with the "biochemical individuality of human beings." In this work, Vogel supported Williams's claim that each individual is a deviant:

What is in fact the "range of the normal"? Isn't there a wide variation of all biochemical processes and functions within the range of the "normal"? [...] Nobody is ever really the same as anybody else in the genetically regulated primary functions of their intermediary metabolism. Perhaps we could even go further and say that in all these functions nobody is completely "normal."

(Vogel 1961, 476–77)

Thus, Friedrich Vogel also described the biochemical genetic analysis of "differences in the range of the normal" as a fundamental requirement for allowing comprehension of the genetic differences in "health and performance." In this task, which numerous Genome Wide Association Studies (GWAS) are currently pursuing, he saw one of the most important future goals of human genetics. Werner

Kalow, whose book *Pharmacogenetics* appeared one year after Vogel's textbook, also described the focus on individual variations as self-evident for pharmacologists (Kalow 1962, 1). In this early phase of pharmacogenetics, he especially stressed the significance of mathematical models for distribution and correlation, which were less established in pharmacology than they were in genetics (Kalow 1962, 2). Despite such forward-looking perspectives, the emerging field of pharmacogenetics also gave signs of pledges that today are widely critically assessed. Vogel's eugenic demand for a prevention of "genetic anomalies" is one of these (Vogel 1961, 629). Remarkably, the research carried out by emerging pharmacogenetics, which took into consideration the differing phenotypical forms in the metabolism, is nowadays regarded mostly as a part of pharmacogenomics, whilst pharmacogenetics currently focusses on DNA sequences (European Medicine Agency 2007, 5).[33]

4 Outlook

Another physician to follow the route set out by Williams for the correlation of biochemical variability with predisposition to disease was Emanuel Cheraskin.[34] In his view, biochemical markers should allow a "predictive medicine" to anticipate diseases before they manifest. As a result, the prospect of the prevention of a not yet manifested illness held out the prospect of more efficient, pain-reducing therapies. Here, Cheraskin, who specialized in nutrition, focused on dietary habits. What is apparent in his case, compared to his predecessors, is his consideration of the disease status in a pre-manifest condition (Cheraskin, Ringsdorf, and Aspray 1969, 121–125).

This shift between the two poles of illness and health was further strengthened by the advent of biomarkers, which have established themselves particularly in cancer research (Keating and Cambrioso 2011). Although publications on biological markers had already appeared (Waalkes et al. 1978, 1685–1703) in conjunction with cancer prognoses in the 1970s, a paper published in 1980 by a scientist working with the oncologist Joseph Paone is regarded as the beginning of the biomarker era (Aaronson 2005, 491–494; Waalkes, Robinson, and Shaper 1980, 59–66). This orientation toward smaller and smaller predictive entities was particularly encouraged in leukemia research in the 1990s, which focused increasingly on molecular markers (Bloomfield, Mrozek and Caligiuri 2006, 3564S–71S). In 2007, the Cancer Biomarker Collaborative was founded in order to promote biomarkers as surrogate endpoints in clinical trials. To date, the U.S. Food and Drug Administration (2015) has listed biomarkers for around 140 medications. While over a third of these drugs come from oncology, drugs from other areas, such as cardiology, psychiatry, or infectious diseases, are increasingly biomarker related.

Over the last three decades, a great deal has taken place, not least on this molecular level, where an increasing number of biomarkers abets more and more abstract models. Once, it was the externally visible that stood in the foreground of medical studies; now, however, the determination of metabolites has been augmented by the determination of DNA and their biochemical function in relation to the coding, transcription, and metabolization of proteins. Human nucleotide sequences have been online since 1992 via the GenBank of the U.S. American National Center for Biotechnology Information. Although not all of the some 16 billion base pairs for human beings have to be considered when it comes to the task of sequencing an individual and interpreting his or her DNA, genetic analyses still rely on an enormous amount of data. Consequently, the medical management of such "big data" is bound more strongly than ever to the development of efficient computers, which reinterpret the confusing variability in gene sequences as stochastic risks via complex algorithms. As a result of the impenetrability of such numerical computations, to a great extent both the physician and the layperson simply have to trust them.

On first sight, present-day PM appears to be in the process of realizing the efforts of its predecessors. Next-generation sequencing and computer-aided evaluations are able to reveal immeasurably more correlations than were imaginable in Brugsch and Lewy's *Die Biologie der Person*. The determination and characterization of genetic changes has been intensified by the establishment of single nucleotide polymorphisms (SNPs). New detection methods, such as the AmpliChip CYP450, which detects variations in the gene regions CYP2D6 and CYP2C19, allow prognoses for the individual metabolism of different medications. However, if we look more closely, a link between the PM of today and the earlier approaches pointing in this direction appears problematic. The sensorially experienced characteristics of a person, which are difficult to operationalize for computational processing, are increasingly neglected. This includes not only symptoms but also a person's biographical background. In the *Personallehre*, both kinetically oriented and psychologically oriented physicians dared to speculate on the formative effect of "biophysical personality building." While the prioritization of the genetic "substructure" of a person had already been laid out by Brugsch (1986, 233), a holistic interpretation of the diverse ways of describing individuals remained more within the reach of human agency due to the lack of interconnectivity of the various approaches gathered in his volumes.

In retrospect, it is evident that both developments described in this chapter shared the tendency to take into account a broadly "normal" variability. In this context, the reference to Hippocrates served as a legitimization strategy. As time passed, it became increasingly harder for the pioneers discussed here to present themselves as the successors of this tradition. Whilst Brugsch continued to highlight his individualizing predecessors in detail, Williams condensed the Hippocratic-Galenic tradition into a few sentences. In the case of Cheraskin, only the name Hippocrates remains. This noticeable trend in the development of PM coincided with a narrowing of focus. Where the interdisciplinary description of behavioral performance held a central position in the *Personallehre*, early pharmacogeneticists had already created a focus on methods of genetics and biochemistry. Williams's devaluation of psychoanalytical explanations is representative of this orientation toward biological material (1956, 198).

The observable loss of the significance of behavior description since then suggests an awareness of less evaluative descriptions. However, the misleadingly applied concept of the person appears disadvantageous here. A reactivation of predetermined interdisciplinary and intercultural reflection at the beginning of the twentieth century, along with the preventative, nutritional approaches of scientists such as Cheraskin or Williams, could be beneficial. With a view to the physical, social and environmental dimensions of human beings—as still healthy, near sick, sick, or recovered—this would be a particularly promising strategy for moving closer to an understanding of the individual, now the center of medicine in the twenty-first century. While Williams and Cheraskin contributed to the notion that risk-associated conditions could be positively influenced by lifestyle choices, Brugsch and contemporaries are particularly instructive with regard to the openness with which they engaged with phenotypic diversity. The heterogenous assessments of individual health gathered in the *Personallehre* highlight the relativity of deviation and therewith the broad frame within which one still could speak of "healthy dyscrasias." It is the intrinsic dynamic of this Galenic concept that inspired Canguilhem's theory of individual normativity (Canguilhem 2011). A wider acceptance of one's own idiosyncrasy may help to come to terms with variation by relativizing the anxiety-ridden risk association studies, which seem to dominate contemporary research perspectives in PM. From a cultural–historical and medical–philosophical perspective, it would be promising to look more closely at the pioneers of PM, who so far have only been selectively explored. A historical reconstruction indicates that recent innovations in PM are aiming more at a refinement of molecular biology and of genetics than at providing a fundamental program. By relativizing the project of PM, examples from the past demonstrate the limits of genetic reductionism and contain a self-critical potential that may dispel fallacies.

Despite the increasingly accurate recording of molecular variability—with very few exceptions, which predominantly come from areas of oncology—the need for an optimized, "tailored" therapy remains hitherto a promise still to be fulfilled.[35] The organization of the correlation between genetic types and phenotypes, influenced by nonhereditary influences (exposome) and by irregularities in transcription and translation, is too complex. Whether in the future the life sciences will be able to "decrypt" this complexity and develop more effective therapies remains to be seen. In the research on PM, it is statistical correlations between genetic sequences and (pre)clinical characteristics that are predominantly in the foreground today. Meanwhile, these algorithmically calculable predictions enjoy a considerable measure of public confidence. The probability of developing cancer early, for example, prompts men and women from certain risk groups toward radical preventative measures such as mastectomy and ovariectomy whilst still in a healthy condition. The different GWAS strengthen Williams's statement that "practically every human being is a deviate in some respects" (Williams 1956, 3), as they associate deviant gene sequences with specific disease risks. In this way, risk groups are treated today more as a stratified collective based on specific SNPs than as individuals with a unique disposition shaped both by their genes as well as their history. It has become clear that the present program of PM tends to ignore essential individual characteristics.

Despite this, the much criticized and supplemented term *personalized medicine* remains—possibly because it has its finger on the pulse of the times. In a global era of a society that is permanently in communication with itself, and in which the borders between the private and public spheres have become porous, individually collected (bio)data can be shared with large communities.[36] This sharing of particularities may have an identity-constituting significance for some people. At the same time, the protection of personal data belongs to the disputed rights of modern society. Despite this difficulty and the so-far limited realization of personalization, "individuality" and "personality" in some cases seem to be little more than convenient labels in the current economic context. In such instances, the break with previous tradition would be all the larger.

It is to be hoped that PM reflects upon the profounder inflexion of the term and the tradition contained within it. If so, it will remain true to its meaningful and continuing task of helping the patient as a vulnerable individual. To respond to this difficult challenge, in addition to increasingly more accurate, high-tech driven "precision medicine," a personal approach is above all necessary—one in which the motives, fears, and history of an individual are considered.

Acknowledgments

The authors would like to thank Giovanni Boniolo, Marco Nathan, and Karl Hughes for their helpful comments and suggestions.

Notes

1 See also, amongst others, Nationale Akademie der Wissenschaften, acatech (2014) and Centre for Technology Assessment TA-SWISS (2014).
2 Status: 11.11.2015. This takes into account the MeSH terms "personalized medicine" and "individualized medicine." Total records for each publication year are taken from https://www.nlm.nih.gov/bsd/licensee/baselinestats.html (accessed December 8th, 2015).
3 In this chapter, we use the wider term *personalized medicine*, which overall is used more frequently—unless past attempts toward the "individualization" of medicine are concerned. In this case, each author's preferred terms are applied.
4 For literature on Theodor Brugsch see also Kaiser and Hübner (1978), Konert (1988), and Brugsch (1986). For the meaning of the term *Personallehre* see Footnote 7.
5 Brugsch (1926, 20) depicted Viola and other European proponents of constitutional medicine, such as Claude Sigaud (1862–1921), as progenitors of his project. For early predecessors of PM in Britain see Tutton (2012).
6 See also Brugsch (1918) and his teacher Kraus (1919). See also Bleker and Schmiedebach (1987).
7 There are no adequate English translations for *Personallehre* and *Individualmedizin*. Brugsch understood it as the "[…] attempt to capture the constitutional core of each person […] in every medical respect" (Brugsch 1926, 20). Thereby, he aimed at an

interdisciplinary expansion of constitutional medicine. Just how inconsistently he dealt with the terminology of the theories of personality, the individual and the constitution, is evident in his concept of constitution, which from one edition of his works to the next is redefined, rejected, or replaced by the concept of the person and then reinstated. See also Brugsch (1922, 43), Brugsch (1926, 20–22), and Brugsch (1927/28, 70).

8   On the constitutional theory of the 1920s and 1930s see Timmermann (2001), Lawrence and Weisz (1998), Hau (2000), and Harrington (1996).

9   In his autobiography (Brugsch 1986, 195), Brugsch described Lewy primarily as an editorial assistant.

10   For a more detailed discussion on standard notions in medicine in the twentieth century see also Gadebusch Bondio (2015, 19–50).

11   See also Hippocrates (2006b, 272–73) and Gadebusch Bondio (2012, 291–311).

12   We used the Kuhn's edition (K) of Galen's works, which will be cited as follows: Galen: Title (K, Vol. No., Book No., Ch. No., Page): Kuhn: Galen Opera Omnia (1821–1833); Galen: De optima corporis nostri constitutione (K, vol. IV, 2, 3, 737–74). Idem: De sanitate tuenda (K, vol. VI, 1, 5, 18–22) and Hippocrates (2006 b, 9).

13   Galen: De methodo medendi (K, X,1–2, 7, 205–6); idem (K, IX, 17, 659–60); idem (K, III, 3, 181–183) and (K, III, 7, 205–10). See also Galen: De sanitate tuenda (K, VI, 1, 2, 7). This idea of a gradual perfecting of conjecture is also central to diagnostics and therapy: Galen: De crisibus (K, IX, 2, 1, 2) and (K, IX, 2, 1, 20); idem: De methodo medendi (K, X, 1–2, 3, 4). See also Gadebusch Bondio and Herrmann (2011, 129–42).

14   For Kretschmer's description scheme see Kretschmer (1925, 5–8).

15   See also Birnbaum (1926, 659–94) and Flatow-Worms (1926, 695–843).

16   Ignaz Kaup worked at the Institute for Hygiene in Munich with Max von Gruber (1853–1927). On Kaup, see also Weindling (1993, 335–37).

17   Following his medical studies, Grote received his bacteriological training at the Robert Koch Institute. In 1914, he joined the Medical University Clinic in Halle, where he qualified as professor in 1918 under Adolf Schmidt, and where he remained until 1924. As professor since 1922, he headed well-known sanatoriums and hospitals. Amongst those areas of research and interest were constitutional theory, regulation pathology, physical therapy, and dietetics. He advocated a "personal" holistic approach, which brought him close to Brugsch's individualized approach (Schulz and Grote 1996, 163–64). See also Grote (1961, 13–39) and Böllinger (2000).

18   See also Grote (1922, 361–77).

19   See also Rieger (2000, 127–29).

20   Like Grote, the anatomist Wilhelm Roux (1850–1924), the internist Hermann Rautmann (1885–1956), the psychiatrist Kurt Hildebrandt (1881–1966), and the philosopher Richard Müller-Freienfels (1882–1949) had either questioned or rejected the concept of the norm. Grote is referring to e.g., Roux (1920), Hildebrandt (1921), and Rautmann (1921). Grote (1922, 369) refers critically to Brugsch's anthropometric investigations. (Brugsch 1918 and 1922).

21   Here Kaup refers, amongst others, to the work of Hans Günther (1921), internist and constitution researcher in Bonn and Leipzig (from 1928). See also the Professorenkatalog der Universität Leipzig (2013).

22   On the time factor in Theodor Brugsch's theory of the individual, see Gadebusch Bondio (2012, 117–38: 128).

23   See also Weindling (1933, 512).

24   Brugsch was married to a Jewess. Despite divorcing and joining a number of National Socialist organizations (National Socialist Motor Corps [NSKK], National Socialist People's Welfare [NSV] and supporting member of the SS), he did not succeed in getting reappointed in Halle (Klee 2005, 78). Paul Weindling (1993, 512) numbers

Brugsch amongst the "liberal professors" who were forced to resign from their posts between 1934 or 1935 and 1937.

25 In 1945, he was appointed professor at the University of Berlin, where he remained as Professor of Internal Medicine and Director of the Clinic for Internal Medicine at the Charité in Berlin until gaining emeritus status in 1957 (Brugsch 1986, 294–96).

26 "Every individual organism that has a distinctive genetic background has distinctive nutritional need which must be met for optimal well-being" (Williams 1956, 167). The term "biochemical individuality" was close to the term "chemical individuality," introduced in 1902 by Archibald E. Garret (1857–1936). See Michl (2015b, 66–68).

27 See also Dern, Beutler, and Alving (1954, 171–76).

28 See Motulsky (1957, 835–37). Motulsky also gave accounts of experiments (on humans and animals) on race, species and population group for the "detection of hereditary biochemical traits that cause drug reactions."

29 Population genetics also remained the central research focus in Vogel's textbook (1961): "Zwei Gebiete sind es vor allem, auf denen in den letzten Jahren besondere Fortschritte erzielt wurden: die biochemische oder, wie man auch sagt, „molekulare" Genetik (die Analyse der Genwirkung) und die Populationsgenetik mit der Analyse der Mutation und der natürlichen Selektion. Hier musste deshalb der Schwerpunkt unserer Darstellung liegen. Gerade bei der Analyse der natürlichen Selektion etwa im Bereich der Hämoglobin-Varianten oder der AB0-Blutgruppen zeigt es sich, wie eng sich diese beiden Arbeitsgebiete miteinander verzahnen."

30 The term "wisdom of the body" has been popularized by Walter B. Cannon's (1932) classic.

31 In 1956, Williams relativized his earlier extrapolation from animals to humans (Williams 1956, 160). However, although admitting the limited validity of his laboratory findings for humans, he still emphasized the "substantial basis" of his observations made on rats and remained convinced of its applicability to human beings. In contrast, pharmacogenticists such as Kalow (1962) highlighted precisely the varying effects of particular drugs in different species. For example, substances such as Procain tend to anesthetize humans, while stimulating other species such as horses.

32 On Vogel's career see also Institut für Humangenetik und Anthropologie (1991).

33 On the inconsistency of definitions see also Centre for Technology Assessment TA-SWISS (2004, 18–20).

34 Cheraskin and Ringsdorf (1971, 511–16), Cheraskin and Ringsdorf (1973), Williams (1961, 325–27).

35 See also Centre for Technology Assessment TA-SWISS (2004, 248) and Gadebusch Bondio and Michl (2012, 117–38: 134–36).

36 On the question of biological identity see also Boniolo and Testa (2011).

References

Allers, R. (1931) "Medizinische Charakterologie." In *Die Biologie der Person, Bd. 2: Allgemeine und psychophysische Konstitution*. Edited by Theodor Brugsch, and Fritz Heinrich Lewy, pp. 511–658. Berlin/Wien: Urban and Schwarzenberg.

Aronson, J.K. (2005) "Biomarkers and surrogate endpoints," *British Journal of Clinical Pharmacology* 59(5), pp. 491–94.

Aschner, B (1928) *Die Krise der Medizin. Konstitutionstherapie als Ausweg*. Stuttgart/ Leipzig/Zürich: Hippokrates.

Bauer, J. (1917) *Die konstitutionelle Disposition zu inneren Krankheiten*. Berlin: Springer.

Bauer, J. (1921) *Allgemeine Konstitutions- und Vererbungslehre*. Berlin: Springer.

Birnbaum, K. (1931) "Die Probleme des biopsychischen Persönlichkeitsaufbaus." In *Die Biologie der Person, Bd. 2: Allgemeine und psychophysische Konstitution*. Edited by Theodor Brugsch, and Fritz Heinrich Lewy, pp. 659–94. Berlin/Wien: Urban and Schwarzenberg.

Bleker, J. and H.-P. Schmiedebach, (eds.) (1987) *Medizin und Krieg: Vom Dilemma der Heilberufe 1865 bis 1985*. Frankfurt am Main: Fischer.

Bloomfield, C.D., K. Mrózek, and M.A. Caligiuri (2006) "Cancer and Leukemia Group B Leukemia Correlative Science Committee: Major Accomplishments and Future Directions," *Clinical Cancer Research* 12, pp. 3564s–3571s.

Böllinger, E.M. (2000) *Leben und Werk des Internisten Louis Radcliffe Grote (1886–1960)*. Diss. med. Universität Leipzig.

Boniolo, G. and T. Giuseppe (2011) "The identity of living beings, epigenetics and the modesty of philosophy," *Erkenntnis* 76(2), pp. 279–298.

Brugsch, T. (1922) *Allgemeine Prognostik oder die Lehre von der Beurteilung des gesunden und kranken Menschen*. 2nd ed. Berlin/Wien: Urban and Schwarzenberg.

Brugsch, T. (1926) Einführung in die Konstitutionslehre. In *Die Biologie der Person. Ein Handbuch der allgemeinen und speziellen Konstitutionslehre in vier Bänden, Bd. 1: Allgemeiner Teil der Personallehre*. Edited by Brugsch, Theodor and Fritz Heinrich Lewy, pp. 1–23. Berlin/Wien: Urban and Schwarzenberg.

Brugsch, T. and F.H. Lewy, (eds.) (1926) *Die Biologie der Person. Ein Handbuch der allgemeinen und speziellen Konstitutionslehre in vier Bänden, Bd. 1: Allgemeiner Teil der Personallehre*. Berlin/Wien: Urban and Schwarzenberg.

Brugsch, T. (1928) Die Klinik. In *Grundlagen und Ziele der Medizin der Gegenwart: Vorträge des Instituts für Geschichte der Medizin an der Universität Leipzig, Bd. 1*, pp. 51–71. Leipzig: G. Thieme.

Brugsch, T. and F.H. Lewy, (eds.) (1929) *Die Biologie der Person, Bd. 4: Soziologie der Person*. Berlin/Wien: Urban and Schwarzenberg.

Brugsch, T. and F.H. Lewy, (eds.) (1930) *Die Biologie der Person, Bd. 3: Organe und Konstitution*. Berlin/Wien: Urban and Schwarzenberg.

Brugsch, T. and F.H. Lewy, (eds.) (1931) *Die Biologie der Person, Bd. 2: Allgemeine somatische und psychophysische Konstitution*. Berlin/Wien: Urban and Schwarzenberg.

Brugsch, T. (1958) "Ganzheit und therapeutischer Aspekt, *Zeitschrift für die gesamte innere Medizin* 13, pp. 932–935.

Brugsch, T. (1986) [1957] *Arzt seit fünf Jahrzehnten. Autobiographie*. Berlin: Verlag der Nation.

Büttner, J. (1997) "Die Herausbildung des Normalwert-Konzeptes im Zusammenhang mit quantitativen diagnostischen Untersuchungen in der Medizin." In *Die Normierung der Gesundheit. Methoden und Verfahren in der Medizin als kulturelle Praxis. Abhandlungen zur Geschichte der Medizin und der Naturwissenschaften, Vol. 82*. Edited by Volker Hess, Rolf Winau, and Heinz Müller-Dietz, 21. Husum: Matthiesen.

Canguilhem, G. (2002) "La philosophie biologique d'Auguste Comte et son influence en France au XIXe siècle." In Georges Canguilhem: *Études d'histoire et de philosophie des sciences*, pp. 61–98. Paris: Vrin.

Canguilhem, G. (2011) *Le normal et le pathologique*. Paris: PUF.

Cannon, W.B. (1932) *The Wisdom of the Body*. New York: W. W. Norton and Company.

Centre for Technology Assessment TA-SWISS (2004) *Pharmakogenetik und Pharmakogenomik*. Zürich: vdf.

Centre for Technology Assessment TA-SWISS (2014) *Personalisierte Medizin*. Zürich: vdf.

Cheraskin, E., W.M. Ringsdorf, and D.W. Aspray (1969) "Cancer proneness profile. A study in ponderal index and blood glucose," *Geriatrics* 24(8), pp. 121–125.

Cheraskin, E. and W.M. Ringsdorf (1971) "Predictive medicine. IV. The gradation concept," *Journal of the American Geriatric Society* 19(6), pp. 511–516.

Cheraskin, E. and W.M. Ringsdorf (1973) *Predictive Medicine. A Study in Strategy.* Mountain View: Pacific Press.

European Medicine Agency (2007) "ICH Topic E15: Definitions for genomic biomarkers, pharmacogenomics, pharmacogenetics, genomic data and sample coding categories." http://www.ema.europa.eu/docs/en_GB/document_library/Scientific_guideline/2009/09/WC500002880.pdf (accessed November 23rd, 2015).

European Science Foundation (2012) *Personalised Medicine for the European Citizen— towards more precise medicine for the diagnosis, treatment and prevention of disease.* Strasbourg: Ireg.

Flatow-Worms, E. (1931) "Handschrift und Charakter." In *Die Biologie der Person, Bd. 2: Allgemeine und psychophysische Konstitution.* Edited by Theodor Brugsch, and Fritz Heinrich Lewy, 695–843. Berlin/Wien: Urban and Schwarzenberg.

Gadebusch Bondio, M. (2015) "Das Individuum – eine Abweichung"... und das Unbehagen der Wissenschaft. In *Norm als Pflicht, Zwang und Traum. Medizinge- schichte im Kontext, Vol. 19.* Edited by E. Brinkschulte and M. Gadebusch Bondio, 19–50. Frankfurt am Main: Peter Lang Verlagsgruppe.

Gadebusch Bondio, M. (2012) "Vom Ringen der Medizin um eine Fehlbarkeitskultur. Epistemologische und ethische Reflexionen." In *Errors and Mistakes. A Cultural History of Fallibility.* Micrologus' Library no. 49. Edited by M. Gadebusch Bondio and A.P. Bagliani, pp. 291–311. Florenz: Sismel.

Gadebusch Bondio, M. and S. Michl (2012) "Von der Medikalisierung des Humanen. Das Individuelle als Herausforderung in der Medizin." In *Konzepte des Humanen.* Edited by M. Gadebusch Bondio and H. Siebenpfeiffer, pp. 117–38. Freiburg: Karl Alber.

Gadebusch Bondio, M., and I.F. Herrmann. (2011) "Ganz persönlich und doch so fremd – Gesundheit in Zeiten der Individualisierten Medizin." In *Was ist Gesundheit? Antworten aus Jahrhunderten.* Edited by K. Bergdolt and I. F. Herrmann, 129–142. Stuttgart: Steiner Verlag.

Galen (1821–1833) *Opera Omnia.* Edited and translated by Carolus Gottlob Kühn. XX Volumes. Leipzig: Lipsiae.

Galen. *De sanitate tuenda* (K, VI, pp. 1–452).

Galen. *De crisibus* (K, IX, 2, pp. 550–768).

Galen. *De methodo medendi* (K, X, 1–2, pp. 1–1021).

Galen. *De optima corporis nostri constitutione* (K, IV, 2, pp. 737–774).

Galen. *De temperamentis* (K, I, pp. 509–694).

Grote, L.R.R. (1918) "Muskeltätigkeit und Blutzucker," *Zentralblatt für innere Medizin* 39, pp. 353–58.

Grote, L.R.R. (1922) "Über den Normbegriff im ärztlichen Denken," *Zeitschrift für die gesamte Anatomie* II Abt. 8, pp. 361–77.

Grote, L.R.R. (1961) *Der Arzt im Angesicht von Leben und Tod. Eine Auswahl aus seinem Werk mit einer biographischen Einführung.* Edited by K.E. Rothschuh, pp. 13–39. Stuttgart: Hippokrates.

Günther, H. (1921) *Grundlagen der biologischen Konstitutionslehre.* Leipzig: Thieme.

Harrington, A. (1996) *Holism in German Culture from Wilhelm II to Hitler.* Princeton: Princeton University Press.

Hau, M. (2000) "The Holistic Gaze in German Medicine, 1890–1930," *Bulletin of the History of Medicine* 74, pp. 495–524.

Hess, V. (1997) *Normierung der Gesundheit. Messende Verfahren der Medizin als kulturelle Praktik um 1900 (Abhandlungen zur Geschichte der Medizin und der Naturwissenschaften, Vol. 82).* Edited by Rolf Winau and Heinz Müller-Dietz. Husum: Matthiesen.

Hildebrandt, K. (1920) *Norm und Entartung des Menschen.* Dresden: Sibyllen-Verlag.

Hippocrates (2006a) "De arte—Über die Kunst." In *Ausgewählte Schriften.* Edited and translated by C. Schubert and W. Leschhorn, pp. 106–129. Düsseldorf/Zürich: Artemis und Winkler.

Hippocrates (2006b) "De vetere medicina—Über die Alte Medizin." In *Ausgewählte Schriften.* Edited and translated by C. Schubert and W. Leschhorn, pp. 272–307. Düsseldorf/Zürich: Artemis und Winkler.

Institut für Humangenetik und Anthropologie, ed. (1991) *Humangenetik in Heidelberg. Das Institut für Humangenetik und Anthropologie von 1962 bis 1990 im Lichte der Habilitationen.* Symposium on March 10th, 1990 for the 65th birthday of Prof. Dr. H.C. Friedrich Vogel. Heidelberg.

Jones, D.S. (2013) "How personalized medicine became genetic, and racial: Werner Kalow and the formations of pharmacogenetics," *Journal of the History of Medicine and Allied Sciences* 68(1), pp. 1–48.

Jori, A. (2003.) *Aristotele.* Mailand: Bruno Mondadori Editore.

Kaiser, W. and H. Hübner (1978) *Theodor Brugsch (1878–1963).* Presented at the Hallesches Brugsch-Symposium, Martin Luther Universität Halle/Wittenberg, Halle an der Saale. Kennan, G.F. (1979) *The Decline of Bismarck's European Order: Franco-Russian Relations, 1875–1890.* Princeton: Princeton University Press.

Kalow, W. (1962) "Pharmacogenetics. Heredity and the response to drugs," *Journal of Pharmaceutical Sciences,* 52(2). Hoboken: Wiley-Blackwelll.

Kalow, W., U.A. Meyer, and R.F. Tindale (2001) *Pharmacogenomics.* New York/Basel: Marcel Dekker, Inc.

Kaup, I. (1922) *Volkshygiene oder selektive Rassenhygiene?* Leipzig: S. Hirzel Verlag.

Kaup, I. (1926) "Bedeutung des Normbegriffs in der Personallehre." In *Ein Handbuch der allgemeinen und speziellen Konstitutionslehre in vier Bänden, Bd. 1: Allgemeiner Teil der Personallehre.* Edited by T. Brugsch and F. H. Lewy, pp. 191–225. Berlin/ Wien: Urban and Schwarzenberg.

Keating, P. and A. Cambrioso (2011) *Cancer on Trial: Oncology as a New Style of Practice.* Chicago: Chicago University Press.

Klee, E. (2005) *Das Personenlexikon zum Dritten Reich. Wer war was vor und nach 1945.* 2nd ed. Frankfurt am Main: S. Fischer Verlag.

Konert, J. (1988) *Theodor Brugsch. Internist und Politiker.* Leipzig: S. Hirzel Verlag.

Kraus, F. (1919) *Die allgemeine und spezielle Pathologie der Person. Klinische Syzygiologie.* Stuttgart: Georg Thieme Verlag.

Kretschmer, E. (1925) *Physique and Character: An Investigation of the Nature of Constitution and of the Theory of Temperament.* London: Kegan Paul, Trench, Trubner.

Kroemer, H.K. (2013) "Personalisierte Medizin—zum Stand der Forschung." In Deutscher Ethikrat: *Personalisierte Medizin—Der Patient als Nutznießer oder als Opfer?* pp. 11–22. http://www.ethikrat.org/dateien/pdf/tagungsdokumentation-personalisierte-medizin.pdf (accessed November 23rd, 2015).

Lawrence, C. and G. Weisz (1998) *Greater than the Parts: Holism in Biomedicine, 1920–1950*. Oxford: Oxford University Press.

Lewy, F.H. (1931) "Experimentelle Untersuchungen zur psychophysischen Typenforschung. I. Die Motorik." In *Die Biologie der Person, Bd. 2: Allgemeine und psychophysische Konstitution*. Edited by T. Brugsch, and F. H. Lewy, pp. 845–858. Berlin/Wien: Urban and Schwarzenberg.

Lindner, M. (1999) *Die Pathologie der Person. Friedrich Kraus' Neubestimmung des Organismus am Beginn des 20. Jahrhunderts*. Berlin/Diepholz: GNT-Verlag.

Michl, S. (2015a) "Inventing Traditions, Raising Expectations. Recent Debates on 'Personalized Medicine.'" In *Individualized Medicine: Ethical, Economical, and Historical Perspectives*. Edited by T. Fischer, M. Langanke, P. Marschall, and S. Michl, pp. 45–60. Cham/Heidelberg/New York/Dodrecht/London: Springer.

Michl, S. (2015b) "The Epistemics of 'Personalized Medicine'. Rebranding Pharmacogenetics." In *Individualized Medicine: Ethical, Economical, and Historical Perspectives*. Edited by T. Fischer, M. Langanke, P. Marschall, and S. Michl, 61–78. Cham/Heidelberg/New York/Dodrecht/London: Springer.

Motulsky, A.G. (1957) "Drug reactions, enzymes and biochemical genetics," *The Journal of the American Medical Association* 165, pp. 835–837.

Motulsky, A.G. (2002) "Pharmacogenetics: A historical account and current status," *Medicina nei secoli* 14(3), pp. 683–703.

Müller-Freienfels, R. (1921) *Die Philosophie der Individualität*. Leipzig: Felix Meiner Verlag.

Nationale Akademie der Wissenschaften Leopoldina, and acatech (2014) *Individualisierte Medizin—Voraussetzungen und Konsequenzen*. Edited by Deutsche Akademie der Technikwissenschaften and Union der deutschen Akademien der Wissenschaften. Halle (Saale).

Paone, J.F., P.T. Waalkes, R.R. Baker, and J.H. Shaper (1980) "Serum UDP-galactosyl transferase as a potential biomarker for breast carcinoma," *Journal of Surgical Oncology* 15, pp. 59–66.

Professorenkatalog der Universität Leipzig. Edited by Lehrstuhl für Neuere und Neueste Geschichte, Historisches Seminar der Universität Leipzig. http://www.uni-leipzig. de/unigeschichte/professorenkatalog/leipzig/Guenther_216.pdf (accessed March 4th, 2013).

Rautmann, H. (1921) *Untersuchungen über die Norm, ihre Bedeutung und Bestimmung. Veröffentlichungen aus der Kriegs- und Konstitutionspathologie*, 2(2). Jena: Gustav Schmidt.

Rieger, S. (2000) "Die Freiheit der Geste und ihre technische Dekodierung." In *Gestik—Figuren des Körpers im Text und Bild. Literatur und Anthropologie* Vol. 8. Edited by M. Egidi, pp. 117–130. Tübingen: Gunter Narr Verlag.

Roux, W. (1920) *Prinzipielle Sonderung von Naturgesetz und Regel, von Wirken und Vorkommen*. Sitzungsberichte der Preußischen Akademie der Wissenschaften, 28. Berlin.

Timmermann, C. (2001) "Constitutional medicine, neoromanticism, and the politics of antimechanism in interwar Germany," *Bulletin of the History of Medicine* 75, pp. 717–739.

Tutton, R. (2012) "Personalizing medicine: Futures present and past," *Social Science & Medicine* 75(10), pp. 1721–1728.

Viola, G. (1932) *La costituzione individuale: dottrina, metodo, tipi morfologici*. Bologna: Cappelli.

Vogel, F. (1959) "Moderne Probleme der Humangenetik," *Ergebnisse. Innere Medizin und Kinderheilkunde* 12, pp. 52–125.

Vogel, F. (1961) *Lehrbuch der allgemeinen Humangenetik*. Berlin/Göttingen/Heidelberg: Springer Verlag.

Vogel, F., and A.G. Motulsky (1997) [1979]. *Human Genetics. Problems and Approaches*. Berlin/Heidelberg/New York: Springer Verlag.

Waalkes T.P., C.W. Gehrke, D.C. Tormey, K.B. Woo, K.C. Buo, J. Snyder and H. Hansen (1978) "Biologic markers in breast carcinoma IV. Serum fucose-protein ratio. Comparison with carcinoembryonic antigen and human chorionic gonadotrophin," *Cancer* 41, pp. 1685–1703.

Weindling, P. (1993) *Health, Race and German Politics between National Unification and Nazism 1870–1945*. Cambridge: Cambridge University Press.

Williams, R.J., J.L. Berry and E. Jr. Beerstecher (1949) "Biochemical individuality. III. Genetotrophic factors in the etiology of alcoholism," *Archives of Biochemistry and Biophysics*. 2, pp. 275–290.

Williams, R.J., J.L. Berry and E. Jr. Beerstecher (1950) "The concept of genotrophic disease," *The Lancet* 255(6599), pp. 287–289.

Williams, R.J. (1956) *Biochemical Individuality. The Basis for the Genetotrophic Concept*. New York: McGraw-Hill.

Williams, R.J. and F.L. Siegel (1961) "'Propetology', A New Branch of Medical Science? Editorial," *The American Journal of Medicine* 3(31), pp. 325–327.

3 From the Concept of Genetic Disease to the Geneticization of Diseases

Analyzing and Solving the Paradox of Contemporary Medical Genetics

Marie Darrason

Abstract

In the sixties, the concept of genetic disease was synonymous with the concept of monogenic disease: a rare inherited Mendelian mutation in one gene leads to one disease. Since then, a double shift has occurred in medical genetics. On the one hand, the concept of genetic disease has extended far beyond the model of monogenic disease, to the point where it can virtually apply to every disease, a process called "the geneticization of diseases." On the other hand, the model of the simple monogenic disease has disintegrated and the distinction between monogenic diseases and polygenic ones has become more and more difficult to draw, to the point where there is no longer a consensual definition for the concept of genetic disease. This is what I call the "paradox of contemporary medical genetics": every disease is considered genetic but no one knows how to define the concept of genetic disease. In this chapter, my aim is threefold. First, I will describe the history and the establishment of this paradox. Second, I will examine three philosophical answers that have been suggested so far in order to solve it and argue they are deeply unsatisfactory. Third, I will suggest a new way to solve the paradox.

1 Introduction

Phenylketonuria, the symptoms of which were described in 1934 by the Norwegian physician Asbjørn Følling, is a metabolic disease due to a deficiency in the liver enzyme phenylalanine hydroxylase. The phenylalanine hydroxylase enzyme is necessary to convert phenylalanine, an essential amino acid found in food, into another amino acid, tyrosine. When this enzyme is mutated, phenylalanine cannot be converted in tyrosine and builds up in the blood, thus exerting a toxic effect on the central nervous system thereby leading to severe mental retardation. However, a simple phenylalanine-free diet, administered from birth, is sufficient to prevent the disease onset. In 1963, Robert Guthrie discovered a simple blood test that could detect this disease, and systemic phenylketonuria (PKU) screening in newborns was quickly made available in many countries, including the U.S. and France. Phenylketonuria was a hereditary disease, with a molecular explanation depending on a single gene (the phenylalanine hydroxylase [PAH] gene, which codes for the phenylalanine hydroxylase enzyme), a screening test, and a treatment; it quickly became a star disease of medical genetics (Lindee 2000; Lindee 2002; Paul 1994; Paul 2000; Paul 2013) in the 1960s. The phenylketonuria model of the concept of genetic disease can thus be described as follows: a Mendelian inherited monogenic mutation causes a deficiency in the corresponding mutated protein, leading to disease onset. This definition of genetic disease is perfectly suited to the central dogma of molecular biology and to the molecular concept of the gene, which developed around the same period. The central dogma of molecular biology states that there is a linear and specific correspondence between a sequence of nucleotides and the amino acids they code for. The molecular concept of the gene identifies the structure of a gene to its function, the code of the sequence of amino acids producing a protein.

However, since the 1960s, when phenylketonuria embodied a paradigmatic example of genetic disease, a double shift has occurred in medical genetics (Melendro-Oliver 2004): more and more diseases have been considered genetic, while the concept of genetic disease has become less and less clear.

Indeed, on the one hand, through several scientific discoveries, the concept of genetic disease has extended far beyond the concept of monogenic disease and has progressively characterized diseases that were considered nongenetic at the time. Several scientific discoveries have played a major part in this evolution. The discovery of susceptibility genes (genes that are associated with the occurrence of a disease but whose presence is not sufficient to cause it) in the 1970s and the discovery of oncogenes and anti-oncogenes (genes whose activation or repression plays a major part in the development of cancer) in the 1980s have drawn attention to the genetics of polygenic common diseases. At the same time, the rise of DNA sequencing and genetic engineering techniques has provided the means for unraveling the genetics of these polygenic common diseases. Indeed, these techniques allowed the development of various methods for identifying allelic variants, that is, for identifying a variation of one or more nucleotides in a gene. Progressively, every disease whose occurrence was associated to an allelic variant has been considered genetic, and nowadays, the concept of genetic disease thus applies to nonhereditary, non-Mendelian, common, complex, polygenic diseases such as cancer, diabetes, hypertension, and even tuberculosis.

On the other hand, several newly discovered mechanisms have disrupted our understanding of monogenic disease and blurred the distinction between simple monogenic diseases and those that are complex and polygenic. To name a few: allelic heterogeneity (several alleles can cause the same disease), genetic heterogeneity (several genes can cause the same disease), phenotypic heterogeneity (the same gene can cause different diseases), or modifier genes (several genes can influence the phenotype of a disease). The physiopathology of phenylketonuria has thus revealed several of these mechanisms (Scriver and Waters 1999; Scriver 1995; Scriver 2007), which have undermined the linear and specific correspondence between a mutation in the PAH gene, the production of a mutated PAH protein, and the occurrence of phenylketonuria, and which have called into question the apparent simplicity of monogenic diseases. At the same time, the discovery of alternative splicing (the same messenger RNA can be spliced into several mature RNA transcripts), transposable elements in the genome, and DNA regulatory sequences located at several hundred thousand base pairs upstream or downstream in the DNA sequence they control have undermined the molecular concept of the gene and the central dogma of molecular biology. All of these complexities convey a much more intricate picture of the relationship between genotype and phenotype and reinforce the resistance to the concept of a monogenic disease.

This is what I call "the paradox of contemporary medical genetics." On the one hand, the concept of genetic disease has extended far beyond the model of monogenic disease, to the point where it can virtually apply to every disease. On the other hand, the model of the simple monogenic disease can be applied to fewer and fewer diseases, and the distinction between monogenic diseases and polygenic ones has become more and more difficult to draw, to the point where there is no longer any consensus in the way we define genetic diseases. To put it differently, every disease is considered genetic, but no one knows what a genetic disease is anymore. There are two main issues at stake in this paradox: an epistemological issue and an ethical one.

The epistemological issue concerns whether it is legitimate to draw a distinction between genetic and nongenetic diseases and whether it is possible to unify the body of disparate knowledge about the genetics of diseases. This issue has been all the more important in the last ten years, as the completion of the Human Genome Project has caused an exponential rise in the number and variety of data acquired in human genetics. Although sequencing the first human genome took more than ten years and cost three billion dollars, it is now possible to complete the same genome sequence in three days for only a thousand dollars. New methods for identifying allelic variants in complex diseases have been elaborated, such as genome-wide association studies that allow the genomes of thousands of individuals to be compared. Acquired data are no longer limited to genomics (the functional study of the genome) but also concern interactions between the genome, the transcriptome (all RNA transcripts from one genome at a given time), and the proteome (all proteins translated from one genome at a given time). Interpreting these data in the postgenomic era and in medical genetics has become a more and more difficult task.

The ethical issue concerns how much genetic exceptionalism is justified and what ethical, legal and social impacts the extension of the concept of genetic disease would have. Genetic exceptionalism involves considering that genetic diseases should be treated separately from an ethical point of view because they raise specific ethical issues that are distinct from those raised by nongenetic diseases (Green and Botkin 2003; Ilkilic 2009; Murray 1997). For example, genetic diseases are often believed to raise some specific issues about prediction and reproduction. As such, in many countries, there are specific legislations about how and when genetic diseases should be detected. Genetic exceptionalism thus justifies the use of a specific legislation to organize the screening and treatment of genetic diseases. If we decide that there is no epistemological difference between genetic disease and nongenetic disease, can we still support genetic exceptionalism? Moreover, what would be the impact of classifying every disease as genetic? What impact would this have, from diagnosis to therapeutics, from individual representations of disease as some "genetic fate" (Dekeuwer 2006) to ethical, legal, and social issues such as medical secrecy, insurance companies, and the exponential development of genetic testing (Kevles and Hood 1992)?

In this manner, the philosophy of medicine has a twofold interest in solving this paradox of contemporary medical genetics. However, I will only address the epistemological issue for two reasons. First, dealing with both points of view would exceed the scope of this work by far. Second, I believe that solving the epistemological issue at stake is primary to solving the ethical issue. In this chapter, I aim to describe the history and the establishment of this paradox and to examine the philosophical answers that have been suggested so far in order to solve it.

First, I will describe the evolution of the Online Mendelian Inheritance on Man database, which is considered to be the reference classification for genetic diseases, in order to analyze the two strands of the paradox of contemporary medical genetics, namely the dissolution of the concept of genetic disease on the one hand and the geneticization of diseases on the other hand.

Second, I will discuss the most commonly used strategy in order to solve the paradox. This strategy usually consists of fighting against the geneticization of diseases while defending the idea that we should redefine the concept of genetic disease in a stricter way. But none of the usually suggested criteria provides a clear-cut redefinition of the concept of genetic disease or a way to distinguish between monogenic and polygenic diseases.

This is why, in conclusion, I claim that solving the paradox of contemporary medical genetics requires adopting the reverse strategy: instead of fighting geneticization and defending the concept of genetic disease, I suggest giving up the concept of genetic disease and taking the geneticization of diseases seriously. To be more specific, I make the hypothesis that the geneticization of diseases may reveal the establishment of regional genetic theories, that is, of neither trivial nor genocentrist explanations of the common causal role of genes in some classes of diseases.

2 Analyzing the Establishment of the Paradox Through the History of the Online Mendelian Inheritance on Man Database

In order to understand the paradox of contemporary medical genetics, it is useful to analyze how the classification of the concept of genetic disease has evolved through its history.

To begin, we should note that there is no official classification of genetic diseases that are acknowledged by worldwide scientific institutions. One might want to turn toward the International Classification of Diseases (ICD-10), but I argue that the ICD cannot be a relevant classification for genetic diseases, given its aims and history. Indeed, the ICD is first and foremost a statistical classification (Bowker and Star, 2000). It includes causes of mortality and diagnoses in morbidity and is primarily used for epidemiological and public-health purposes. As a consequence, genetic diseases are widely dispersed across several chapters of the ICD-10 (for instance, in the chapter on congenital diseases and chromosomal abnormalities but also in every anatomoclinical category). They do not constitute a specific etiological category, or an anatomoclinical category, or even an operational category within the ICD-10.

For this reason, I choose to analyze the Online Mendelian Inheritance on Man (OMIM) database. Created in 1967 in its paper version by Victor McKusick, "the father of medical genetics" (Collins 2008; Valle 2008), under the title "Mendelian Inheritance on Man," the catalog was originally dedicated to the nosology and nosography of three types of monogenic disorders: autosomal dominant phenotypes, autosomal recessive phenotypes, and X-linked phenotypes (McKusick 2007).

Over the years, the catalog went through several rounds of revision, a process overseen by the John Hopkins Institute. Since 1986, OMIM has become an online database that is now widely acknowledged by medical geneticists and is considered by most as the standard classification of genetic diseases (Amberger, Bocchini, and Hamosh 2011). Since its creation, I identify two main shifts in OMIM, which both seem especially relevant to understanding how our understanding of genetic diseases has evolved.

2.1 The Calling into Question of Mendelian Inheritance and the Extension of the Classification to Non-Mendelian Diseases

On the one hand, the database has progressively extended to include non-Mendelian diseases, while the very definition of Mendelian inheritance itself has been challenged. This change is reflected by the evolution of the numbering system in OMIM. In OMIM, each entry is given a six-digit number, the first number corresponding to a category of phenotypes (see Figure 3.1). Until 1994, only Mendelian phenotypes were referenced in OMIM (categories 1, 2, and 3, corresponding to "autosomal dominant phenotypes," "autosomal recessive phenotypes," and "X-linked phenotypes," respectively). In 1994, new categories were created in order to include non-Mendelian diseases (such as Y-linked and mitochondrial loci or phenotypes). Moreover, a new category called "autosomal loci" or "phenotypes" was created in order to replace the previous categories 1 and 2. This substitution of categories acknowledges that dominance and recessivity (which are very important concepts of Mendelian inheritance) are relative rather than absolute notions. More precisely, whether a given trait is dominant or recessive depends on the phenotypic level at which the trait is observed

1	Autosomal dominant loci or phenotypes (before 1994)
2	Autosomal recessive loci or phenotypes (before 1994)
3	X-linked loci or phenotypes
4	Y-linked loci or phenotypes (category created in 1994)
5	Mitochondrial loci or phenotypes (category created in 1994)
6	Autosomal loci or phenotypes (category created in 1994)

Figure 3.1 Evolution of the numbering system of OMIM. In OMIM, each phenotype is given a six-digit number. The first number designates the category to which the given phenotype belongs. In 1994, three new categories were created to acknowledge that dominance and recessivity are relative concepts and include non-Mendelian diseases.

(McKusick 2007). For instance, sickle-cell anemia is usually considered a recessive trait at the genotypic level: most of the time, a heterozygous individual (HbS/HbA) does not display any symptoms of the disease. But at the level of proteins, a heterozygous genotype leads to the production of a proportion of abnormal hemoglobin, thus explaining why, under certain circumstances (such as high altitude, the course of an infection, etc.), some heterozygous individuals can display mild symptoms of sickle-cell anemia, as if the trait was dominant. The quite simple model of Mendelian inheritance has thus been challenged at its core.

2.2 From a Classification of Diseases to a Classification of Allelic Variants

At the same time, the aim of the database is no longer restricted to the classification and the nomenclature of diseases: OMIM does not only list phenotypes, it also lists every allelic variant involved in a given disease. This major change is reflected by the change of names of the catalog, previously called "the catalog of Mendelian disorders" and called "the catalog of human genes and genetic disorders" since 1994. The purpose of this separation between the classification of phenotypes and the classification of allelic variants was to take into account three major issues for the concept of monogenic diseases (one allele, one gene, one phenotype, one disease): *genetic heterogeneity* (one phenotype, several genes); *allelic heterogeneity* (several alleles, one disease); and *phenotypic heterogeneity* or *pleiotropy* (one gene, several phenotypes). These three phenomena have completely called into question the apparent simplicity of the concept of monogenic disease, as is shown in the case of phenylketonuria. Even if the PAH gene is considered as the determinant of the disease, the disease exhibits allelic heterogeneity (more than 500 distinct mutations in the PAH gene can be involved), genetic heterogeneity (other genes, such as the BH4 gene can cause the disease), and phenotypic heterogeneity (the same mutation can lead to different phenotypes).

Eventually, Mendelian disorders came to be considered complex diseases (Badano and Katsanis 2002), the previously sharp frontier between monogenic

Mendelian disorders and polygenic Mendelian disorders has blurred (Dipple and McCabe 2000a; Dipple and McCabe 2000b), and every allelic variant that is involved in a given disease is included in OMIM, leading to the idea of a "genetic continuum" between monogenic and polygenic diseases:

> The distinction between "single" gene disorders ("simple" Mendelian diseases) and multiple gene disorders (complex traits) has become less clear as it has been recognized to be artificial and a representation of human perception rather than biological reality. We would argue that for a so-called single gene disorder, there is one gene that may be primarily responsible for the pathogenesis, with one or more independently inherited modifier genes that influence the phenotype. On the other hand, for a complex trait, the primacy of any individual gene is not perceptible, and the interaction of two or more independently inherited pairs of alleles, most likely influenced by additional modifier genes, results in the disease. *The consequence of this conceptual framework is that there is no such thing as a "single" gene disorder.* In other words, there is no obvious clear distinction between simple Mendelian and complex traits: *genetic diseases represent a continuum* with diminishing influence from a single primary gene influenced by modifier genes, to increasingly shared influence by multiple genes.
>
> (Dipple and McCabe 2000a, 47; my italics)

This frontier between monogenic diseases and polygenic diseases is all the more blurred now that more and more allelic variants are considered to be involved in many common disorders such as diabetes, arterial hypertension or schizophrenia. In other words, the evolution of OMIM from a classification of diseases to a classification of allelic variants serves a double purpose: it aims at integrating the complexities of Mendelian disorders, while acknowledging that common disorders (which had until then been considered nongenetic) are also influenced by genetic factors.

2.3 The Paradox of Contemporary Medical Genetics

What can we learn from the history of OMIM? Since 1994, two major shifts have happened: the online database has been extended to include non-Mendelian diseases, and it has evolved from being a classification of diseases and phenotypes to a classification of allelic variants. I argue that these two major shifts in the history of OMIM clearly demonstrate that the concept of genetic disease has been in crisis; it has extended to the point where it virtually encompasses every disease in some kind of "genetic continuum of disease" (every disease is genetic but some disease are considered "more genetic than others"). That is what I call the paradox of contemporary medical genetics: every disease[1] is considered genetic but there is no consensual definition of the concept of genetic disease.

This ambiguity is reflected by the term *geneticization of disease*. This term is mostly used in a neutral way to describe the fact that every disease is linked to some allelic variant. However, it was originally defined by Abby Lippman as "an ongoing process by which differences between individuals are reduced to their DNA codes, with most physical and behavioural diseases defined at least in part as genetic in origin. It refers as well to the process by which intervention employing genetic technologies are adopted to manage problems of health" (Lippman 1994, 13). Thus, the term *geneticization* originally involves an inherently negative connotation (Lippman 1991; Lippman 1994), since it aims at denouncing an illegitimate extension of the concept of genetic disease. In the following, I will use the term *geneticization* in its most neutral sense: "One solution is to adopt a 'stripped down' definition of geneticization, stating simply that, in medicine, geneticization takes place when a condition is linked to a specific stretch of DNA" (Hedgecoe 2002, 8). However, the ambiguities of this term underline the question at the heart of the paradox: is the extension of the concept of genetic disease legitimate, and what does it mean to say that every disease is genetic?

3 Framing the Paradox as an Instance of the Causal Selection Problem and Failing To Solve It

So, how do philosophers deal with these two intricate issues, namely the definition of the concept of genetic disease and the geneticization of diseases? A common strategy for addressing these questions, shared by several philosophers (Hesslow 1984; Gifford 1990; Smith 2007; Hull 1979), is to begin by approaching the project of defining the concept of genetic disease as an instance of the causal selection problem. The causal selection problem consists in picking out the main cause of an event occurring in a multicausal context and in determining in virtue of which criterion a cause is considered as the main cause of this event (Hesslow 1984; Hesslow 1988; Mackie 1965; Sober 1988). In his well-known paper, Mackie takes the example of a fire in a house: in virtue of which criterion do the experts conclude that an electrical short-circuit is the cause of the fire? The fire could never have broken out without some inflammable material or without some air, and there would not have been an electrical short-circuit if Mary had repaired the electrical wire in the pantry. In other words, this means that when we say that "A is the cause of B," we consider A as the most important cause of B, thus choosing to ignore some other causes. The causal selection problem therefore consists in determining how and why we can pick a cause as the most important cause of a phenomenon.

Applied to the problem of genetic disease, this means that labeling a disease *genetic* would imply that genes are the most important cause in disease explanation:

> It is evident that in most cases of genetic disease the term's application reflects a choice having been made among the causal factors to emphasize the genetic component and deemphasize the environmental component.
> (Hull 1979, 78)

There are genetic and nongenetic factors that are causally relevant to every trait, a fact recognized by virtually all commentators on the concept of genetic disease [...]. So the real issue in deciding that something is a genetic disease, is whether the causal factors that are genetic are the most important causes. I will call this the "selection problem." How do we decide whether genetic factors or environmental factors are more important in the production of various diseases?

(Magnus 2004, 234)

If the concept of genetic disease is understood in the context of the causal selection problem, geneticization can then be understood as an expansion of the concept of genetic disease to all diseases: the geneticization of diseases would amount to stating that genes are the most important causes for every disease. In this case, geneticization would amount to an acceptance of genocentrism—the belief that genes are the most important causal factor in explaining any biological phenomenon. Genocentrism, however, has already been heavily criticized (Hull 1979; Gifford 1990; Magnus 2004; Smith 2007). I will not review here the numerous arguments against genocentrism. It is enough to say that genocentrism seems to be both scientifically unjustified and ethically questionable. Since geneticization, on this causal selection understanding of the term, can be identified with genocentrism, it follows that geneticization cannot be an acceptable approach to disease. These philosophers, therefore (Hesslow 1984; Gifford 1990; Smith 2007; Hull 1979), reject geneticization on the one hand and, on the other hand, attempt to give a more restricted account of genetic disease that would address the causal selection problem without leading to the pervasive geneticization of disease:

It is crucial, therefore, that causal selection always be done on the basis of an explicit criterion which has been critically evaluated. If we identify a disease as genetic, we should be able to give a clear account of precisely what this entails and why such a description gains more through concise transmission of important causal information than it loses through incompleteness. Unfortunately, this is almost never the case in current descriptions of genetic traits. Typically, there is no explicit criterion at all. When criteria are indicated or can be inferred, they are often so simplistic that the explanations they produce distort more than they inform.

(Smith 2007, 84)

Thus, these philosophers (Hesslow 1984; Gifford 1990; Smith 2007; Hull 1979) usually attempt to define the concept of genetic disease strictly in an attempt to distinguish between diseases that are "true" genetic diseases (usually the Mendelian monogenic diseases), where genes can be said to be the most important cause of the disease; and diseases where gene–environment interactions are more difficult to assess (usually the polygenic disorders).

At this point, if the general two-step method for resolving the paradox of contemporary medical genetics is common (1. Reject geneticization of diseases as genocentrism and 2. Redefine the concept of genetic disease in a more strict way), I identify three main approaches to justify the causal priority of genes in the "true" genetic diseases. I will now describe these three variations of the common strategy to solve the paradox of contemporary medical genetics and then demonstrate how each of these approaches fails to solve the paradox.

3.1 The Cause–Condition Approach

I call the first approach the "cause–condition approach." Its aim is to define genetic diseases as diseases for which genetic factors can be considered as the cause of the disease, while other environmental determinants are considered as simple background conditions (Wulff 1984). In order to distinguish cases where genetic factors are the cause of the disease from cases where genetic factors are simple background conditions, different criteria have been suggested, such as necessity, sufficiency, and manipulability. However, every suggested criterion fails to justify this distinction between causes and background conditions.

Indeed, necessity can be quickly excluded as a good criterion for at least two reasons. First, precisely since the dissolution of the classical concept of genetic disease, there are now many examples of so-called monogenic diseases for which penetrance is incomplete and for which having the gene does not necessarily mean you will have the disease. Second, even if there were some clear-cut cases of such diseases, we would still need to justify in which sense having this genetic cause is more important for the disease to occur than other necessary causes. For example, in the case of sickle-cell disease, the disease occurs only if a certain threshold of mutated protein is reached, and such a threshold does not depend only on the mutation of the gene that codes for this protein. This protein threshold is as much a necessary condition as the genetic mutation, so why should the genetic cause be considered the most important one?

The manipulability account, according to which a disease is genetic if the gene is the most manipulable cause, is also quite easy to discard. First, since "genetic therapy" is still a rarity, we should discuss whether the manipulability account is about manipulability in theory or in practice. Second, even if we admit that genetic therapy will become a widely used cure, and thus consider that the gene will be a manipulable cause of disease, it does not ensure that it will be the most manipulable cause, as the case of phenylketonuria shows.

Finally, the sufficiency account clearly raises some difficult issues for cases such as phenylketonuria: the PAH gene cannot be considered to be sufficient for phenylketonuria to occur, since the disease would never happen in a phenylalanine-free environment. In this kind of example, the only way to consider that the PAH gene is a cause while the phenylalanine-free environment is a background condition would be to consider that the PAH gene is sufficient not *per se*, but relatively to a certain population of contrast living in a fixed environment, thus considered as *a ceteris paribus* condition. This kind of difficulty—that is, the need to consider genetic traits in a population rather than genetic diseases in an individual—naturally leads to the second approach, which I call the "differentialist approach."

3.2 The Differentialist Approach

The differentialist approach does not seek to explain the distinction between genetic and nongenetic diseases, but rather to explain how genes are responsible for the differences between healthy and sick populations (Gifford 1990; Smith 1992; Smith 2001; Smith 2007; Waters 2007). This strategy does not define a concept of genetic disease *per se*; it consists in defining genetic diseases relatively to a given population of contrast where genetic differences are considered to be the cause of phenotypic differences. The issue at stake in this strategy is then to find nonarbitrary criteria to justify the choice of the population of contrast. There are more or less detailed accounts of this approach; Kelly Smith describes at length one of the most complex versions of it, which he called his "epidemiological approach" (Smith 2007).

However, regardless of the complexity of the criteria chosen to justify the choice of the population of contrast, the differentialist approach meets the same shortcomings over and over again. The first difficulty is that every differentialist approach forbids any assertion on individual cases unless some causal information at the population level is imported. In other words, when looking at a specific population, I will be able to say that, for example, in this population, this given gene explains the differences between healthy and sick people and that this specific disease is genetic relatively to this specific population living in this given environment. However, I could not tell anything about a specific patient without relying on population-level information.

The second difficulty is a direct consequence of the differentialist approach: adopting a differentialist approach means that there is no genetic disease *per se*, only genetic diseases in a given population. In this kind of account, lactose intolerance would be considered a genetic disease in the U.S., where milk consumption is frequent, while the genetic mutation is quite rare. However, lactose intolerance would be considered an environmental disease in Asia, where the situation is the reverse. The difficulty that I am pointing out here is not that there is no absolute distinction between genetic and nongenetic diseases. The problem is that there is no absolute criterion to justify choosing this population over this one as a population of contrast. Let's take an example suggested by David Magnus: let's assume that half a village experiences abdominal pain, diarrhea, and severe dehydration. In order to understand what causes these symptoms, medical experts are sent there, and they discover that the water has been tainted. It could seem pretty obvious that the cause of the disease is the parasite that has tainted the water. However, let's assume that everyone in the village drinks the same water and that only half the population of the village develops the disease; in such a case, there is every reason to consider that the environment is the *ceteris paribus* condition and that there is a gene that makes one half of the village more sensitive to the parasite than the other. So, depending on how we choose our population of contrast in this example, the disease could be considered genetic or not (Magnus 2004). To put it differently, there is no way to embrace the differentialist approach without concluding that there is no absolute way to distinguish between genetic and nongenetic diseases and thus no way to resolve the causal selection problem. Defining a disease as genetic relatively to a population means that it is necessary to find nonarbitrary criteria in order to justify the choice of the population of contrast. Since such criteria remain to be found, the differentialist approach naturally leads to the third strategy, which I call the "pragmatic approach."

3.3 The Pragmatic Approach

"The pragmatic approach" (Gannett 1999) goes beyond the differentialist approach; rather than seeking to justify the choice of a population of contrast, it admits that such a choice is dependent on the context of explanation. Thus, this approach acknowledges that the concept of genetic disease cannot be defined, even if genetic explanations can be relevant depending on the context in which we are interested in this kind of explanation:

> To argue, as I have, that traits are designated "genetic" for pragmatic reasons is not to deny that genes are causally efficacious agents. We can speak sensibly about genetic causes and their effects, using either deterministic or probabilistic language, provided we recognize that we do so only relative to a particular set of background conditions, a specific population, and the present state of knowledge. What I do deny is that terms such as "genetic trait," "genetic disease," and "genes for," are objective, if we understand "objective" to mean devoid of pragmatic content. I contend that how the cause–condition distinction is drawn, what population is selected, and which paths of research are followed, are choices that are influenced by the aims, interests, and orientations of those who make them. By appreciating the pragmatic dimensions of genetic explanations, we are forced to recognize their contingency and the need to interrogate the desires that shape the focus on genetic causes.
>
> (Gannett 1999, 370–371)

In other words, the pragmatic approach provides a practical solution to the causal selection problem (the solution always depends on our explanatory context and interests). But it also means that the distinction between genetic disease and nongenetic disease is never absolute, not even relative to a given population, but relative to our epistemic interests. In a given context, for example, for a research center that is trying to do some fundraising, labeling a disease (let's say lung cancer) "genetic" will be interesting, while in other contexts, such as when a state promotes some health messages about tobacco, labeling the same disease nongenetic would help to insist on individual responsibility toward health and disease.

4 Shifting Paradigms: From Geneticization to the Hypothesis of a Genetic Theory

4.1 The Concept of Genetic Disease Should Be Abandoned

The common strategy behind the three approaches discussed in Section 3 was to first denounce the geneticization of diseases as an illegitimate genocentrist extension of the concept of genetic disease, and then to redefine the concept of genetic disease in a stricter way in order to firmly distinguish between genetic and nongenetic diseases. However, I argued in the previous section that each of these approaches fails to provide a satisfactory account of the concept of genetic disease. I suspect that this repeated failure is linked to an outdated attachment to the concept of genetic disease and to the way philosophers frame the paradox of contemporary medical genetics as an instance of the causal selection problem. In other words, since the common strategy fails to reject the geneticization of diseases and to establish a firm delineation between genetic and nongenetic diseases, why not adopt the reverse strategy and let go of the concept of genetic disease?

Indeed, let's discuss why we would need the concept of genetic disease in the first place. In Section 1, I discussed at length how the simple and strong model of monogenic disease has progressively vanished and is now completely outdated from a scientific point of view. I also claimed that, from a philosophical point of view, there is no way to defend a concept of genetic disease *per se*. Some might worry about the social and ethical consequences of letting go of the concept of genetic diseases; for example, the concept of "genetic disease" has clearly been pivotal for raising money or for public awareness. However, in such cases, many other concepts, such as "orphan diseases," for example, can probably do as well and may avoid some confusion carried by the term "genetic disease." Indeed, other philosophers underline that the focus on a gene-centered approach of bioethical discourse could be an impediment to a more fine-grained evaluation of the issues raised by the geneticization of diseases and their impact on health and disease management (Kakuk 2008). For such authors, the bioethical discourse is mistaken in framing the philosophical issues raised by genetic information in our society by relying on the monogenic concept of genetic disease; this would amount to ignoring that most diseases are the product of a complex interaction between genes and environment, raising different issues, such as how to make ethical choices under scientific uncertainty or how to pick up the relevant information in our genome, for example. Thus, there is no social or ethical argument that should prevent us from abandoning the concept of genetic disease.

4.2 The Causal Selection Problem Should Be Abandoned

If we abandon the concept of genetic disease, this means that the entire strategy that consists in framing the paradox of contemporary medical genetics as an instance of the problem of causal selection is misguided. Why do philosophers seem to be so interested in solving the causal selection problem when considering the involvement of genes in diseases?

First, I believe that solving the causal selection problem always seems appealing in medicine because of the implicit belief that identifying the main cause provides a direct way to manipulate and find a treatment for the disease. However, solving the causal selection problem in disease explanation, that is, identifying a cause considered as the most important one in a multicausal event, may not prove as useful in medicine as one may think, for at least three reasons. First, the supposedly main cause of a disease is not necessarily the best therapeutic target. For example, in the case of phenylketonuria, intervening on the gene is less efficient (and much more complicated) than prescribing a phenylalanine-free diet. Second, the same disease can have multiple causes that are independent from each other. For example, most cases of gastric or duodenal ulcers are caused by a bacterium called *Helicobacter pylori*. But some cases are the result of a simple excess of gastric acid secretion. So, for the same disease, the cause of the disease can differ from a patient to another. Third, the same disease can have multiple and interacting causes, in which case, curing the main cause does not mean that the other factors have been eradicated. To get back to the example of the gastric ulcer, it is believed nowadays that there is a link between *Helicobacter pylori* and an excessive secretion of gastric acidity. This means that in some cases, both causes play a role in the occurrence of the disease and that getting rid of either cause is not sufficient to get rid of the disease. This is why a gastric ulcer is nowadays treated both by a combination of antibiotics against *Helicobacter pylori* and by proton pump inhibitors against excessive acid secretion.

Moreover, I argue that solving the causal selection problem would still not provide a full explanation of disease. Knowing that the PAH gene causes phenylketonuria does not tell us why some patients with the same mutations display mild symptoms or how to regulate the phenylalanine-free diet in patients. In the end, solving the causal selection problem tends to oversimplify the causal matrix of disease that is reduced to a binary opposition between genes and environment. This oversimplification is all the more problematic given that genes and environment are often poorly defined. On the gene side, chromosome abnormalities that you may find in Down's syndrome and that imply the deletion or the addition of thousands of genes are often put on the same level as allelic mutations in one allelic variant. As for the environment, it often designates everything which is not the gene, including the internal environment as well as the external one, epigenetics, socioeconomic background, stochastic cellular process, and so on and so forth. Not only is this binary opposition crude in the depiction of each component, but it also completely denies the existence and the importance of the complex interactions between different genes (gene–gene interactions) or between genes and environment (gene–environment interactions) (Tabery 2007; Tabery 2009). For all these reasons, I strongly suggest that the causal selection problem should be abandoned.

4.3   Taking Geneticization Seriously: The Hypothesis of a Genetic Theory

I thus propose the reverse strategy: first, I suggest that we stop trying to solve the causal selection problem in disease explanation. Second, since there seems to be little point in saving the concept of genetic disease, perhaps one can salvage the concept of geneticization. Of course, I do not mean "geneticization" understood as essentially equivalent to genocentrism, which, as I noted, is subject to significant objections. I suggest a meaningful interpretation of geneticization that bypasses the issues of causal selection. Rather than interpreting geneticization as a misguided expansion of the concept of genetic disease, I propose that geneticization be understood as an epistemic reorganization of our disparate knowledge about the genetic side of diseases. To put it differently, I make the hypothesis that the geneticization of diseases in the contemporary biomedical literature could reveal the progressive development at play of one or several explanations for the common role of genes in diseases, and I call such explanations "genetic theories of disease."

I have suggested elsewhere that such general explanations for the role of genes in diseases could take several forms (Darrason 2013). Among the spectrum of possible genetic theories, two particular kinds of general genetic theories can be considered: what I call "a genetic *theory of diseases*" that would be a heterogeneous set of regional genetic theories, each one being specific of a disease class (Figure 3.2); or what I call "a genetic theory of *Disease*" that would rely on a definition of the common role of genes in disease and that might possibly renew the way we classify diseases (Figure 3.3).

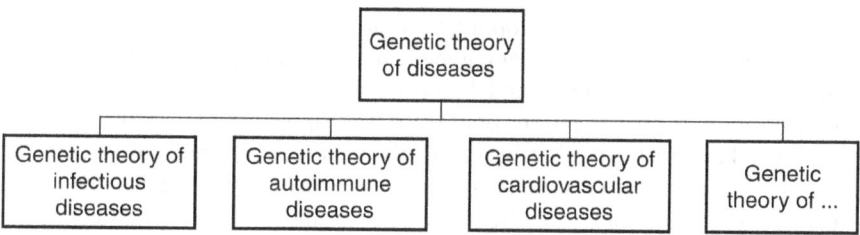

Figure 3.2 Typical representation of a genetic theory of diseases. The genetic theory of diseases is a set of regional genetic theories. For each category of diseases, there is a specific genetic theory with specific mechanisms. Genetic mechanisms may differ for each class of diseases. This kind of theory does not change the way we classify diseases (Darrason 2013, 339).

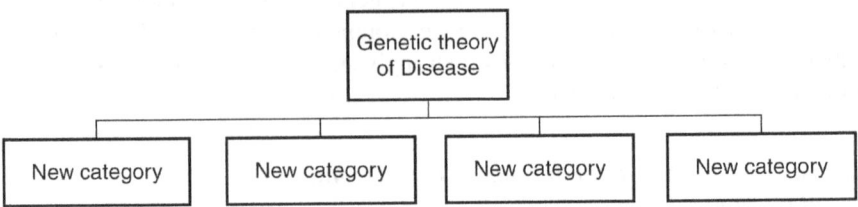

Figure 3.3 Typical representation of a genetic theory of Disease. In a genetic theory of Disease, we may expect a genetic definition of Disease in general. Depending on this definition, some classificatory principles would appear and these principles would likely renew the way we classify diseases (Darrason 2013).

In the first kind of genetic theory, namely the genetic *theory of diseases*, every disease class, as we nowadays know them, would admit its own genetic theory; for example, cardiovascular diseases would have their own genetic theory about the common role of genes in cardiovascular diseases. The main principles or mechanisms at play in each genetic theory could differ or not for each disease class. In the second kind, the genetic *theory of Disease*, a general definition of the role of genes in disease could be suggested. In such a theory, important changes in the way we classify diseases may be expected, since the molecular definition of the role of genes in disease in general could completely dissolve our anatomoclinical division of diseases. While this description of two particular forms of genetic theories may seem quite speculative, I argue that at least one example of each of these genetic theories can be found in the contemporary biomedical literature.

On the one hand, the genetic theory of infectious diseases that has been developed by two French geneticists, Jean-Louis Casanova and Laurent Abel, may be considered as an example of a genetic regional theory of diseases (Abel and Casanova 2000; Casanova and Abel 2013; Alcaïs, Abel, and Casanova 2010; Casanova and Abel 2007). Indeed, its aim is to explain the clinical intervariability to infectious diseases, that is, why only a small fraction of individuals that are infected by *Mycobacterium tuberculosis* actually develops the disease. They identify four common genetic mechanisms (Mendelian predisposition to multiple infections, Mendelian predisposition to one infection, major gene/resistance to one infection, polygenic predisposition to one or multiple infections) that can be at play in and explain the development of infectious diseases (Darrason 2013). Obviously, in such a theory, there is no place for genocentrism: genes cannot be the main cause or the most important cause of diseases that required the intervention of an external agent (the vector of the infection). But Abel and Casanova offer a general background to understand the common yet diverse role of genes in the explanation of a specific disease class, infectious diseases, thus providing us with an example of a successful genetic theory of diseases (on this theme, see also Liu, Love, and Travisano, this volume).

On the other hand, network medicine is a new research field, developed by the proponents of systems medicine and biology (Barabási, Gulbahce, and Loscalzo 2011; Loscalzo, Kohane, and Barabasi 2007; Loscalzo and Barabási 2011). It offers some interesting perspectives for a genetic theory of disease (Darrason, 2015), since it relies on a general definition of disease which can be instantiated for virtually every single disease: "A disorder then represents the perturbation or breakdown of a specific functional module caused by variation in one or more of the components producing recognizable developmental and/or physiological abnormalities" (Goh et al. 2007, 8688). Each disease is then defined by a specific disease module, composed of four main types of factors: the primary disease genome, the secondary disease genome, the intermediate phenotype, and the environmental determinants. In the example of phenylketonuria, the primary genome would be the PAH gene, which codes for the phenylalanine hydroxylase enzyme. The secondary genome would include modifier genes such as the BH4 gene. The intermediate phenotype would include all the intermediate physiopathological states, such as phenylalanine accumulation, tyrosin deficiency, demyelinisation, and so on and so forth. The environmental determinants would be diet types, intellectual stimulation, and so on. In such a theory, not only it is possible to understand the role of a given gene in a given disease through its role in the disease module, but it is also possible to understand how diseases are not distinct entities but have intertwined physiopathology and modules, depending on the genes they share. It is precisely the aim of the diseaseome, a tool developed by the proponents of network medicine, in order to represent two networks, which are the two faces of the same coin (Goh et al. 2007). On the one hand, the human disease network represents diseases as nodes and two diseases are connected to each other if they share the same gene in their physiopathology. On the other hand, the human gene network represents genes as nodes, and two genes are connected to each other if they are involved in the same disease. Such an analysis can explain how obesity and diabetes are intertwined entities by underlining the genetic components they share. Thus, if network medicine is still far from an exhaustive and detailed genetic theory of disease, it is definitely a big step in this direction.

5 Conclusion

While the concept of genetic disease was well defined in the 1960s and was clearly distinct from nongenetic diseases, the last 50 years have witnessed two major shifts in medical genetics. On the one hand, the concept of monogenic disease was called into question, while on the other hand, many allelic variants were discovered to be involved in common disorders. This results in what I have called the "paradox of contemporary medical genetics": every disease is considered as genetic, but no one knows how to define a genetic disease.

To address both the dissolution of the concept of genetic disease and the geneticization of diseases, most philosophers framed this paradox as an instance of the causal selection problem: they denounce the geneticization of diseases as a misguided gene-centered extension of the concept of genetic disease, and they attempt to find a restricted definition of genetic disease for which the gene is the main cause of the disease. However, none of them achieves a firm distinction between genetic and nongenetic diseases nor provides a definition of the concept of genetic disease *per se*. In light of these failures, I suggest a different way of tackling the paradox: instead of fighting the geneticization of diseases and defending the concept of genetic disease, I strongly advocate that we choose the reverse strategy, and I make the hypothesis that the geneticization of diseases may reveal the current elaboration of genetic theories of diseases—that is, of unified explanations of the involvement of genes in diseases—in the contemporary biomedical literature. I identify two specific forms of such theories, namely a genetic theory of diseases and a genetic theory of disease, and I point out that examples of such theories, namely the genetic theory of infectious diseases and network medicine, can already be found in the contemporary biomedical literature. I thus strongly believe that the future of molecular medicine lies in the development of a plurality of genetic theories.

Acknowledgments

This chapter heavily relies on the first part of my PhD dissertation, titled "Is there a genetic theory of disease?" and written under the supervision of Jean Gayon at Université Paris 1 Panthéon Sorbonne. I am therefore very grateful to my dissertation committee for helping me to improve my initial work on this topic. I also want to thank the members of my research department, the IHPST (Institut d'Histoire et de Philosophie des Sciences et des Techniques). Finally, I want to address my warmest thanks to both editors (Giovanni Boniolo and Marco Nathan) for allowing me to take part in this volume and for their very enthusiastic and helpful comments.

Note

1   To be perfectly clear, for practical purposes, the term *disease* here refers to every disease compiled in the International Classification of Diseases in its latest version. This therefore includes some controversial borderline cases, such as injuries or poisoning cases that are, however, considered diseases in the ICD-10. Some authors may object that genes cannot play a role in such diseases, and that therefore, I cannot argue that "every disease is genetic." There are three ways to answer this objection. First, I could say that in order to state the paradox, I do not need to state that every disease is considered genetic—I merely have to demonstrate that most diseases are considered so. Second, even in borderline cases, I believe we could discuss the implications of genes as difference-makers in the occurrence of the disease. For example, for the same kind of fall, it is possible that patient A does not have a broken arm while patient B does, because patient B has genes that make his bones frailer than patient A's bones, or vice-versa. Similarly, poisoning could be more or less lethal depending on some individual's genetic make-up. The third way to answer this objection is to acknowledge that my second answer is a trivial interpretation of the geneticization of diseases,

one that is made most of the time in the biomedical literature but that I do not defend *per se*. I do believe that genes play a role in every disease and that this common role can be formalized (see Section 4 of this paper), but I do not believe that this role is going to be the same from one disease to another.

References

Abel, L., and J.L. Casanova (2000) "Genetic predisposition to clinical tuberculosis: Bridging the gap between simple and complex inheritance," *American Journal of Human Genetics* 67(2), p. 274.

Alcaïs, A., L. Abel, and J.L. Casanova (2010) "Human genetics of infectious diseases." In *Vogel and Motulsky's Human Genetics—Problems and Approaches*, edited by M. Speicher, S.E. Antonarakis, and A.G. Motulsky, Fourth Edition, pp. 403–16. Heidelberg; New York: Springer.

Amberger, J., C. Bocchini, and A. Hamosh (2011) "A new face and new challenges for Online Mendelian Inheritance in Man (OMIM®)," *Human Mutation* 32(5), pp. 564–67. doi:10.1002/humu.21466.

Badano, J.L., and N. Katsanis (2002) "Beyond Mendel: An evolving view of human genetic disease transmission," *Nature Reviews Genetics* 3(10), pp. 779–89.

Barabási, A.L., N. Gulbahce, and J. Loscalzo (2011) "Network medicine: A network-based approach to human disease," *Nature Reviews Genetics* 12(1), pp. 56–68. doi:10.1038/nrg2918.

Bowker, G.C., and S.L. Star (2000) *Sorting Things Out: Classification and Its Consequences*. Cambridge, Mass.: MIT Press.

Casanova, J.L., and L. Abel (2007) "Human genetics of infectious diseases: A unified theory," *EMBO Journal* 26(4), pp. 915–22.

Casanova, J.L., and L. Abel (2013) "The genetic theory of infectious diseases: A brief history and selected illustrations," *Annual Review of Genomics and Human Genetics* 14(1), pp. 215–43. doi:10.1146/annurev-genom-091212-153448.

Collins, F.S. (2008) "Victor A. McKusick (1921–2008)," *Science* 321(5891), pp. 925–25.

Darrason, M. (2013) "Unifying diseases from a genetic point of view: The example of the genetic theory of infectious diseases," *Theoretical Medicine and Bioethics* 34(4), pp. 327–344. doi:10.1007/s11017-013-9260-6.

Darrason, M. (2015) "Mechanistic and topological explanations in medicine: The case of medical genetics and network medicine," *Synthese*. doi:10.1007/s11229-015-0983-y.

Dekeuwer, C. (2006) "Liberté de choix et destins individuels. Examen des concepts et des problèmes éthiques impliqués par la médecine prédictive." Doctorat de philosophie, Paris, France: Université Panthéon-Sorbonne.

Dipple, K.M., and E.R.B. McCabe (2000a) "Modifier genes convert 'simple' Mendelian disorders to complex traits." *Molecular Genetics and Metabolism* 71(1–2), pp. 43–50. doi:10.1006/mgme.2000.3052.

Dipple, K.M., and E.R.B. McCabe (2000b) "Phenotypes of patients with 'simple' Mendelian disorders are complex traits: Thresholds, modifiers, and systems dynamics," *American Journal of Human Genetics* 66(6), pp. 1729–35.

Gannett, L. (1999) "What's in a cause?: The pragmatic dimensions of genetic explanations," *Biology and Philosophy* 14(3), pp. 349–73.

Gifford, F. (1990) "Genetic traits," *Biology and Philosophy* 5(3), pp. 327–47.

Goh, K.I., M.E. Cusick, D. Valle, B. Childs, M. Vidal, and A.L. Barabási. 2007. "The human disease network," *Proceedings of the National Academy of Sciences* 104(21), pp. 8685–90.

Green, M.J., and J.R. Botkin (2003) "'Genetic exceptionalism' in medicine: Clarifying the differences between genetic and nongenetic tests," *Annals of Internal Medicine* 138(7), pp. 571–75.

Hedgecoe, A.M. (2002). "Reinventing diabetes: Classification, division and the geneticization of disease," *New Genetics and Society* 21(1), pp. 7–27.

Hesslow, G. (1984) "What is a genetic disease?" In *Health, Disease, and Causal Explanations in Medicine*, edited by L. Nordenfelt and B.I.B. Lindahl, 183–93. Philosophy and Medicine; v. 16. Dordrecht, Pays-Bas: D. Reidel publ. Co.

Hesslow, G. (1988) "The problem of causal selection." In *Contemporary Science and Natural Explanation: Commonsense Conceptions of Causality*, edited by D.J. Hilton, pp. 11–31. Brighton: Harverster Press.

Hull, R.T. (1979) "Why 'genetic disease'?" In *Genetic Counseling: Facts, Values, and Norms*, by National Foundation, edited by A.M. Capron, M.Lappé, R.F. Murray, T.M. Powledge, S.B. Twiss, and D. Bergsma. Birth Defects Original Article Series, v. 15, no. 2. New York: A. R. Liss.

Ilkilic, I. (2009) "Coming to grips with genetic exceptionalism: Roots and reach of an explanatory model," *Medicine Studies* 1(2), pp. 131–42. doi:10.1007/s12376-009-0015-7.

Kakuk, P. (2008) "Gene concepts and genethics: Beyond exceptionalism," *Science and Engineering Ethics* 14(3), pp. 357–75. doi:10.1007/s11948-008-9056-7.

Kevles, D.J. and L. Hood (1992) *The Code of Codes: Scientific and Social Issues in the Human Genome Project*. Harvard: Harvard University Press.

Lindee, M.S. (2000) "Genetic disease since 1945," *Nature Reviews Genetics* 1(3), pp. 236–41. doi:10.1038/35042097.

Lindee, M.S. (2002) "Genetic disease in the 1960s: A structural revolution," *American Journal of Medical Genetics* 115(2), pp. 75–82. doi:10.1002/ajmg.10541.

Lippman, A. (1991) "Prenatal genetic testing and screening: Constructing needs and reinforcing inequalities," *American Journal of Law and Medicine* 17(1–2), pp. 11–50.

Lippman, A. (1994) "The genetic construction of prenatal testing: Choice, consent or conformity for women," In *Women and Prenatal Testing: Facing the Challenges of Genetic Testing*, edited by K.H. Rosenberg and E.J. Thomson, pp. 9–34. Columbus: Ohio State University.

Loscalzo, J., and A.L. Barabási (2011) "Systems biology and the future of medicine," *Wiley Interdisciplinary Reviews: Systems Biology and Medicine* 3(6), pp. 619–27.

Loscalzo, J., I. Kohane, and A.L. Barabási (2007) "Human disease classification in the postgenomic era: A complex systems approach to human pathobiology," *Molecular Systems Biology* 3(1), pp. 1–11.

Mackie, J.L. (1965) "Causes and conditions," *American Philosophical Quarterly* 2(4), pp. 245–64.

Magnus, D. (2004) "The concept of genetic disease." In *Health, Disease, and Illness: Concepts in Medicine*, edited by Arthur L. Caplan, James J. McCartney, and Dominic A. Sisti, pp. 233–42. Washington, D.C.: Georgetown University Press.

McKusick, V.A. (2007) "Mendelian inheritance in man and its online version, OMIM," *American Journal of Human Genetics* 80(4), p. 588.

Melendro-Oliver, S. (2004) "Shifting concepts of genetic disease," *Science Studies* 17(1), pp. 20–33.

Murray, T.H. (1997) "Genetic exceptionalism and 'future diaries': Is genetic information different from other medical information?" In *Genetic Secrets: Protecting Privacy and Confidentiality in the Genetic Era*, Yale University Press, pp. 60–76. New Haven.

Paul, D.B. (1994) "Toward a realistic assessment of PKU screening." In *PSA: Proceedings of the Biennial Meeting of the Philosophy of Science Association*, pp. 322–28.

Paul, D.B. (2000) "PKU and procreative liberty: Historical and ethical considerations." In *Ethical Issues in Health Care on the Frontiers of the Twenty-First Century*, edited by S. Wear, J. J. Bono, G. Logue, and A. McEvoy, 65, pp. 171–90. Dordrecht: Kluwer Academic Publishers.

Paul, D.B. (2013) *The PKU Paradox: A Short History of a Genetic Disease*. Johns Hopkins Biographies of Disease. Baltimore: Johns Hopkins University Press.

Scriver, C.R. (1995) "Whatever happened to PKU?" *Clinical Biochemistry* 28(2), pp. 137–44.

Scriver, C.R. (2007) "The PAH gene, phenylketonuria, and a paradigm shift," *Human Mutation* 28 (9), pp. 831–45.

Scriver, C.R., and P. J. Waters (1999) "Monogenic traits are not simple: Lessons from phenylketonuria," *Trends in Genetics* 15(7), pp. 267–72.

Smith, K.C. (1992) "The new problem of genetics: A response to Gifford," *Biology and Philosophy* 7(3), pp. 331–48.

Smith, K.C. (2001) "A disease by any other name: Musings on the concept of a genetic disease," *Medicine, Health Care and Philosophy* 4(1), pp. 19–30.

Smith, K.C. (2007) "Towards an adequate account of genetic disease." In *Establishing Medical Reality—Essays in the Metaphysics and Epistemology of Biomedical Science*, edited by H. Kincaid and J. McKitrick, pp. 83–110. Dordrecht, The Netherlands: Springer.

Sober, E. (1988) "Apportioning causal responsibility," *The Journal of Philosophy* 85(6), pp. 303–18.

Tabery, J. (2007) "Biometric and developmental gene–environment interactions: Looking back, moving forward," *Development and Psychopathology* 19(04), pp. 961–76. doi:10.1017/S0954579407000478.

Tabery, J. (2009) "From a genetic predisposition to an interactive predisposition: Rethinking the ethical implications of screening for gene–environment interactions," *Journal of Medicine and Philosophy* 34(1), pp. 27–48. doi:10.1093/jmp/jhn039.

Valle, D. (2008) "Victor Almon McKusick, MD, 1921–2008, In Memoriam," *American Journal of Human Genetics* 83(3), p. 301.

Waters, C.K. (2007) "Causes that make a difference," *Journal of Philosophy* 104(11), pp. 551–79.

Wulff, H. R. (1984) "Comments on Hesslow's 'What is a genetic disease?'" In *Health, Disease, and Causal Explanations in Medicine*, edited by L. Nordenfelt and B.I.B. Lindahl, pp. 195–97. Dordrecht, Pays-Bas: D. Reidel publ. Co.

Part II
Explanation

4   Molecular Complexity

Why Has Psychiatry Not Been Revolutionized by Genomics (Yet)?

Maël Lemoine

Abstract

This chapter looks into the results of Genome-Wide Associations Studies, one of the leading trends in molecular psychiatry. The main reason for their limited achievements to-date is precisely what they have been designed to handle: complexity. Yet this much used term needs further clarification in the context of Genome-Wide Association Studies. A distinction should be made between 1) unreliable phenotypes, a problem that may eventually solve itself with a molecular approach, 2) complex inheritance processes, and 3) complex mechanisms, in the forms of genomic, etiological, and pathophysiological complexity. This situation is of interest to the study of several traditional questions in the philosophy of medicine.

1 Introduction

Medicine has been revolutionized by molecular biology. In particular, since the 2000s, genome-wide association studies (GWASs) have cast light on biological pathways in many diseases. GWASs consist in investigating all genes in the genomes of a group of diseased people, as compared with healthy people's genomes, in search for a statistically significant association with particular diseases. For instance, one of the very first GWASs discovered that people with likely mutations in two regions called *rs380390* and *rs10272438* are seven times more likely to have age-related macular degeneration (ARMD) than people without them (Klein et al. 2005). One of these regions contains a gene encoding, an already known immunoregulator called complement factor H (CFH), which in turn could be shown to be involved in ARMD.

Psychiatry is not lagging behind in this enthusiastic embrace of genomics. One of the leading journals in this domain, published by the Nature Publishing Group, bears the unambiguous name of *Molecular Psychiatry*. GWASs in psychiatry have been published in the dozens. Molecular psychopharmacology, metabolomics, and transcriptomics now provide the main tools of investigation for a domain of medicine, *mental disorders* (MD), long considered to be specific because of ontological as well methodological problems befalling the realm of the "mental." Despite this conversion to molecular biology, doubts still override facts, dissent still poisons consensus and expectations overcome track record in psychiatry. This has led to the question: "Why has psychiatric research not yet been revolutionized by molecular biology?" (Crow 2011)

In tackling this question, this paper focuses on the methods and results of GWASs in psychiatry (§2). Indeed, for GWASs of any condition to be a success, the following conditions must be met:

1   There is a reliably identifiable phenotype, that is, the observable features of a disease.
2   It is heritable.
3   It is frequent enough in the population.
4   It is underpinned by a not too-complex and sufficiently shared mechanism.

Each of these conditions in turn raises specific concerns for GWASs in psychiatry because of *complexity*. Phenotypes of mental disorders are known to be clinically "complex," which bears the risk of heterogeneous samples: actually, this kind of complexity might just be an illusion created by the clinical approach to mental disorders (§3). The heritability of mental disorders is also known to involve a "complex" process: as opposed to simple Mendelian diseases, the heritability of MDs is probably due to many genes with small effects. Although GWASs should help detect them, much of the explanation of heritability is missing. It is a theoretical question why, of all pathological conditions, MDs should be characterized by such complex heritability (§4). On the whole, these two sorts of complexity already undermine the statistical power of GWASs, but a third sort of complexity, that is, the complexity of mechanisms, probably makes things insuperable if mental disorders really have a higher degree of complexity than GWASs are powered to reveal. At the levels of the genome itself, of triggering events and of underlying mechanisms, many different genes may be involved in a causal network at some point in the onset of a mental disorder (§5). Hence, my conclusion about the power of GWASs in psychiatry is rather pessimistic. This is fuel for many traditional debates in the philosophy of medicine.

2 An Overview of the Achievements of GWASs in Psychiatry

GWASs generally (but not exclusively) investigate genetic markers called *single-nucleotide polymorphisms* (SNPs).[1] SNPs are single nucleotides presenting different forms in at least 1 percent of the population (Reece et al. 2013). Thus, if an SNP is associated with the occurrence of a disease on the one hand, and with a coding allele on the other hand, it is highly likely that this SNP signals a coding allele involved in the disease or the vulnerability thereof (Reece et al. 2013). GWASs commonly consist in a blind comparison of about 1,000,000 SNPs present in a population. If the frequency of any one SNP, that is, a single nucleotide change, is significantly different in a group affected with a given disease than in a group not affected with it, the conclusion is drawn that this SNP may signal a gene involved in the onset of the disease (Clarke et al. 2011). This is called a case/control study.[2]

In psychiatry, GWASs have had disappointing results so far. In the case of major depressive disorder (MDD), the largest GWAS to date has failed to find any significant association with any gene (Levinson et al. 2014), although the GWAS catalog mentioned 14 of them with *p*-value $\leq$ 5.0 x 10^{-8} (http://www.ebi.ac.uk/gwas/) in August 2015. In the case of bipolar disorder (BD), no gene associated has been discovered according to the GWAS catalog as of May 2014, although BD has long been known to run in families, heritability being assessed as high as ~90% (Kieseppä et al. 2004). GWASs have, however, succeeded in rejecting previous hypotheses and suggesting new ones (Barnett and Smoller 2009; Craddock and Sklar 2013; Ferreira et al. 2008; Psychiatric GWAS Consortium Bipolar Disorder Working Group 2011). In the case of schizophrenia, the schizophrenia forum registered in April 2008 1,291 published studies of 690 candidate genes (www.schizophreniaforum.org/res/sczgene/default.asp). In 2009, the Psychiatric GWAS consortium cited a meta-analysis retaining only four candidate genes (Psychiatric GWAS Consortium 2009). Two years later, the schizophrenia group in this consortium mentioned five loci (Schizophrenia Psychiatric Genome-Wide Association Study (GWAS) Consortium 2011). In May 2014, 39 were mentioned by the GWAS catalog. It is a confusing picture of poor replication of results, a problem discussed by (Schaffner forthcoming).

In the face of these poor results so far, the consensus has been that more numerous cohorts are needed (Craddock et al. 2008; Psychiatric GWAS Consortium 2009; Ripke et al. 2013). Levinson and colleagues draw an analogy: in the case of schizophrenia, 9,000 cases have been necessary for 5 SNPs to appear, and 35,000 for 108 of them (most of which have not been confirmed). In the case of mood disorders, a much more prevalent and much less heritable series of conditions, an extrapolation justifies the expectation of 75,000–100,000 cases before the inflection point is reached, that is, the point of the curve where there is linear increase between the number of cases and the number of significantly associated SNPs (Levinson et al. 2014; Ripke et al. 2013).

Because the prior, theoretical estimates of the number of cases required were lower (Risch and Merikangas 1996), GWASs were typically designed with several hundreds of cases at most. The difficulty of discovering new genes involved in fairly heritable conditions has obviously come as a surprise, and it was generally blamed on the unexpectedly high *complexity* of mental disorders. But what does this complexity consist in? Three different arguments are often made. The first one identifies complexity with diagnostic heterogeneity: if one either fails to signal cases (false negative) or wrongly signals cases (false positive), the statistical power of the study decreases sharply (Ripke et al. 2013). The reason for so many detection errors is *phenotypic complexity*: many phenotypes related to one genotype, and many "phenocopies" unrelated to a given genotype (§3). The second argument questions estimates of heritability and invokes *developmental complexity*: theoretical models that match mental disorders state that such "complex" diseases cannot be explained by highly penetrant susceptibility genes but only by many with small to large effect sizes (Craddock et al. 1995; Wang et al. 2005) (§4). The third argument invokes the "complexity" of the underlying mechanisms of mental disorders (§5). The rest of the paper looks into these three sorts of complexity in turn.

3   Phenotypic Complexity, or Why There are No Reliably Identifiable Phenotypes in Psychiatry

GWASs consist in systematic comparison of cases *versus* controls. People are assigned to either group on the basis of known, observable features associated with the disease. Further, nonobservable features are investigated: are they associated with the observable features? Non-observable features are genetic. Observable features are *phenotypic*. Hence, the determination of reliable phenotypes is the critical first step of GWASs.

The problem with phenotypes in psychiatry is their purported complexity, as testified by the so-called polythetic nature of psychiatric diagnoses—any mixture of at least *n* symptoms out of a list makes the diagnosis at the price of a high heterogeneity of the mental disorder (American Psychiatric Association 2013). In this sense, mental disorders are complex because there may be multiple biological realizations corresponding to the same descriptive predicates in clinical psychiatry, and possibly multiple clinical manifestations of the same biological disorder. Philosophers have tackled the issue in the light of the philosophical treatment of "natural kinds" (Haslam 2000; Zachar 2000; Cooper 2005; Kendler et al. 2011; Zachar 2014). This has also fueled debates about the ontological status of disease entities and the reality of MDs (Reznek 1987, 1991; Hacking 2000; Simon 2011). These are general questions scientists are aware of but leave unanswered as ones that may eventually solve themselves with scientific progress.

But phenotypic complexity raises a specific problem for GWASs. At bottom, the problem is the isomorphism of two different approaches to the same mental disorder, one clinical and the other molecular. Indeed, they do not necessarily decompose the same system into the same elementary components. We may have molecules, neurons, and neuronal maps on the one hand; and beliefs, emotions, and behaviors on the other hand. This gives rise to what Wimsatt calls "descriptive complexity" (Wimsatt 2007). As many different manifestations of the "same" disorder are underpinned by many different biological processes, the number of genes possibly involved in any mental disorder increases so fast that there is no theoretical limit to the number of patients necessary for candidate genes to emerge.

This is the general question addressed in this section. GWASs should classify cases *versus* noncases on the basis of *phenotypes*, which are supposed to be isomorphic with *genotypes*. But they classify cases on the basis of *diagnostic criteria* that define *clinical forms* of a mental disorder. Are clinical forms of a mental disorder *phenotypes*? In medicine, they are generally considered phenotypes. But, strictly speaking, a clinical form of a disease is the realization of one of the alternative sets of criteria by which this disease can be detected with maximum reliability, that is, sensitivity and specificity, given the current state of knowledge and the accuracy of clinical judgment. In other words, clinical forms contain more than just phenotypes: there is much noise in the signal from phenotypes that clinical forms of a disease imperfectly capture.[3]

The obvious source of noise is clinical judgment itself. In *judging* whether a drosophila's eyes are black or red, judgment itself can generally be considered a perfect rendition of the phenotype. For various reasons, judgment is a variable in reports of wheezing, purpura, or irritable mood. In the case of psychiatric disorders, most traits, such as anhedonia or delusion, obtain through a highly sophisticated interpretation process by a trained psychiatrist. Although the standardization of criteria may improve reproducibility, it does not make the disease entity a phenotype rather than a construct of clinical judgment, albeit one widely shared in the community of psychiatrists. For these reasons, interprets of GWASs in psychiatry have often invoked the search for biomarkers as the key to unlocking the door to robust results.[4]

Yet biomarkers raise exactly the same problem as clinical judgment when it comes to capturing phenotypes in medicine. For instance, in *determining* whether type-2 diabetes is associated with a genotype, elevated fasting glycemia or HbA1c is resorted to. High HbA1c is considered more reliable than hyperglycemia. Yet neither *is* the corresponding phenotype, that is, the pathophysiological process underlying type-2 diabetes, but only a trait frequently *associated with* the phenotype. Such biomarkers are generally defined as surrogate markers, that is, not necessarily a measurement of the pathophysiological process itself. Ideally, biomarkers should be a direct expression of this process, not just of its symptoms and consequences (Ritsner 2009). In depression, for instance, they should reflect the core process of depression rather than, say, an indicator of bad sleep or fatigue, that is, common symptoms or even consequences of depression. Indeed, most such symptoms are likely not to be specific. In psychiatry, where the pathophysiological process is not even localized but only hypothesized, it is difficult to assess whether any biomarker is associated with it.

As an illustration of the problem, Table 1 provides a list of the most well-known and currently investigated biomarkers of MDD. Biomarkers useful in clinics should be reliable detection tools, that is, avoiding false positives and false negatives to a certain extent. Schneider and Prvulovic assess that sensitivity and specificity should both be over 80 percent (Schneider and Prvulovic 2013).[5] This is a sensible standard (although admittedly nothing more than a conventional one). Yet biomarkers are rarely specific enough in psychiatry. For instance, brain derived neurotrophic factor (BDNF) is a marker of adult neurogenesis. An interesting hypothesis is that it is low in depression. In a recent meta-analysis, ca. 700 persons, half of them healthy and half of them depressed, were tested for levels of BDNF in serum (Sen et al. 2008). It seems to show high sensitivity but low specificity: as a matter of fact, BDNF is also low in bipolar disorder, schizophrenia, bulimia and anorexia nervosa, autism, Alzheimer's disease, Huntington's disease, and many other somatic conditions. Either it is sensitive to far too many potential physiological changes to be useful for diagnosis—just as heart rate or blood pressure is—or it reveals a widely shared endophenotype of mental disorders. Like adenocorticotropic hormone (ACTH), it could also simply reveal a biologically common reaction to any mental disorder—it is not surprising that one is chronically anxious or unable to be "open to life" with depression, schizophrenia, autism, or most other mental disorders.

Strictly speaking, it does not follow from this that mental disorders are complex but only that their mechanisms are unknown, and as a consequence, that their phenotypes are unidentifiable. There might be simple mechanisms and highly sensitive/specific biomarkers of at least some mental disorders. The current approach to them might generate artificial diversity and create the illusion of complexity. Or, they can be complex indeed: there is no way to know.

Take a second example to illustrate the consequences of this situation: ACTH (cortisol/CRH in Table 4.1). The hypothalamic–pituitary–adrenocortical (HPA) system regulates stress by a feedback process involving cortisol. Dexamethasone is a surrogate for cortisol. In subjects where the feedback process functions normally, a dexamethasone shot elicits a downregulation of the HPA axis, as assessed by a reduction in the production of ACTH, a mediator between the inhibition and the production centers of cortisol. A nonsuppression of the ACTH production, or any other marker of the activity of the HPA axis, is interpreted as the inability of this system to downregulate. This test, the dexamethasone suppression test (DST), is standard for Cushing's syndrome, a neuroendocrine disorder associated with depressive symptoms. It has also been investigated since the 1960s as an interesting test in MDD. Indeed, the dexamethasone-CRH test is positive in up to 90 percent of depressed patients depending on age and sex (Heuser et al. 1994), but also on the depression profile, such as melancholic, endogenous, familial, and psychotic (Nelson and Davis 1997; Lamers et al. 2013). Eighty-eight percent of patients with anxious retarded depression have higher plasma levels of arginine vasopressin (AVP), a hormone associated with the HPA axis (de Winter et al. 2003). Yet, this test is supposed to have an overall sensitivity of just 50 percent, meaning that half of clinically depressed patients respond normally to the DST. If we considered a positive reaction to the DST to be a necessary condition for the diagnosis of MDD, half of clinically depressed patients would not be "depressed" any more. Alternatively, the DST would specify a subtype of depression or at least would be consistent with a specific clinical presentation of it, or the presence of one symptom. In any case, if one biological process really underlies all cases of depression, or at least most of them, and if the clinical diagnosis is reasonably sensitive and specific, ACTH under the DST cannot be considered a relevant biomarker of depression.

Table 4.1 A list of biomarkers of major depressive disorder, after Schneider and Prvulovic (2013). More detailed explanations in (Schneider et al. 2011)

Biochemical/molecular markers
  BDNF
  HPA-Axis:
    AVP
    Cortisol response to DEX/CRH stimulation
"Core"-CSF markers
    T-τ
    P-τ
    APP and APLP2
    Ab1-42
Neuroimaging findings
  Structural findings (volumes)
    ACC
    OFC
    Hippocampus
    Putamen
    Caudate nucleus
    Basal ganglia
    SGPFC
  DTI
    FA in prefrontal, callosal and medial temporal regions
  Functional imaging markers: no result
  Nuclear imaging markers:
  CBF in the amygdala, orbital cortex, and medial thalamus
  CBF in the dorsomedial/dorsal-anterolateral PFC and subgenual PFC
  HT1A BP across large areas of the brain
  HT1A BP in the medial temporal cortex, in the dorsal raphe nucleus of elderly depressed subjects, and in the sgACC, pgACC, lateral–orbital, and mesotemporal cortices of postpartum MDD cases
Neurochemical imaging markers
  Glutamate and glutamine (Glx) and GABA in frontal WM
  Glx in the anterior cingulate cortex and left DLPFC
  Glu in occipital cortex
  NAA in frontal cortex and in subcortical regions
  Creatine and phosphocreatine
  γ-aminobutyric acid in occipital cortex

On the other hand, if one abandons one of these two hypotheses, either MDD is biologically diverse, or it is clinically ill-defined, or both.

The upshot of this section is that GWASs hinge on the tractability of phenotypic complexity in psychiatry to build case/control studies. While biologically oriented psychiatrists are eager to discover biological mechanisms of mental disorders, they also are, by necessity, conservative of the behavioral/ mental symptoms signaling these mental disorders. After all, GWASs consist in a *phenotype-to-genotype approach* starting with phenotypes. Replacement of these symptoms by biomarkers would hopefully reduce phenotypic diversity, but it is impossible to drop them altogether and necessary to smoothly translate them into molecular features. This necessity is a potential source of artificial complexity. Thus, either molecular psychiatry is ready to stipulate new definitions of disorders of the mental, or it is doomed to deal with much too many observable, but parasitic, traits.

Three final remarks should nevertheless qualify this somewhat skeptical conclusion. The first is that, far from having been investigated over and over, the question of biomarkers of MDD remains an open one. For instance, recent plasma metabonomics tests have shown a sensitivity of 92.8 percent and a specificity of 82.3 percent, according to Zheng et al. (2012). Second, both latent factor analysis and latent component analysis of *biomarkers* (instead of *clinical variables*) could yield interesting results but have not been investigated. Indeed, I have found only one old cluster analysis applied to biomarkers (Staner et al. 1994). The third remark is that, even if justified, reservations about necessary reliance on clinical forms of disease are likely to be, but by no means are necessarily, fatal to the possible results of GWASs using them: noise may be filtered enough, or not important enough, for significant results to obtain.

4　Developmental Complexity, Or Why Supposedly Highly Heritable Phenotypes in Psychiatry Cannot Be Associated With Genes

There is an altogether different sense in which MDs are considered "complex" diseases: along with phenotypical complexity, MDs are developmentally complex. In this sense, they contrast with genetically simple diseases, that is, Mendelian diseases. Indeed, both types of complexity are linked. In biology, a phenotype is defined as the observable result of the combined influence of genes and environment leading to diversity of the manifestations of the same "genetic" trait. The second condition for a GWAS to be a success is that the so-called phenotypic features under study are associated with genetic features, not just with environmental factors. Some diseases do not fulfill this condition, such as carbon monoxide poisoning or a sunlight-impaired retina. In the most general but trivial sense, there always are genes involved in any pathological condition: genes are relevant to the fact that we have brains, muscles, blood, and lungs; how they function; and consequently that CO is poisonous, and so on. But this is the case for *everybody*, diseased or not. What GWASs are interested in is therefore not what is associated with genes in all humans, but rather in some of them. In other terms, we are interested in differences or *variance*, defined as "the average of the squares of deviations from the arithmetic mean" (Bryant 1960), and noted σ^2.

Philosophers have tackled the issue of difference making in biology in general and in molecular biology in particular (Waters 2007; Nathan 2012). The broad question of the genetic and environmental parts of behavioral traits has also been examined in compelling philosophical contributions (Schaffner forthcoming; Griffiths and Machery 2008; Griffiths and Tabery 2008; Tabery 2014). Not much has been written on the way this question has been raised in GWASs in particular.

In classical genetic epidemiology, genetic "determination" is assessed through the ratio of genotypic variance on phenotypic variance, which expresses the proportion of the phenotypic variance that can be attributed to genotypic variance:

$$H^2 = \sigma_G^2/\sigma_P^2$$

This is heritability broadly construed, that is, a measurement of the relative influence[6] of genes and environment in the variation of an observable trait.[6] It can be assessed in (classical) genetic epidemiology by means of twin studies, adoption studies, family studies, and population studies. These studies largely predate the molecular era of GWASs and will not be investigated further here (Lynch and Walsh 1998; Falconer and Mackay 1995; Visscher et al. 2008; Schaffner forthcoming).

The main point is that before any GWAS is undertaken, an assessment of heritability justifies which conditions to study and gives a sense of how great the influence of certain genetic variations may be. Importantly, what is expected to be heritable in a disease based on genetic epidemiological studies is typically much higher than what can be actually attributed to specific parts of the genome in a GWAS. This is referred to as the problem of missing heritability (Manolio et al. 2009). In the case of schizophrenia, for instance, it has been estimated that although the pond of genes has been drained dry through GWASs, only 2 percent of the heritability of the disease, estimated at 90 percent, has been explained (Crow 2011).

Two main explanations have questioned the hypotheses underlying GWASs. The first questions the so-called *linkage disequilibrium* (LD) hypothesis, according to which the physically closer coding regions of the genome are, the less likely they are to be separated due to crossing over. This is based on the interpretation of the observation that alleles are not linked randomly. Here what may be questioned is the number of SNPs (usually 1m nucleotides out of 3.3bn). The second is the so-called *common disease/common variant* (CDCV) hypothesis, according to which diseases investigated through GWASs are frequent and due to frequent variations in the genome (Risch and Merikangas 1996; Reich and Lander 2001; for a synthetic discussion on this hypothesis, see Visscher et al. 2012, 10). Some have suggested a common disease-rare variant hypothesis in the case of schizophrenia (McClellan et al. 2007; McClellan and King 2010). Another hypothesis goes that when an SNP is associated with a disease, it is associated not with high LD with one common causal variant, but with low LD with neighboring rare causal variant(s) with a joint effect, a possibility dubbed "synthetic associations" (Dickson et al. 2010). However, most doubt that either situation happens often enough to explain much of the disappointment with

GWASs (Anderson et al. 2011; Wray et al. 2011; Visscher et al. 2012; Sullivan et al. 2012).

The upshot of this discussion is the general agreement that polygenic, complex diseases are the joint result of many alleles with very different frequency, from very rare to very common (Visscher et al. 2012; Sullivan et al. 2012). Me Mental disorders are generally considered to rank among the highest in the monogenic/simple diseases–polygenic/complex diseases spectrum. This might suffice to explain why the genetics of most mental disorders is more difficult to study than the genetics of most somatic disorders. The question is rarely investigated further in the scientific literature. Yet a new interesting philosophical question is raised: it is not a question of *whether* MDs are heritable—a question already investigated at length in the aftermath of classical psychiatric epidemiology—but a question of *why*, most of all, even the most heritable of mental disorders do not yield much results when they are submitted to systematic genetic screening such as those performed in GWASs. If the scientifically accepted answer is: "because they are particularly polygenic and complex," then an explanation must be provided as to why they are polygenic and complex.

At least three reasons for this polygenicity and complexity of mental disorders in particular have been advanced. First, it is sometimes argued that mental disorders impair reproduction to a greater extent, and consequently, they would most likely be strongly selected against were they due to genes with large effects. Therefore, the effects must be small, and given that they are nevertheless highly heritable, mental disorders must be particularly polygenic disorders (Craddock et al. 2008). Second, it is possible that the heritability of most MDs is overestimated. As a matter of fact, even when one does not admit that social forces such as normative constraints on behaviors have a stronger influence on MDs than on somatic disorders, it is more difficult to identify their existence and their nature. What part of the same MDs suffered by adopted twins raised in different families should be attributed to being submitted to the same social influence? What is a "stressful" environment? For instance, why exactly are demands about school performance differently stressful, as compared to abuse, for children aged 6–12? Here the problem is not with the complexity of GxE interactions (Tabery 2014), but more specifically with assessing the direction and respective weights of various causal factors. Third, diagnostic heterogeneity, which is also greater in the case of mental disorders, as said above, necessarily impacts the genetic heterogeneity of the population under study and amplifies the assessed complexity of the disorder.

5 The Molecular Complexity of Mental Disorders

In a final—and more fundamental—sense, mental disorders are deemed *complex* diseases because of specific features of the "mechanisms" of the disease at the molecular level. There is no question that many other diseases are mechanistically complex in the same sense. This last section is an attempt to clarify what, if anything, makes mental disorders particularly complex.

There is a first type of mechanistic complexity, emphasized by philosophers, which characterizes organized systems, that is, systems "having multiple parts that stand in nonsimple relations" as Mitchell puts it (Mitchell 2003). Diverse functions attributed to parts make for mechanistic complexity. Specific forms are Bechtel and Williamson's "functionally integrated systems," with cascading and feedback effects and nonlinearity (Bechtel and Richardson 2010); and Wimsatt's "interactional complexity," assessed as the general quantity of strong causal interaction with other systems (Wimsatt 2007). Although they raise real challenges to reductionist strategies prevalent in molecular biology, it is trivial to say that mental disorders are complex in this general sense. Indeed, all diseases are complex in that sense, as all diseases are organic processes; yet "complex" is meant here as a contrastive term. On the other hand, there is no reason to think that Mendelian diseases are more simply organized than polygenic diseases. What is complex in the latter case is not the mechanism of the disease. It is the gene × environment (G × E) mechanism of inheritance of the disease, as emphasized in the previous section. Moreover, there is also complexity in the causal network between the genome and the mechanism of the disease.

Indeed, GWASs are early studies of the complexity of the genomic causal network. It is not a decomposition/localization approach, which is yet to come. That means that "complexity" still is a more basic notion than the one used in physiology, biology, and traditional molecular biology. Eventually, it may turn out to become "complexity" in the sense of organizational complexity, as prophesied by network medicine (Loscalzo et al. 2007; Barabási et al. 2011). Yet, for the moment, what is clearly meant by complexity in the context of GWASs is that "many" genes, not just one, are likely to be implied in the genome–mechanisms network. More precisely, the complexity of the mechanism of a disease such as a mental disorder refers both to *I)* the *length* of the causal pathway that explains the occurrence of the disease and its manifestations, and *II)* to *alternative* parts of this causal pathway (e.g., if the disease is characterized by a causal chain A→···→C→D→E→···→G, alternative causal pathways may fill in the blanks). In the former case, the number of genes involved is proportional to the number of stages in the mechanism of the disease, whereas in the latter case the number of genes involved is proportional to the number of alternative pathways in the disease. Of course, both types of complexity are not mutually exclusive: on the contrary, the longer the causal chain, the greater the number of possible mechanistic alternatives.

The general abstract schema scientists are trying to fill in is a long chain of causes with possible alternative pathways at given points. In medicine, it is traditional (and useful) to distinguish three general stages: dispositional, etiological, and pathophysiological. The first generally explains vulnerability with genes and developmental environment, the second deals with triggering events, and the third deals with the underlying causal process sustaining the disease (Belzung and Lemoine 2011). In psychology and psychiatry, this is called the diathesis-stress model (Panksepp 2004). Complexity takes a different turn for each in the case of molecular explanations of mental disorders.

At the level of genes, *genomic complexity* refers to the fact that genes may be associated with the disease only in alternative combinations. Any gene might yield a significant or detectable effect only under a series of conditions, such as: allele A_1 is a factor of vulnerability only in the presence of alleles $\{A_3, A_{57}, a_{43}, a_{101}\}$ or $\{a_3, a_{57}, A_{101}, a_{127}\}$, or... and so on, a phenomenon referred to as epistasis. Such a set of combinations would dramatically increase the size of the cohorts needed to provide results, to the effect of "dwarfing" "readily available computational power" (Richardson 2015). Besides, it may have consequences at any of the afore-mentioned levels of complexity and, possibly for some genes, at several levels at the same time, given the ubiquity of many molecules in the organism. The only reason why genomic complexity is made worse in the case of MDs as compared to most somatic disorders, is that it multiplies with pathophysiological and etiological complexity, which are worse in the case of MDs.

As a matter of fact, at the level of triggering events, *etiological complexity* indicates alternative combinations of *entries* to one pathophysiological pathway to symptoms. Depression, we have strong evidence to say, may be triggered by events as diverse as bereavement of a loved one, a symbolic loss of social status, and the onset of a chronic disease, which probably all trigger it via the prefrontal cortex; unpredictable mild chronic stress and sexual assault, via the amygdala; a tumor in the pituitary gland, a shot of dexamethasone and depriva-tion of sleep, via the HPA axis of stress; influenza, autoimmune diseases, and cancer chemotherapy, via neurogenesis in the hippocampus. The list also con-tains documented links between depression and atheroma, stroke, Alzheimer's, Parkinson's, and so on. Even if there were only one pathophysiological causal pathway to the symptoms of depression, say, deregulated hippocampal neuro-genesis, genetic heterogeneity could be relevant in explaining vulnerabilities in the specific causal pathways that may trigger this pathophysiological process—here, a specific vulnerability to influenza, autoimmune disease, or chemotherapy. Two reasons make MDs very likely to be etiologically complex diseases. The first is their link to cognitive functions. If the general function of a cognitive system is to translate an almost unlimited repertoire of stimuli into a limited repertoire of behaviors, the first step is to interpret stimuli. There are infinitely many particular cognitive processes of interpretation leading to the conclusion that one is worthless or that people are hostile. There may be biological vulner-abilities specific to each of them. The second reason why MDs are likely to be etiologically complex diseases is the fact that they are diseases of the most integrated system in the organism, the central nervous system: any dysfunction in any mechanism may trigger the same repertoire of pathological responses.

On the contrary, at the level of pathophysiology, *pathophysiological complexity* may be thought of, in the case of mental disorders, as a consequence of highly unspecific emotional and cognitive responses to environmental challenge. As compared to a highly evolved system such as the immune system, which is capable of very specific responses to environmental challenge, the cognitive system tends to answer in the same limited ways each time it is overcome by stress, whatever the nature. Yet, if the symptomatic response is indeed poorly specific, a hypothesis defended in Section 2 of the present paper, it means that there are possibly many pathophysiological pathways leading to the same limited set of responses—*symptoms*. These alternative pathophysiological mechanisms contribute to the complexity of mental disorders. The number of genes associated with these symptoms is proportional to the number of alternative pathways leading to the symptoms. Although a symptom like anhedonia would supposedly translate into one biological process performed by one and the same biological mechanism, say, a shortage of dopamine in the reward circuit in the brain, this process could be the last part of many alternative chains of events, which are more or less likely to start depending on various dispositions; genetic vulnerabilities to, say, a shortage of serotonin in the whole brain, downregulation of cortisol, upregulation of hippocampal neurogenesis, control of neuroinflammation, and so on, are all *ipso facto* genetic vulnerabilities to anhedonia. Each of them may be explained by its own type of genetic vulnerability and involve its own set of genes in the explanation of this vulnerability (Sullivan et al. 2012), which complicates detection in GWASs.

Complexity is increased in GWASs by the fact that it is not known at which level—pathophysiological, etiological, or genomic—a given gene may be associated with a part of one of the many mechanisms possibly involved. The resulting picture is that GWASs are attempts at detecting genes involved in an unknown mechanism at an unknown level. The silver lining of this cloud is data-driven hypotheses: indeed, in GWASs, it is not necessary to proceed inductively from any hypothesis about the pathophysiology, that is, the mechanism involved in the disease, to proteins and then to candidate genes. It is possible to study the disease right at the level of genes and then to make hypotheses about the pathophysiology of the disease. However, this does not mean that one could dispense with having any hypothesis about who belongs to the group of diseased individuals, that is, a prior definition of the disease. Yet, what is common and what is not in a mental disorder, and where, if any, is the causal nexus that unifies all cases of any given mental disorder? Besides, to take all this data into account and make sense out of it, it might be necessary to develop complex theoretical models, like those emerging in network medicine (Loscalzo et al. 2007; Barabási et al. 2011; Goh et al. 2007) or in stochastic dynamical systems approaches (Rolls and Deco 2011).

The necessity of theoretical models in order to make sense of molecular information has often been pointed out by philosophers (O'Malley and Dupré 2005; Leonelli 2012; Keating and Cambrosio 2012; O'Malley and Soyer 2012). In the beginning, what is given is molecules in great numbers and some of their interactions. In the end, what is expected is a synoptic, that is, simplified, abstracted away and thereby understandable, picture of what is going on. Yet it is not only an end in itself, one that would define theoretical science—knowledge for its own sake—as opposed to practical or applied science, which could afford to remain "blind." Researchers look for understandable models in order to relate parts of the universe together, further model their relations, and avoid directly studying the "whole" itself.

The problem is not therefore that there is no signal to detect in molecular approaches in psychiatry. It is rather that a psychiatric diagnosis may be too baroque a detector to provide any reliable information for the purpose of molecular biology.

6 Conclusion

The question that sparked the discussion was: why has psychiatric research not yet been revolutionized by molecular biology? This chapter examined why psychiatry has not been revolutionized by genomics and, more specifically, by GWASs. GWASs rely on the notion of SNPs, that is, tags likely to be associated with coding regions and covering the whole genome. Their statistical association with common diseases, on the other hand, can be more or less strong, and statistical power may not be sufficient to establish it. GWASs have so far been successful in 1) confirming the role of a gene in the onset of a disease, 2) discovering new candidate genes, 3) suggesting new molecular pathways to diseases, 4) suggesting new molecular pathways to new candidate biomarkers or treatments, and 5) challenging the received nosology. None of these achievements has so far been reached in psychiatric research on schizophrenia, BD, and MDD, the three most investigated MDs in molecular psychiatry. Four main causes to that situation are that 1) MDs are not, and cannot just be translated into, reliably identifiable phenotypes; 2) their heritability is particularly difficult to assess; 3) the size of groups needed to reach statistical power and thus detect effects may be difficult, if not impossible, to get; and 4) the entrenchment of genomic, etiological, pathophysiological, and diagnostic complexity in a condition whose mechanisms have not yet been investigated with sufficient results to limit possibilities might well hinder any further progress with GWASs. I am skeptical about reason (1) being associated with "phenotypic complexity," and rather blame it on "descriptive complexity," that is, an epistemological problem rather than an ontological one. According to this paper's conclusion, (3) should be blamed not only on inheritance complexity (2) but also on mechanistic complexity (4). In other terms, genomics requires that enough of the molecular details of a condition are known before providing more. Whereas identifying mechanisms is the ultimate goal, the first step in the molecularization of mental disorders has to be statistical and probabilistic models. The fear is that molecular psychiatry cannot progress much further. Strictly speaking, there is no way to know it, but to wait and see... and bet in the meanwhile.

Such a situation is of interest to several more traditional questions in the philosophy of medicine and in the philosophy of psychiatry. First, it may be either an illustration or a challenge for those (referred to above) tending to think of nosological problems in psychiatry in terms of natural kinds: could Cooper's classic analysis of the problem stand in the face of a more systematic, biological investigation? Second, it constitutes a relevant case study of the question of what disease entities are, and fuel for the project of clarifying usual distinctions between signs and biomarkers, symptoms, syndrome, and disease. Third, it lays a bridge between discussions of heritability in the philosophy of biology and in the philosophy of medicine. As suggested above, it is a generally unexplored question for philosophers not to assess whether it makes sense to say that mental disorders are heritable but to explain why systematic screening of genes cannot account for estimates of heritability. At last, I would like to emphasize the interest of investigating such problems into biological details for those interested in the naturalism/normativism debate about the concept of disease. As a professed naturalist, I consider it crucial to assess why the naturalization of mental disorders in GWASs has been a failure so far. An interesting, yet controversial, hypothesis would be that although *diseases* as we know them fail to be separately naturalized, the process of building a general theory of pathophysiological processes, mental disorders included, is on its way.

Notes

1 In psychiatry, common tags of other sorts, like *copy number variations* (CNVs), have only been shown to be associated with a rare form of schizophrenia frequent in the so-called velo-cardio-facial syndrome (Murphy 2002), but not, for instance, with mood disorders (Levinson et al. 2014).

2 The existence of an association between allele A and disease is assessed through a standard χ^2 statistical test. Different methods prevail depending on the penetrance model adopted. *Penetrance* γ is defined as the risk of disease in a given individual carrying allele A. It may be *multiplicative* if the risk of disease with two alleles AA is γ^2, *additive* if it is 2γ, *dominant* if it is γ in both Aa and AA, or *recessive* if it is γ in AA only. In the case of complex diseases typically explored in GWASs, the most likely hypothesis is that effects are additive. The *strength of association* between gene and disease, in turn, is assessed by the *allelic risk ratio* γ^*, that is, the ratio of the proportions of presence of allele A to absence of allele A in cases relative to noncases. This ratio approximately equals penetrance γ of allele A. This has been theoretically demonstrated, provided that we consider penetrance to be multiplicative, and under some other general conditions; for details, see Clarke et al. (2011).

3 The invocation of biomarkers and the notion of *endophenotypes* of mental disorders are supposed to improve the situation (Murphy 2006). Endophenotypes refer to more distal endpoints than the symptoms, more likely to be attached to the same mechanism in heterogeneous tableaus than to the same tableau resulting from different mechanisms. They are postulated on the basis of the clinical or theoretical judgment that many different associations of symptoms should realize the same disease entity. Yet it remains to be seen how exactly to establish endophenotypes.

4 The NiH defines a biomarker as "a characteristic that is objectively measured and evaluated as an indicator of normal biologic processes, pathogenic processes, or pharmacologic responses to a therapeutic intervention" (http://www.genomicglossaries.com/CONTENT/Biomarkers.asp).

5 Remember that sensitivity of 80 percent means that 8 cases of a disease out of 10 will be detected, and that specificity of 80 percent means that 8 out of 10 diagnosed cases really are cases of this disease.

6   There is also a narrower sense of heritability that does not take into account interactions between alleles at the same locus σ_D^2 and at different loci σ_I^2, because these are not transmitted but result from the interaction of alleles inherited from both parents, and only refers to additive genetic effects: $h^2 = \sigma_A^2/\sigma_P^2$.

References

American Psychiatric Association (2013) *Diagnostic and Statistical Manual of Mental Disorders: Dsm-5* Édition: 5th edition, Washington, D.C.: American Psychiatric Publishing.

Anderson, C.A., N. Soranzo, E. Zeggini, J.C. Barrett (2011) "Synthetic associations are unlikely to account for many common disease genome-wide association signals," *PLoS biology*, 9(1), p.e1000580.

Barabási, A.-L., N. Gulbahce, and J. Loscalzo (2011) "Network medicine: A network-based approach to human disease," *Nature Reviews. Genetics*, 12(1), pp. 56–68.

Barnett, J.H. and J.W. Smoller (2009) "The genetics of bipolar disorder," *Neuroscience*, 164(1), pp. 331–343.

Bechtel, W. and R.C. Richardson (2010) *Discovering Complexity—Decomposition and Localization as Strategies in Scientific Research,* Reissue, Cambridge, Mass: MIT Press.

Belzung, C. and M. Lemoine (2011) "Criteria of validity for animal models of psychiatric disorders: Focus on anxiety disorders and depression," *Biology of Mood and Anxiety Disorders*, 1(1), p. 9.

Bryant, E.C. (1960) *Statistical Analysis*. McGraw-Hill Book Company, New York.

Clarke, G.M., C.A. Anderson, F.H. Pettersson, L.R. Cardon, A.P. Morris, and K.T. Zondervan (2011) "Basic statistical analysis in genetic case-control studies," *Nature Protocols*, 6(2), pp. 121–133.

Cooper, R. (2005) *Classifying Madness: A Philosophical Examination of the Diagnostic And Statistical Manual of Mental Disorders*, Springer.

Craddock, N. and P. Sklar (2013) "Genetics of bipolar disorder," *Lancet (London, England)*, 381(9878), pp. 1654–1662.

Craddock, N., M.C. O'Donovan, and M.J. Owen (2008) "Genome-wide association studies in psychiatry: Lessons from early studies of non-psychiatric and psychiatric phenotypes," *Molecular Psychiatry*, 13(7), pp. 649–653.

Craddock, N., V. Khodel, P. Van Eerdewegh, and T. Reich, (1995) "Mathematical limits of multilocus models: The genetic transmission of bipolar disorder," *American Journal of Human Genetics*, 57(3), pp. 690–702.

Crow, T.J. (2011) "The missing genes: what happened to the heritability of psychiatric disorders?" *Molecular Psychiatry*, 16(4), pp. 362–364.

Dickson, S.P., K. Wang, I. Krantz, H. Hakonarson, and D.B. Goldstein (2010) "Rare variants create synthetic genome-wide associations," *PLoS biology*, 8(1), p. e1000294.

Falconer, P.D.S. and P.T.F.C. Mackay (1995) *Introduction to Quantitative Genetics* 4th ed., Harlow: Longman.

Ferreira, M.A.R. et al. (2008) "Collaborative genome-wide association analysis supports a role for ANK3 and CACNA1C in bipolar disorder," *Nature Genetics*, 40(9), pp. 1056–1058.

Goh, K.-I., M.E. Cusick, D. Valle, B. Childs, M. Vidal, and A.-L. Barabási (2007) "The human disease network," *Proceedings of the National Academy of Sciences of the United States of America*, 104(21), pp. 8685–8690.

Griffiths, P.E. and Machery, E. (2008) "Innateness, Canalization, and 'Biologicizing the Mind'," *Philosophical Psychology*, 21(3), pp. 397–414.

Griffiths, P.E. and Tabery, J. (2008) "Behavioral genetics and development: Historical and conceptual causes of controversy," *New Ideas in Psychology*, 26(3), pp. 332–352.

Hacking, I. (2000) *The Social Construction of What?* Harvard University Press.

Haslam, N. (2000) "Psychiatric categories as natural kinds: Essentialist thinking about mental disorder," *Social Research*, 67(4), pp. 1031–1058.

Heuser, I., A. Yassouridis, and F. Holsboer, (1994) "The combined dexamethasone/CRH test: a refined laboratory test for psychiatric disorders," *Journal of Psychiatric Research*, 28(4), pp. 341–356.

Keating, P. and A. Cambrosio (2012) "Too many numbers: Microarrays in clinical cancer research," *Studies in History and Philosophy of Science Part C: Studies in History and Philosophy of Biological and Biomedical Sciences*, 43(1), pp. 37–51.

Kendler, K.S., P. Zachar, and C. Craver (2011) "What kinds of things are psychiatric disorders?" *Psychological Medicine*, 41(6), pp. 1143–1150.

Kieseppä, T., T. Partonen, J. Haukka, J. Kaprio, and J. Lönnqvist (2004) "High concordance of bipolar I disorder in a nationwide sample of twins," *The American Journal of Psychiatry*, 161(10), pp. 1814–1821.

Klein, R.J. et al. (2005) "Complement Factor H polymorphism in age-related macular degeneration," *Science (New York, N.Y.)*, 308(5720), pp. 385–389.

Lamers, F. et al. (2013) "Evidence for a differential role of HPA-axis function, inflammation and metabolic syndrome in melancholic versus atypical depression," *Molecular Psychiatry*, 18(6), pp. 692–699.

Leonelli, S. (2012) "Introduction: Making sense of data-driven research in the biological and biomedical sciences," *Studies in History and Philosophy of Science Part C: Studies in History and Philosophy of Biological and Biomedical Sciences*, 43(1), pp. 1–3.

Levinson, D.F., S. Mostafavi, Y. Milaneschi, M. Rivera, S. Ripke, N.R. Wray, and P.F. Sullivan (2014) "Genetic studies of major depressive disorder: Why are there no genome-wide association study findings and what can we do about it?" *Biological Psychiatry*, 76(7), pp. 510–512.

Loscalzo, J., I. Kohane, and A.-L. Barabási (2007) "Human disease classification in the postgenomic era: a complex systems approach to human pathobiology," *Molecular Systems Biology*, 3, p. 124.

Lynch, M. and B. Walsh (1998) *Genetics and Analysis of Quantitative Traits*, Sunderland, Mass: Sinauer Associates Inc.,U.S., U.S.

McClellan, J. and M.-C. King (2010) "Genetic heterogeneity in human disease," *Cell*, 141(2), pp. 210–217.

McClellan, J.M., E. Susser, and M.-C. King (2007) "Schizophrenia: A common disease caused by multiple rare alleles," *The British Journal of Psychiatry: The Journal of Mental Science*, 190, pp. 194–199.

Manolio, T.A. et al. (2009) "Finding the missing heritability of complex diseases," *Nature*, 461(7265), pp. 747–753.

Mitchell, S.D. (2003) *Biological Complexity and Integrative Pluralism,* Cambridge University Press, Cambridge, Mass.

Murphy, D. (2006) *Psychiatry in the Scientific Image*, MIT Press.

Murphy, K.C. (2002) "Schizophrenia and velo-cardio-facial syndrome," *Lancet (London, England)*, 359(9304), pp. 426–430.

Nathan, M.J. (2012) "The varieties of molecular explanation," *Philosophy of Science*, 79(2), pp. 233–254.

Nelson, J.C. and J.M. Davis (1997) "DST studies in psychotic depression: A meta-analysis," *The American Journal of Psychiatry*, 154(11), pp. 1497–1503.

O'Malley, M.A. and J. Dupré (2005) "Fundamental issues in systems biology," *BioEssays*, 27(12), pp. 1270–1276.

O'Malley, M.A. and O.S. Soyer (2012) "The roles of integration in molecular systems biology," *Studies in History and Philosophy of Science Part C: Studies in History and Philosophy of Biological and Biomedical Sciences*, 43(1), pp. 58–68.

Panksepp, J. ed. (2004) *Biological Psychiatry*, Wiley-Liss., Hoboken, NJ.

Psychiatric GWAS Consortium (2009) "Genomewide association studies: History, rationale and prospects for psychiatric disorders," *The American Journal of Psychiatry*, 166(5), pp. 540–556.

Psychiatric GWAS Consortium Bipolar Disorder Working Group (2011) "Large-scale genome-wide association analysis of bipolar disorder identifies a new susceptibility locus near ODZ4," *Nature Genetics*, 43(10), pp. 977–983.

Reece, J.B., L.A. Urry, M.L. Cain, S.A. Wasserman, P.V. Minorsky, and R.B. Jackson (2013) *Campbell Biology*, 10th ed. Boston: Benjamin Cummings.

Reich, D.E. and E.S. Lander (2001). "On the allelic spectrum of human disease," *Trends in Genetics: TIG*, 17(9), pp. 502–510.

Reznek, L. (1987) *The Nature of Disease*, London, New York: Routledge and Kegan Paul.

Reznek, L. (1991) *The Philosophical Defence of Psychiatry*, London: Routledge.

Richardson, S.S. (2015) *Postgenomics: Perspectives on Biology after the Genome*, Duke University Press Books.

Ripke, S. et al. (2013) "A mega-analysis of genome-wide association studies for major depressive disorder," *Molecular Psychiatry*, 18(4), pp. 497–511.

Risch, N. and K. Merikangas (1996) "The future of genetic studies of complex human diseases," *Science (New York, N.Y.)*, 273(5281), pp. 1516–1517.

Ritsner, M.S. (ed). (2009) *The Handbook of Neuropsychiatric Biomarkers, Endophenotypes and Genes: Volume I: Neuropsychological Endophenotypes and Biomarkers*, 2009 edition, Dordrecht: Springer.

Rolls, E.T. and G. Deco (2011) "A computational neuroscience approach to schizophrenia and its onset," *Neuroscience and Biobehavioral Reviews*, 35(8), pp. 1644–1653.

Schaffner, K.F. (forthcoming) *Behaving: What's Genetic and What's Not, and Why Should We Care?* New York: Oxford University Press.

Schizophrenia Psychiatric Genome-Wide Association Study (GWAS) Consortium (2011) "Genome-wide association study identifies five new schizophrenia loci," *Nature Genetics*, 43(10), pp. 969–976.

Schneider, B. et al. (2011) "Biomarkers for major depression and its delineation from neurodegenerative disorders," *Progress in Neurobiology*, 95(4), pp. 703–717.

Schneider, B. and D. Prvulovic (2013) "Novel biomarkers in major depression," *Current Opinion in Psychiatry*, 26(1), pp. 47–53.

Sen, S., R. Duman, and G. Sanacora (2008) "Serum brain-derived neurotrophic factor, depression, and antidepressant medications: Meta-analyses and implications," *Biological Psychiatry*, 64(6), pp. 527–532.

Simon, J.R. (2011) "Medical ontology." In *Philosophy of Medicine*. Elsevier.

Staner, L., P. Linkowski, and J. Mendlewicz (1994) "Biological markers as classifiers for depression: A multivariate study," *Progress in Neuro-Psychopharmacology and Biological Psychiatry*, 18(5), pp. 899–914.

Sullivan, P.F., M. Daly, and M. O'Donovan (2012) "Genetic architectures of 46 psychiatric disorders: The emerging picture and its implications," *Nature Reviews Genetics*, 13, pp. 537–551.

Tabery, J. (2014) *Beyond Versus: The Struggle to Understand the Interaction of Nature and Nurture*, MIT Press.

Visscher, P.M., W.G. Hill, and N.R. Wray (2008) "Heritability in the genomics era—concepts and misconceptions," *Nature Reviews. Genetics*, 9(4), pp. 255–266.

Visscher, P.M., M.E. Goddard, E.M. Derks, and N.R. Wray (2012) "Evidence-based psychiatric genetics, AKA the false dichotomy between common and rare variant hypotheses," *Molecular Psychiatry*, 17(5), pp. 474–485.

Visscher, P.M., M.A. Brown, M.I. McCarthy, and J. Yang (2012) "Five years of GWAS discovery," *American Journal of Human Genetics*, 90(1), pp. 7–24.

Wang, W.Y.S., B.J. Barratt, D.G. Clayton, and J.A. Todd (2005) "Genome-wide association studies: Theoretical and practical concerns," *Nature Reviews. Genetics*, 6(2), pp. 109–118.

Waters, C.K. (2007) "Causes that make a difference," *Journal of Philosophy*, 104(11), pp. 551–579.

Wimsatt, W.C. (2007) *Re-engineering Philosophy for Limited Beings: Piecewise Approximations to Reality*, Cambridge, Mass.: Harvard University Press.

de Winter, R.F.P. et al. (2003) "Anxious-retarded depression: Relation with plasma vasopressin and cortisol," *Neuropsychopharmacology: Official Publication of the American College of Neuropsychopharmacology*, 28(1), pp. 140–147.

Wray, N.R., S.M. Purcell, and P.M. Visscher (2011) "Synthetic associations created by rare variants do not explain most GWAS results," *PLoS biology*, 9(1), p. e1000579.

Zachar, P. (2014) *A Metaphysics of Psychopathology* 1st ed., Cambridge, Masschusetts: MIT Press.

Zachar, P. (2000) "Psychiatric disorders are not natural kinds," *Philosophy, Psychiatry, and Psychology*, 7(3), pp. 167–182.

Zheng, P. et al. (2012) "Plasma metabonomics as a novel diagnostic approach for major depressive disorder," *Journal of Proteome Research*, 11(3), pp. 1741–1748.

5 How Cancer Spreads

Reconceptualizing a Disease

Katherine E. Liu, Alan C. Love, and Michael Travisano

Abstract

Despite tremendous advances in cancer research, a stubborn gap exists between these advances and successful treatments that reduce mortality. One strategic way to address this gap is to model cancer as an infectious disease that we give ourselves. This conceptual maneuver shifts attention from cellular proliferation and tumor growth (how cancer grows) to cellular motility and metastasis (how cancer spreads), and emphasizes properties of cancerous cells that are responsible for the majority of deaths. We use the case of cystic fibrosis as an analogy to show the value of conceptualizing a genetic disease that is recalcitrant to treatment as an infectious disease. One consequence of modeling cancer in this manner is a more direct engagement with the pathological features of cancer's biology and, therefore, it has increased promise for identifying novel clinical applications—the primary goal of translational medicine.

1 Taking Translation Seriously

It is a matter of life or death. Despite 20 years of decreasing death rates in the U.S., cancer remains the second leading cause of death and is expected to overtake heart disease as the leading cause relatively soon (Siegel, Miller, and Jemal 2015). The irony is that cancer researchers have been incredibly productive in terms of publishing research papers and having new drugs approved. In 2012 alone, over 100,000 cancer-related papers were published worldwide. Between 2002 and 2014, the U.S. Food and Drug Administration approved 71 drugs for the treatment of cancers. However, the median gain in survival for those drugs was a mere 2.1 months (Fojo, Mailankody, and Lo 2014). In the U.S., 1.6 million new cases of cancer are diagnosed each year and lead to over 500,000 deaths. A patient's 5-year survival rate is tightly related to how early a cancer is diagnosed. Some cancers, if caught at an early stage, have a 90–100 percent 5-year survival rate (e.g., breast and skin cancers), whereas other cancers have a 5-year survival rate of 2–4% if diagnosed post-metastasis (e.g., pancreatic and lung cancers). What accounts for this stubborn gap between the tremendous advances in cancer biology and a continuing failure to successfully treat cancers beyond the earliest stages of detection?

"Translational research" focuses on the conversion of basic biology discoveries into clinically useful applications. This endeavor has many success stories, including the polio vaccine and antiretroviral treatments for HIV. In these cases, researchers secured an understanding of the relevant biology contributing to disease pathology, which facilitated clinical application. A robust understanding of the polio virus and our immune system has made it possible to almost eradicate polio globally (The Polio Eradication Initiative 2016). Similarly, an understanding of retrovirus biology and our immune system led to the development of HIV drug regimens that have greatly increased survival rates for HIV patients (Djawe et al. 2015), as well as decreased transmission (Cohen et al. 2011). Why has this kind of success not happened for most forms of cancer?

One potential success story in cancer biology is human papillomavirus (HPV). Nearly 100 percent of all cervical cancers are caused by HPV (Walboomers et al. 1999). Thus, if we could eliminate the virus, then we would stop the cancer. A study published in 2013 estimated that within four years of initiating a program of HPV vaccination, there had been a 56 percent decrease in HPV prevalence in 14–19 year old females despite only 34 percent of females being vaccinated (Markowitz et al. 2013). Longer-term studies have shown protection against HPV strains for 8–9 years after vaccination (Ferris et al. 2014; Naud et al. 2014). Decreasing HPV prevalence has translated directly into decreasing rates of cervical cancer. Though HPV is not yet eradicated, higher rates of vaccination foreshadow the eventual elimination of this cancer.

There is one parallel in cancer biology to the example of HIV treatments. Chronic myeloid leukemia (CML) patients who receive treatment with Imatinib (Gleevec) have an estimated 5-year survival rate of 89 percent (Druker et al. 2006) and a 10-year survival rate of 84 percent (Kalmanti et al. 2015). CML results from a specific chromosomal translocation, which leads to a fusion gene that produces an abnormally active tyrosine kinase (Lugo et al. 1990).[1] Because the resulting kinase is structured slightly differently than other kinases, researchers can precisely target and thus intervene on kinase functionality (Druker et al. 1996). This intervention specificity allows for the treatment's high efficacy. However, the most effective approaches to treating the majority of other cancers are chemotherapy, radiation, and surgery. Although successful in some cases, these treatments are often accompanied by debilitating side effects. Why haven't researchers achieved more effective and less harmful treatment regimens for cancer? These treatments do not engage with features of the biology of cancer related to its causing detrimental health outcomes (i.e., its pathology). Chemotherapy and radiation target dividing cells *generally*, not cancer cells *specifically*.[2] Noncancerous cells are targeted equally in these treatments, and quiescent cancerous cells can evade the toxins, especially after metastasis.[3] The most successful targeted treatments available beyond Imatinib that engage with aspects of the pathology of cancer, such as herceptin for ~15–20% of breast cancers (Ignatiadis et al. 2009), are combined with chemotherapy, which accounts for much of their efficacy (Moja et al. 2012; O'Sullivan et al. 2015; Perez et al. 2014). Researchers must concentrate more attention on how cancer kills in order to achieve successful translations into effective treatments.

How might we facilitate research that can secure a better understanding of the pathology of cancer? How can engagement with relevant aspects of the biology be spurred so as to generate clinically useful treatment regimens, thereby closing the gap between advances in cancer biology and failures to successfully treat cancers that kill? One way to productively address these questions is to rethink how we understand cancer as a disease. Cancer is currently modeled as a *genetic* disease—a pathology derived from a mutation or other alteration of genomic material (including epigenetic changes). This encourages researchers to focus on identifying the differential expression of particular genes or look for antibodies specific to proteins displayed by cancerous cells (Brennan and Wild 2015). Although not strictly wrong, this conception of cancer misses central aspects of its pathology. Cancer does involve the differential expression of particular genes, but it also involves the differential behavior of cells through complex causal sequences in the contexts of tumor growth and various dimensions of metastasis. Tumor growth dynamics have been studied intensely, acknowledging the importance of unregulated cellular proliferation, but less attention has been given to metastasis (the spread of cancer to other parts of the body), which is what actually kills people in the majority of cases.

Shifting more attention to the dimensions of metastasis (e.g., epithelial–mesenchymal transition, extravasation) has potential for fostering novel clinical applications, though only if we engage with the distinctive biological properties of metastatic cancers (Massagué and Obenauf 2016). A central element of this biology is cellular motility: how cancerous cells invade new environments through transformations of developmental processes. With cellular motility in the foreground, a new conceptualization of cancer emerges that engages directly with aspects of the biology that contribute to its pathology. Instead of primarily scrutinizing how cancer grows (in a tumor), we also must investigate *how cancer spreads* (via metastasis).[4] This spreading from one site to another within a host is analogous to an infection. Reconceptualizing cancer as an infectious disease *that we give to ourselves* suggests novel paths of research that have the potential to translate into successful mitigation strategies.

Our argument is structured as follows. First, we review the aims of cancer translational research and look at how well current conceptions of cancer match those aims. Next, we discuss the common distinction between genetic and infectious diseases and show how this breaks down in the context of cystic fibrosis. Cystic fibrosis is both a genetic and an infectious disease, and its pathological facets require taking the latter into account explicitly. This case study provides a segue to modeling cancer as an infection in addition to thinking about it as a genetic disease. After detailing this conception, we explore potential payoffs for cancer translational research, especially the application of approaches from epidemiology and ecology. We conclude that reflecting on how cancer is conceptualized in research can have direct consequences for investigation aimed at the discovery of new clinical interventions.

2 Idealized Models and Cancer Translational Research

A scientific community often has several distinct conceptualizations or models of the natural phenomena it attempts to investigate, manipulate, and explain. These are guided by a variety of assumptions embedded in, and sometimes constitutive of, the pertinent group of researchers. Multiple conceptualizations of natural phenomena can be maintained in investigative communities, such as treating physical matter as discrete particles, rigid bodies, or continuous masses (Wilson 2013). Biologists treat the interior space of a cell as if it was relatively empty even though intracellular space is known to be crowded (Ellis 2001). Idealization—a reasoning strategy that scientists use to describe or model that purposefully departs from features known to be present in nature (Jones 2005; Weisberg 2007)—plays a central role in these conceptions by facilitating the investigation of specific properties and making it difficult, if not impossible, to study others. For example, matter is composed of atoms and therefore not strictly continuous. Likewise, sometimes biologists ignore variation in the timing of developmental events to produce normal stages that facilitate studying ontogeny (Love 2010). Idealized models always combine distinctive strengths alongside latent weaknesses. Thus, idealizations should be closely matched with the aims of inquiry and coordinated with particular properties of interest for any natural phenomenon being investigated (see Plutynski, this volume).

With the introduction of the translational research initiative at the U.S. National Institutes of Health (NIH) in the 2000s, then-director Elias Zerhouni encouraged the development of "novel approaches" that can "be truly transforming for human health" (Zerhouni 2005, p. 1621). This led to the NIH Roadmap for Medical Research (Zerhouni 2003) and a reorganization of the NIH in 2012 to accommodate the addition of the National Center for Advancing Translational Science (Wadman 2011). The aims of translational research revolve around the identification and application of novel clinical treatments that benefit human health. In terms of cancer, this initiative has led to programs like The Cancer Genome Atlas to track key molecular changes that occur in various cancers (National Cancer Institute 2016a), and the formation of Specialized Programs of Research Excellence to serve as collaborative interdisciplinary teams with projects "that will result in new and diverse approaches to the prevention, early detection, diagnosis and treatment of human cancers" (National Cancer Institute 2016b).

These programs tend to presume a particular model of cancer as a natural phenomenon: cancer as a *genetic* disease caused by alterations in the molecular structure and function of the genome (Brennan and Wild 2015). As a consequence, researchers focus on those properties that are salient in this conception, such as the differential expression of particular genes or DNA sequence variants that serve as molecular signatures of tumor formation and growth. They also concentrate on those properties for a particular time in the sequence of cancer progression: early detection and prevention.

However, with respect to the goals of translational research, there are many reasons not to rely on early detection and prevention alone in efforts to eliminate cancer. For one, many tumors (or what is diagnosed as "cancer"[5]) will grow but remain localized, inflicting no harm on the individual. Identification of a growing tumor conjoined with the fear of its spreading can lead to unnecessary surgeries, such as thyroidectomies, mastectomies, and prostatectomies (Scudellari 2015). In the case of breast cancer, early treatment via chemotherapy, radiotherapy, or mastectomy does not reduce the risk of breast cancer–specific mortality; patients who undergo early treatment of ductal carcinoma *in situ* (stage 0 breast cancer) are just as likely to die of breast cancer as women in general society (Narod et al. 2015). Additionally, some cancers are identified via molecular proxies (cf. Nathan, this volume), such as elevated levels of prostate-specific antigen (PSA) in screening for prostate cancer, which can encourage more invasive tests, like prostate biopsies (Prasad, Lenzer, and Newman 2015). In these types of cases, attempts to detect cancer early are more likely to harm than to help the individual due to the high rate of false positives. Furthermore, early screening can some-times lead to reduced specific mortality (e.g., dying from prostate cancer), but not reduced overall mortality (e.g., dying from something else just as early or earlier).

Our intention is not to discourage attending to early detection and cancer prevention in clinical settings. However, even if screening and detection worked better, they would never catch all cases early. We still need a way to treat can-cers that have moved beyond the early stages of tumor growth and into the dimensions of metastasis (Massagué and Obenauf 2016). And since metastasis is the pathological aspect of cancer—essentially the causative factor of death—we need approaches that enhance the ability of researchers to understand, detect, prevent, and treat these later stages of cancer. We need a conceptualization of cancer that facilitates investigating specific properties that make it pathological; we need a model that is more closely matched with the aims of inquiry (i.e., translational research). Only then will we be positioned to better identify novel and effective treatment regimens for the metastatic dimensions of cancer.

Metastasis is remarkably different from tumorigenesis. Early in the process, unregulated and uncontrolled tumor growth are the greatest concerns (*how cancer grows*). Later, the transformation, movement, and recolonization of cancerous cells move to the foreground (*how cancer spreads*). Tumorigenesis and metastasis exhibit different environmental contexts, thus suggesting distinct ecological dynamics as well as different cellular properties, thereby highlighting different developmental processes. However, if the aims of inquiry are translational, and pathological aspects of cancer are concentrated largely in the metastatic phase (Massagué and Obenauf 2016), then we should reconceptualize cancer accordingly to facilitate investigation of those properties pertaining to metastasis, such as cell motility. An idealized con-ception of cancer that makes properties appearing at these later temporal stages of cancer progression more salient is desirable. An analogy from epidemiology is apt: treat metastasis as a kind of infection whereby cells of a single individual propagate out to new locations in the body. Instead of focusing primarily on molecular genetic markers of tumorigenesis, we should also investigate the properties underlying the infectious propagation. The advantages of this type of idealization are more readily apparent in an altogether different case: cystic fibrosis.

3 Genetic *and* Infectious Diseases: Illumination from Cystic Fibrosis

Many, if not most, diseases are now conceptualized in terms of molecular deficiencies and labeled "genetic" diseases (Darrason, this volume). This is most notable for those displaying a Mendelian pattern of inheritance, such as Phenylketonuria, due to a deficiency in phenylalanine hydroxylase (Blau, van Spronsen, and Levy 2010), or Huntington's disease, resulting from an autosomally dominant mutation in the *huntingtin* gene (Walker 2007). Diseases showing more complex, non-Mendelian inheritance patterns (van Heyningen and Yeyati 2004) are also conceptualized in terms of alterations in normal gene expression or some other molecular activity (e.g., abnormal protein folding). This conceptualization focuses on intrinsic properties of the individual and is reflected in our language: we "have" genetic diseases. By way of contrast, a different category of diseases consists of those associated with specific microbial agents, such as *Mycobacterium tuberculosis*. Patterns of inheritance are not in view for most infectious diseases; instead, the concern is about their *transmission*. This conceptualization focuses on the properties of microbes extrinsic to the individual: we "get" or "catch" infectious diseases. Therefore, standard categorizations of disease typically distinguish genetic diseases from infectious diseases. However, there are some diseases that blur the line.

Cystic fibrosis (CF) is characterized by thick mucus buildup that restricts the function of the lungs and airways. It originates due to mutations in a single gene, *cystic fibrosis transmembrane conductance regulator (CFTR)*, which produces a transmembrane channel protein that regulates chloride ion transport. In cells that generate water-based secretions (e.g., mucus, sweat, saliva, or tears), the regulation of chloride ions is critical for moving water in and out of tissues. Because of this causal etiology, and the fact that it displays an autosomally recessive Mendelian inheritance pattern, CF is routinely labeled a *genetic* disease. However, the pathological aspects of CF do not derive primarily from a buildup of mucus. CF is also characterized by chronic infections due to multiple microbial species, especially *Pseudomonas aeruginosa* and *Burkholderia cepacia*. These infections are transmissible from patient to patient and comprise an essential part of the negative health effects associated with CF (Sun et al. 1995; Fothergill, Walshaw, and Winstanley 2012). Thus, CF is also an *infectious* disease (Lyczak, Cannon, and Pier 2002; Sibley, Rabin, and Surette 2006).

If the goal is to adequately address CF as a disease, then conceptualizing it only, or even primarily, as a genetic disease will lead to clinical frustration. CF has been stubbornly difficult to treat for a genetic disease where we have pinpointed the specific gene and that displays a relatively simple inheritance pattern; gene therapies continue to be disappointing (Alton et al. 2015). Importantly, this does not mean that conceptualizing CF as a genetic disease will prevent all research advances into aspects of its basic biology. Conceptualizations are idealized models in the sense of purposefully neglecting known features of natural phenomena in order to facilitate investigation of specific properties. Researchers know that bacterial infections are a key component of the disease etiology of CF, but the genetic disease conceptualization ignores these for the purpose of focusing on molecular genetic mechanisms that break down in the early stages of the condition. Models should be closely matched with the aims of inquiry and coordinated with particular properties of interest for a natural phenomenon being investigated. Conceptualizing CF as an infectious disease is an idealization that intentionally neglects the breakdown in chloride ion transport regulation due to a mutated transmembrane protein. However, it is an idealization that facilitates the investigation of specific properties of CF related to its pathology.

The realization that CF's complex pathogenicity involves infection has yielded multiple advances in treatment. CF was originally described as a pathological condition of the pancreas ("steatorrhea"; Andersen 1938; Parmelee 1935). Poor pancreatic function often resulted in death due to malnutrition for many patients, but it also would lead to other conditions, such as diabetes (O'Sullivan and Freedman 2009). CF was long considered a pediatric disease because a majority of individuals with the condition died in their first year of life. At this time, the latest observed age of death was 14.5 years, but the average age of death was 1 year (Parmelee 1935).

By the 1950s, the characterization of CF had expanded to encompass general exocrine gland dysfunction, which included pancreatic deficiencies, susceptibility to chronic bronchitis, and increased electrolytes in the sweat (Andersen 1958). Because pancreatic dysfunction could be treated with pancreatin substitution therapy, the usual cause of death in those affected was respiratory infection. By then, mortality was not occurring until between 3 and 5 years of age, and many were living through their teenage years. The introduction of penicillin was a key part of this improved lifespan (Andersen 1958). Now patients live well into adulthood; the median survival age more than doubled between 1969 and 2001, from 14 years to 30 years (Döring and Hoiby 2004). It has been estimated that patients born in 2000 will live into their 50s (Dodge et al. 2007). This is due to particular antibiotics, such as sulfonamides, tetracyclines, and carbenicillin, which came into use from the 1940s through the 1970s (Fernald and Boat 1987). Patients with specific infection types (*P. aeruginosa, B. cepacia* complex, or MRSA infections) are encouraged to be seen in separate clinics or at separate times than other patients to reduce the chances of cross-infection (Kerem et al. 2005).

Conceptualizing CF as an infectious disease has encouraged increased attention to the evolutionary dynamics of its microbial pathogens. Many researchers have analyzed the long-term evolution of *P. aeruginosa* in respiratory infections to understand how it can adapt to the dynamic lung environment of CF patients. A recent study looked at the immediate impact of various treatments on *P. aeruginosa* in a 34-year old female patient over a 1-year period (Diaz Caballero et al. 2015). Sputum samples were obtained longitudinally and correlated with the prescribed antibiotic treatments (12 samples over a 1-year period). It was discovered that *P. aeruginosa* populations evolved quickly in stressful environments (e.g., in response to antimicrobial treatments). There were genomic signs of recurrent selective sweeps and many indicators of parallel adaptation at loci for antibiotic resistance. However, the advantage of these alleles was dependent on specific antibiotic environments; an allele was beneficial during certain antibiotic treatments but disadvantageous in others. Understanding short-term evolutionary dynamics is critical to deciphering the relationships between treatment regimes and clinical success. Given that treatment regimes for CF usually include long-term antibiotics to prevent or control infections (along with various methods for loosening mucus build up), conceptualizing CF as an infectious disease helps to accent how microbes evolve in the host's lungs and thereby intensify the pathology of CF. Identifying and characterizing these features increases the potential for translating research findings into effective clinical treatments.

The trajectory of research on CF shows the differential value of distinct disease conceptualizations. CF can be modeled as either genetic or infectious, and each idealization serves the purpose of securing a better understanding of the relevant features of its basic biology. Since each conceptualization combines distinctive strengths (in terms of the specific properties in view) and latent weakness (in terms of those features neglected or ignored), the important question is not whether one conceptualization is better than the other but rather what end or goal they serve. Given that the pathological dimensions of CF arise later in the disease progression due to evolutionarily dynamic populations of microbes, a conceptualization of CF as infectious is currently better suited to identifying clinically relevant and effective treatments. Recent research amply supports this claim.

Three key lessons emerge from recognizing that we can model CF as either a genetic or infectious disease. The first is that a conceptualization should be well matched to the aims of inquiry. If the aims of inquiry are translational, then the disease conceptualization should make salient those properties that contribute to pathology. Knowing these features of the basic biology for a particular disease are most relevant to finding effective clinical treatments. Second, the case of CF (and what we have already noted for cancer) suggests that conceptualizations of a disease depend on where one concentrates in its temporal progression. The initial stages of CF are better construed as a genetic disease, where the buildup of mucus is a function of mutations in *CFTR*. The later stages of CF are better construed as an infectious disease, in which different bacterial communities invade under the conditions of excess mucus and progressively become better adapted to the host's immune defenses (and therefore more pathological). The evolutionary dynamics that yield this progressive adaptation are germane for fine tuning treatment regimens to increase their effectiveness. Finally, shifting from a model of CF as a genetic disease to CF as an infectious disease alters how we view causal responsibility within the disease progression. At the outset, consistent with modeling cancer as a genetic disease, the responsibility is located in the deficient molecular component, which leads to the mucus buildup in the lungs. However, later in the temporal progression of CF, the responsibility is located in the bacterial species invading the environment of the lungs and ultimately producing pathogenic effects in the host.

The three lessons—matching conceptualizations to aims of inquiry, focusing on particular segments in the temporal progression of a disease, and isolating where causal responsibility lies for pathology—can be encapsulated as we segue to explicitly conceptualizing cancer as an infectious disease. For cancer translational research, a conceptualization of the disease should be matched to finding effective clinical treatments beyond the earliest stages of detection, which means later stages of its progression (i.e., metastatic dimensions), where the relevant causal responsibility lies with motile cells invading and adapting to new areas of the body rather than with mutations that initiate tumor formation by upsetting cell cycle regulation.

4 Conceptualizing Cancer as an Infection

Cancer is often thought to be the classic example of a genetic disease. Cancers start with a mutation (either spontaneous or induced), followed by unrestricted proliferation to generate a mass of cells (i.e., a tumor), and then, with variable frequency, transition to disseminating cancerous cells to other locations in the body (i.e., metastasis), which usually results in death. Because a cancer starts and ends in an individual who harbors a pertinent genetic mutation, it seems odd to think of cancer as an infectious disease. You (seemingly) cannot "catch" cancer. The pathology is not transmitted from one individual to another, at least not typically. There are cases of transmissible cancer, such as in clams or dogs (Metzger et al. 2015; Murchison 2009). Perhaps the most famous of these has been observed in Tasmanian devils.

The Tasmanian devil is facing extinction due to devil tumor facial disease (DTFD). DTFD is transmitted cellularly though facial bites while fighting. The cells grow rapidly, spreading throughout the face and neck and then metastasize through the lymph nodes to reach distant organs (Loh et al. 2006). Within months, the cancer's spread leads to death. DTFD shows all of the conventional signs of cancer: unrestricted growth of cells that spread throughout the body (or metastasize) and lead to death. However, molecular analyses of multiple individuals have demonstrated that the cancer is clonal and monophyletic, arising from a single outbreak of Schwann cell origin. Therefore, it is not initiated genetically by a mutation in the affected individual. The cancer is transmitted as an allograft (i.e., as a distinct tissue); cancerous cells between individuals exhibit more similarity than the cancer does to the cells of its host (Murchison et al. 2010). The transmissibility of DFTD is associated with a decreased immune response. Since the tumor is a graft from a different individual, it should be recognized as foreign by the immune system, but it is not (Siddle et al. 2007). Again, this behavior tightly parallels what we would expect for an infectious disease.

Despite their intrinsically fascinating status, DFTD and other transmissible cancers are usually considered exceptions to the rule. In part, this is because the majority of cancer cases thus far studied do not involve any transmission of the pathology from one individual to another. However, if we set aside the criterion of transmission from one "distinct" individual to another and concentrate on how cancer spatially spreads through the various dimensions of metastasis, cancer is quite analogous to a transmissible disease that moves from one area or tissue in an individual to another. Subsequent to tumor origination and growth, it behaves like an infectious disease that we give ourselves, or that one part of an individual gives to other parts. This conceptualization concentrates our attention on how cancerous cells move from one location to another, from the original tumor to establishment elsewhere in the body (Massagué and Obenauf 2016), which is how cancer primarily kills individuals.

Recall our three lessons from the case of CF. First, conceptualizations of a disease should be suited to finding effective clinical treatments by drawing attention to specific properties that contribute to a disease's pathology. Second, a conceptualization is keyed to a particular segment of the temporal progression of a disease. Third, any conceptualization highlights particular types of causal responsibility for a disease. Conceptualizing cancer as an infectious disease focuses on the later stages of its progression (i.e., metastatic dimensions), rather than earlier stages where a molecular characterization of unregulated cellular proliferation in terms of genetic mutations is apt. These later stages of cancer's progression are where the specific, pathological properties of the disease are largely confined. Metastasis is the primary killer, not tumor growth. And causal responsibility attaches to the property of cell motility whereby tumor cells detach and invade new environments (i.e., other locations in the body) through a variety of developmental transformations (e.g., epithelial–mesenchymal transition or intravasation), as well as adapt to different conditions within the body (Massagué and Obenauf 2016). An idealized model of cancer as an infectious disease appears well suited to the aims of cancer translational research.

Once this conceptualization of cancer is in view, it encourages us to scrutinize in more detail the specific properties made salient in the model. For example, cancerous cells exhibit distinct evolutionary dynamics in the host individual in the later stages of progression (i.e., the metastatic dimensions), in part to evade the immune system, which is being encountered in different ways as these cells migrate into new environments of the body. The fallible engagement of the immune system helps to account for the existence of remission phases, similar to what we see for other infectious diseases (e.g., HIV), as well as resistance to treatments. Although the cancerous cells that evolve to evade the host immune system or become resistant to treatment are not microbes (as in CF), the cellular behavior is remarkably similar.

The first similarity is that both cancer and infectious diseases have underlying clonal evolutionary dynamics that are specific to that individual and depend on a variety of local circumstances. For example, there are episodes of periodic selection and clonal interference that regularly affect populations of cancerous cells and infectious agents (Nowell 1976; Diaz Caballero et al. 2015), but these exhibit characteristics unique to each particular patient, such as how rapidly their pathological effects become manifested due to individual differences in immune response. A second similarity is that there will be convergent evolution of cancerous cells and infectious agents as different functional niches are filled, leading to the appearance of homogeneity across and between cancers and infectious diseases, especially on longer time scales and for analyses at higher levels of organization. This convergence is what yields typical ranges of disease progression for particular forms of cancers and transmissible infections. However, historical contingency and differences in the local ecologies of each individual will lead to variation in disease progression and thus the need for different treatment plans between individuals. The order and frequency of treatments are important for understanding progression and predicting the evolution of a cancer

(Greaves and Maley 2012; Diaz Caballero et al. 2015). Different treatments will stress a cancer in different ways, leading to differential growth and variegated propagation among various clones in the population. Thus, analyses at lower levels of organization and across shorter time scales will often reveal heterogeneity that is relevant to effective treatments, especially for the dimensions of metastasis (Massagué and Obenauf 2016).

A third similarity between cancer and infectious diseases is that both depend on decreased immune system function for disease persistence and progression. In the case of CF, as populations of *P. aeruginosa* adapt to the lungs of an affected individual, immune system function decreases, which allows for the persistence of the bacterial infection and nurtures its evolution. With DFTD, the contracted cancerous cells can only persist in the presence of a depressed immune system. Although mutations occur frequently in somatic cells, cancers do not occur as a necessary result because the immune system recognizes cells harboring these mutations as abnormal and eliminates them. Cancerous cells are able to evade the immune system and proliferate. This principle is what immunotherapies aim to leverage: increase the activity of the immune system so as to better detect and destroy abnormal and potentially cancerous cells. However, less attention has been given to how the immune system might detect and destroy circulating tumor cells post-metastasis. For example, epithelial–mesenchymal transition may not be required for metastasis in lung cancers, but when this transition does occur, there is an increase in drug resistance (Fischer et al. 2015). Modeling cancer as an infectious disease points us toward immunotherapies, which are receiving increasing attention (see below, Section 5), though we might have arrived there earlier.[6] Drawing an analogy between CF and cancer illustrates the value of conceptualizing cancer as an infection; it directs our attention to the relevant biological features for identifying novel clinical applications.

5 Payoff: Conceptual Reflection Leads to Research that Saves Lives

The reconceptualization of cancer as an infectious disease reminds us starkly of the primary source of cancer mortality—metastasis. This reorients the usual success story told about cancer research. Every January, the American Cancer Society, the Centers for Disease Control and Prevention, the National Cancer Institute, and the North American Association of Central Cancer Registries jointly release the latest statistics surrounding cancer. These annual reports show that death rates continue to decline, as they have for the last 20 years (see, e.g., Siegel, Miller, and Jemal 2015). However, these reports lump together all cases of the same type of cancer. For example, the 2015 report shows that lung cancer survival rates are improving. This is almost exclusively because of early detection through screening rather than because of better treatments. The 5-year survival rate for lung cancer if diagnosed early is over 50 percent, but the 5-year survival rate for lung cancer if diagnosed post-metastasis is 2 percent. Unfortunately, more than half of all cases are not diagnosed until the cancer has metastasized. While new molecular methods for early detection certainly count as contributions to translational medicine, the pathogenicity of metastasis is sobering; the need to achieve an understanding of metastasis that will translate into more effective clinical treatments becomes readily apparent.

That an idealized model of cancer as an infectious disease can change how research programs are interpreted is perhaps not surprising. For example, with cancer modeled as an infectious disease, immunotherapies move into the spotlight for addressing metastasis. If the idealized model changes how research programs proceed, then the proposal is more substantial. We think this is the payoff of conceptual reflection on how we think about cancer as a natural phenomenon; in a very real sense, it could lead to saved lives. Research that concentrates on achieving a better understanding of cancer's sources of mortality as a consequence of this reconceptualization is poised to make a big clinical difference. Before isolating these effective treatments, changes in the research programs need to unfold. We see at least two domains where conceptualizing cancer as an infectious disease could help redirect aspects of cancer translational research: epidemiology and ecology.

Cancer epidemiologists typically study trends in incidence and mortality rates with the goal of identifying risk factors that can be used in screening and prevention practices. This is how tobacco and asbestos were identified as risk factors for cancer. Similarly, epidemiologists track the spread of diseases via molecular markers, such as new subtypes of influenza that are measured globally on an annual basis (cf. Russo and Vineis, this volume). Identification of subtypes is critical for influenza vaccine development; even a partial misprediction of the annual type results in suboptimal vaccines (Flannery et al. 2015; Ohmit et al. 2014; Pebody et al. 2013). Tuberculosis is another example; strains of *Mycobacterium tuberculosis* can be identified to determine the proportion of cases due to new infections versus reactivation of previous infections. Thus, researchers can identify which strains are more or less able to spread through populations (Foxman and Riley 2001). These examples highlight three familiar aspects associated with infectious disease: pathogen infectiousness, host susceptibility, and host–pathogen interactions. Influenza is one of the most infectious pathogens but is readily cleared by otherwise healthy individuals. *M. tuberculosis* is far less infectious, but chronic and lethal infections are commonplace without antibiotic treatment. In both cases, there is substantial variation in host susceptibility among individuals.

While some of these insights have been incorporated into cancer research, this has mostly occurred in a piecemeal fashion. Cancer treatment outcomes are negatively correlated with the degree of metastatic spread, which has encouraged attempts to systematize the stages of cancer (Bülzebruck et al. 1992; Bundred 2001). Ideally, such a systemization would provide better approaches to screening, estimates of survival, and therapy. However, debates about approaches to all three of these are extremely contentious (Hari et al. 2013; Miller et al. 2014; Pace and Keating 2014). This is largely because epidemiological perspectives are not well integrated into cancer medicine. Although there are now very sophisticated mathematical models (Kam, Rejniak, and Anderson 2012), their clinical relevance has been modest. The explicit incorporation of mathematical models and epidemiological perspectives into translational medicine more broadly should provide clarity about the sources of cancer mortality (e.g., through more precise delineations of resistance to treatment at distinct stages). Recent studies have identified genetic variation among cancer cells as a possible indicator of cancer aggressiveness and persistence (Lauren et al. 2006; Park et al. 2010). Epidemiological studies of many infectious diseases over the past half-century show similar correlations (Bloom 1979; Demerec 1948; Webster et al. 1992). If cancer had been conceptualized as an infectious disease, this could have provided a faster pathway to discovering potential strategies of treatment. More generally, cancer translational research has not focused on the origin of genetic variation beyond assuming it arises from mutation (Burrell et al. 2013). Genetic variation—a pivotal element of infectious disease epidemiology—is the ultimate source of cancer recurrence after therapy.

Researchers have long recognized that individuals vary in their cancer susceptibility and survivorship, but the emphasis was primarily on mutations in specific genes or behavioral and environmental factors that promote differences in cancer initiation and spread (Danaei et al. 2005; Nigro et al. 1989). Until recently, there was much less interest in the host environment. Individuals vary in their susceptibility to infectious diseases, and these differences have provided insights for understanding disease mechanisms and facilitating the development of treatments (e.g., partial immunity to HIV). Perhaps of greater value is the recognition that the infectiousness of pathogens and susceptibility of hosts can *interact*. Some individuals are relatively unsusceptible to pathogens that are lethal to others. Longitudinal and comparative observations for cancer incidence and progression have provided clues to individualized treatments, but identifying differential susceptibility of distinct tissues within the same individual merits further investigation. Comparisons within and between individuals with specific cancers that progressed to metastasis in preferred locations with those individuals whose tumor (of the same type) stayed benign or exhibited different patterns of metastasis could illuminate the nature of these interactions. Scrutinizing instances where progression is anticipated but not observed (especially in cases of spontaneous remission) is likely to generate translational insights. Epidemiological approaches premised on studying cancer as an infectious disease will improve our understanding of the dynamics of its spread and yield insights relevant to effective treatment regimes that reduce mortality beyond early detection.

A focus on epidemiology naturally presages the relevance of ecological perspectives to the study of cancer as an infectious disease. The probability of transmission (or propensity to spread), contact rates between individuals (or between parts of the body), and the susceptibility of the next host (or organ system) are relevant to infectious disease and metastatic cancers alike. Although epidemiological models are powerful approaches to formulating hypotheses about patterns of contagion and pertinent risk factors, the actual mechanisms involve ecological interactions (Merlo et al. 2006). For example, cells (or clusters of cells) constantly break off from tumors and move through the circulatory system, but only a small fraction of those clusters successfully seed new tumors at a distant location (Aceto et al. 2014; Massagué and Obenauf 2016). "Seed versus soil" views of metastasis suggest that cells (the seeds) must match characteristics of the new environment to be occupied (the soil) in order for metastases to be successful. Models that include relevant ecological interactions provide mechanistic details about both matches and mismatches to explain phenomena like preferred metastatic destinations.

The application of concepts from infectious disease biology can help in identifying the primary characteristics of environments that permit metastasis. For example, what properties are pertinent to the susceptibility of the new host? Which tissues are more likely to allow for the successful planting of metastases, and why? In parallel with the infectious disease examples discussed earlier, susceptible tissues are likely found in systems that have lessened immune responses. Organ systems with stronger immune responses will recognize cancerous cells and eliminate them. In cases of organotropic metastasis, where there are special affinities between the tissue of tumor origination and likely sites of spread (e.g., lung, liver, and brain), general immune reactions to cancerous cells may be subverted by a distinct molecular indicator, such as exosomes containing distinctive suites of integrins (Hoshino et al. 2015). Knowing these "ecological" interactions of host immune responses (or lack thereof) is critical for developing novel clinical interventions.

Although cancer research has increasingly focused on the role of the immune system, especially with the rise of immunotherapy (Restifo, Dudley, and Rosenberg 2012), progress on this front could have been achieved earlier and in a less haphazard fashion by recognizing general ecological principles that apply to the infectious aspects of cancer. For example, what are the contact rates between cancer cells and different regions of the body? What kind of contact is the most important for successful metastasis (e.g., intermittent, repeated, or constant)? Environments with lots of cell movement permit mass-action dynamics where all cells have access to resources, which allows for the invasion of cheaters or cells that exploit resources without contributing. On the other hand, environments that are spatially structured severely limit the ability of cheaters to invade (Chao and Levin 1981; Escalante et al. 2015; Greig and Travisano 2004; Greig and Travisano 2008; Travisano and Velicer 2004). Thus, we would expect tissues with more spatial structure to be more resistant to the spread of cancer, whereas tissues with less spatial structure should be more permissive to metastasis. Addressing these ecological components is essential for fighting infectious diseases and therefore must be included in mitigating the spread of cancers as well.

Ecological perspectives on cancer that incorporate interactive dynamics and environmental context are not new (see. e.g., Gatenby, Brown, and Vincent 2009; Kareva 2015; Merlo et al. 2006). Nowell's well-known clonal evolution model had ecological elements, even if not explicitly stated (Nowell 1976). Most of these perspectives come in one of two forms. First, there are reviews that attempt to apply as many ecological (and evolutionary) concepts as possible to cancer as a system. Everything ranging from competition (for space and resources) and niche construction (changes in the microenvironment) to predation (via the immune system) and ecological succession (progression through the stages of cancer) are discussed (Kareva 2015; Merlo et al. 2006). The second form is to highlight similarities with already familiar systems, such as drawing analogies between cancer and invasive species or pests (Gatenby 2009; Gatenby, Brown, and Vincent 2009), or the evolution of cancer and the evolution of a new species (Kaznatcheev 2014). Our goal is not to add more ecological concepts or analogous systems to the list but to narrow the list to idealized models that will lead to novel translational results. This means applying select concepts to cancer biology so as to understand why and how metastasis occurs. That cells moving through the circulatory system are similar to organisms migrating is interesting, but can an analogy to migration yield novel treatments? The discussion of mass-action dynamics and cheating suggests an affirmative answer, though more work is needed to elucidate the details. As with any analogy, cancer stages and ecological succession have many similarities, but also many differences. Comparing metastasis to ecological succession highlights that they share a sequence of relatively predictable events (Kareva 2015). However, the predicted progression would be linear and end when the climax or stable state is reached, whereas the processes relevant to cancer are more cyclical and involve diverse environments (Divoli et al. 2011).[7] An idealized model based on the similarities should be matched to the aims of inquiry. If the aim is to understand the pathology of metastasis, ecological succession might not be the best analogy for identifying effective clinical treatments.

6 Conclusion

Cancer translational research is making notable progress, but that progress is almost entirely through the vehicle of prevention and early detection from screening practices. There is a gap between these advances in cancer biology and our ability to treat cancers beyond the earliest stages of detection. We have argued that one way to address this gap is through reconceptualizing cancer as an infectious disease that we give ourselves. Most research programs model cancer as a genetic disease and focus on the early stages of how cancer grows. Both of these conceptualizations are idealized models that facilitate the investigation of some properties and elide the study of others. If the aims of inquiry are guided by the demands of identifying effective clinical treatments (i.e., translational research), then an idealized model that accents the pathological properties of cancer is a necessity. These pathological properties are almost universally located in the metastatic dimensions of cancer.

Support for switching from a genetic disease conceptualization to an infectious disease conceptualization derives from the case of CF, where increasing scrutiny of microbial infections in the later stages of disease progression has yielded significant advances in patient treatment. Additionally, the dynamics of transmissible cancers, such as DFTD, assist us in reorienting our conception of cancer as infectious despite most cases of cancer being confined to a single individual. Cancerous cells spread to new locations in an individual's body just as a pathogen can spread to new hosts. This conceptual move fixes our attention on the relevant biological features of metastatic cancers that should help translational research meet its goals. This includes novel angles for cancer research programs, such as the application of epidemiological and ecological approaches, which are both sources for inspiration in finding shared patterns of transmission dynamics and isolating mechanistic principles that will predict the direction and rate of metastasis, thereby nurturing novel clinical applications. It is noteworthy that this exemplifies the repeated call for cross-disciplinary approaches to cancer (e.g., Ogden 2015).

In matters of life and death, conceptual reflection is usually deemed a luxury. However, stepping back to evaluate how our idealized models do or do not contribute to the goals of translational medicine may be indispensable for identifying a suite of new treatment regimens that effectively deal with metastatic cancers. Only then will we be positioned strategically to understand the pathogenic properties of cancer's biology and thereby close the gap between major research advances and a paucity of treatments that save lives.

Notes

1   In noncancerous individuals, the genes *c-abl* and *brc* are separated spatially on different chromosomes. CML occurs when the two chromosomes break and fuse together at the location of the two genes forming the *brc-abl* gene. This new version of the gene produces an enzyme (a tyrosine kinase) that is constitutively active, leading to excess differentiation of blood stem cells into white blood cells (Sawyers 1999).
2   Immunotherapies are an attempt to secure a general treatment that will target cancerous cells by harnessing the specificity of our adaptive immune system.
3   Some chemotherapies might target cancerous cells more than noncancerous cells because the former upregulate normal processes. However, because these processes occur in normal cells, they also will be affected. For example, the chemotherapy drug cisplatin induces apoptosis by interfering with DNA repair, which operates in normal cells but is more active in cancerous cells (Dasari and Bernard 2014).
4   It is important to emphasize that many biologists work on cell motility and cancer metastasis. Our argument is that modeling cancer in a particular way encourages investigating some properties rather than others (see below, Section 2). Conceptualizing cancer as an infectious disease shifts attention away from tumor origination toward cancerous cell propagation. This does not devalue research focused on finding ways to eliminate nonmetastatic tumors.
5   *Cancer* and *tumor* are frequently used interchangeably, but the label *tumor* is reserved for uncontrolled growth, and *cancer* refers to the spread of the cells to other locations (National Cancer Institute 2015). The term *cancer* covers both solid (e.g., breast and prostate cancers) and liquid cancers (e.g., leukemias and lymphomas). Our discussion focuses only on solid cancers.
6   The "father of present day immunotherapy," William B. Coley, treated inoperable sarcoma by injection of bacteria in the early 1900s. This treatment was based on the observation that cancer regressions were associated with bacterial infections called erysipelas (Wiemann and Starnes 1994).
7   Kareva admits the analogy does not hold exactly because the patient usually dies when the climax state is reached, but this is a different application of ecological succession than our view suggests.

References

Aceto, N., A. Bardia, D.T. Miyamoto, M.C. Donaldson, B.S. Wittner, J.A. Spencer, M. Yu, A. Pely, A. Engstrom, H. Zhu, B.W. Brannigan, R. Kapur, S.L. Stott, T. Shioda,

S. Ramaswamy, D.T. Ting, C.P. Lin, M. Toner, D.A. Haber, and S. Maheswaran. (2014) "Circulating tumor cell clusters are oligoclonal precursors of breast cancer metastasis," *Cell* 158 (5), pp. 1110–1122.

Alton, E.W.F.W., D.K. Armstrong, D. Ashby, K.J. Bayfield, D. Bilton, E.V. Bloomfield, A.C. Boyd, J. Brand, R. Buchan, R. Calcedo, P. Carvelli, M. Chan, S.H. Cheng, D.D.S. Collie, S. Cunningham, H.E. Davidson, G. Davies, J.C. Davies, L.A. Davies, M.H. Dewar, A. Doherty, J. Donovan, N.S. Dwyer, H.I. Elgmati, R.F. Featherstone, J. Gavino, S. Gea-Sorli, D.M. Geddes, J.S.R. Gibson, D.R. Gill, A.P. Greening, U. Griesenbach, D.M. Hansell, K. Harman, T.E. Higgins, S.L. Hodges, S.C. Hyde, L. Hyndman, J.A. Innes, J. Jacob, N. Jones, B.F. Keogh, M.P. Limberis, P. Lloyd-Evans, A.W. Maclean, M.C. Manvell, D. McCormick, M. McGovern, G. McLachlan, C. Meng, M.A. Montero, H. Milligan, L.J. Moyce, G.D. Murray, A.G. Nicholson, T. Osadolor, J. Parra-Leion, D.J. Porteous, I.A. Pringle, E.K. Punch, K.N. Smith, N. Soussi, S. Soussi, E.J. Spearing, B.J. Stevenson, S.G. Sumner-Jones, M. Turkkila, R.P. Ureta, M.D. Waller, M.Y. Wasowicz, J.M. Wilson, and P. Wolstenholme-Hogg. (2015) "Repeated nebulisation of non-viral CFTR gene therapy in patients with cystic fibrosis: A randomised, double-blind, placebo-controlled, Phase 2b trial," *Lancet Respiratory Medicine* 3 (15), pp. 684–691.

Andersen, D.H. (1938) "Cystic fibrosis of the pancreas and its relation to celiac disease: A clinical and pathologic study," *The American Journal of Diseases of Children* 56, pp. 344–399.

— (1958) "Cystic fibrosis of the pancreas," *Journal of Chronic Diseases* 7 (1), pp. 58–90.

Blau, N., F.J. van Spronsen, and H.L. Levy (2010) "Phenylketonuria," *The Lancet* 376 (9750), pp. 1417–1427.

Bloom, B.R. (1979) "Games parasites play: How parasites evade immune surveillance," *Nature* 279 (3), pp. 21–26.

Brennan, P., and C.P. Wild (2015) "Genomics of cancer and a new era for cancer prevention," *PLoS Genetics* 11 (11), e1005522.

Bülzebruck, H., R. Bopp, P. Drings, E. Bauer, S. Krysa, G. Probst, G. Van Kaick, K.M. Müller, and I. Vogt-Moykopf (1992) "New aspects in the staging of lung cancer," *Cancer* 70 (5), pp. 1102–1110.

Bundred, N.J. (2001) "Prognostic and predictive factors in breast cancer," *Cancer Treatment Reviews* 27, pp. 137–142.

Burrell, R.A., N. McGranahan, J. Bartek, and C. Swanton (2013) "The causes and consequences of genetic heterogeneity in cancer evolution," *Nature* 501 (7467), pp. 338–345.

Chao, L., and B.R. Levin (1981) "Structured habitats and the evolution of anticompetitor toxins in bacteria," *Proceedings of the National Academy of Sciences of the United States of America* 78 (10), pp. 6324–6328.

Cohen, M.S., Y.Q. Chen, M. McCauley, T. Gamble, M.C. Hosseinipour, N. Kumarasamy, J.G. Hakim, J. Kumwenda, B. Grinsztejn, J.H.S. Pilotto, S.V. Godbole, S. Mehendale, S. Chariyalertsak, B.R. Santos, K.H. Mayer, I.F. Hoffman, S.H. Eshleman, E. Piwowar-Manning, L. Wang, J. Makhema, L.A. Mills, G. deBruyn, I. Sanne, J. Eron, J. Gallant, D. Havlir, S. Swindells, H. Ribaudo, V. Elharrar, D. Burns, T.E. Taha, K. Nielsen-Saines, D. Celentano, M. Essex, and T.R. Fleming. (2011) "Prevention of HIV-1 infection with early antiretroviral therapy," *The New England Journal of Medicine* 365 (6), pp. 2187–2198.

Danaei, G., S. Vander Hoorn, A.D. Lopez, C.J.L. Murray, and M. Ezzati (2005) "Causes of cancer in the world: Comparative risk assessment of nine behavioural and environmental risk factors," *The Lancet* 366 (9499), pp. 1784–1793.

Dasari, S., and P. Bernard (2014) "Cisplatin in cancer therapy: Molecular mechanisms of action," *European Journal of Pharmacology* 740, pp. 364–378.

Demerec, M. (1948) "Origin of bacterial resistance to antibiotics," *Journal of Bacteriology* 56 (1), pp. 63–74.

Diaz Caballero, J., S.T. Clark, B. Coburn, Y. Zhang, P.W. Wang, S.L. Donaldson, D.E. Tullis, Y.C.W. Yau, V.J. Waters, D.M. Hwang, and D.S. Guttman (2015) "Selective sweeps and parallel pathoadaptation drive *Pseudomonas aeruginosa* evolution in the cystic fibrosis lung," *mBio* 6 (5), e00981–15.

Divoli, A., E.A. Mendonça, J.A. Evans, and A. Rzhetsky (2011) "Conflicting biomedical assumptions for mathematical modeling: The case of cancer metastasis," *PLoS Computational Biology* 7 (10), e1002132.

Djawe, K., K. Buchacz, L. Hsu, M.-J. Chen, R.M. Selik, C. Rose, T. Williams, J.T. Brooks, and S. Schwarcz (2015) "Mortality risk after AIDS-defining opportunistic illness among HIV-infected persons—San Francisco, 1981–2012," *The Journal of Infectious Disease* 212, pp. 1366–1375.

Dodge, J.A., P.A. Lewis, M. Stanton, and J. Wilsher (2007) "Cystic fibrosis mortality and survival in the UK: 1947–2003," *European Respiratory Journal* 29 (3), pp. 522–526.

Döring, G., and N. Hoiby (2004) "Early intervention and prevention of lung disease in cystic fibrosis: A European consensus," *Journal of Cystic Fibrosis* 3 (2), pp. 67–91.

Druker, B.J., F. Guilhot, S. G. O'Brien, I. Gathmann, H. Kantargian, N. Gattermann, M.W.N. Deininger, R.T. Silver, J.M. Goldman, R.M. Stone, F. Cervantes, A. Hochhaus, B.L. Powell, J.L. Gabrilove, P. Rousselot, J. Reiffers, J.J. Cornelissen, T. Hughes, H. Agis, T. Fischer, G. Verhoef, J. Shepherd, G. Saglio, A. Gratwohl, J.L. Nielsen, J.P. Radich, B. Simonsson, K. Taylor, M. Baccarani, C. So, L. Letvak, and R.A. Larson. (2006) "Five-year follow-up of patients receiving Imatinib for chronic myeloid leukemia," *New England Journal of Medicine* 355, pp. 2408–2417.

Druker, B.J., S. Tamura, E. Buchdunger, S. Ohno, G.M. Segal, S. Fanning, J. Zimmermann, and N.B. Lydon (1996) "Effects of a selective inhibitor of the Abl tyrosine kinase on the growth of Bcr-Abl positive cells," *Nature Medicine* 2 (5), pp. 561–566.

Ellis, R.J. (2001) "Macromolecular crowding: Obvious but underappreciated," *Trends in Biochemical Sciences* 26 (10), pp. 597–604.

Escalante, A.E., M. Rebolleda-Gomez, M. Benitez, and M. Travisano (2015) "Ecological perspectives on synthetic biology: Insights from microbial population biology," *Frontiers in Microbiology* 6 (143), doi:10.3389/fmicb.2015.00143.

Fernald, G.W., and T.F. Boat (1987) "Cystic fibrosis: Overview," *Seminars in Roentgenology* XXII (2), pp. 87–96.

Ferris, D., R. Samakoses, S.L. Block, E. Lazcano-Ponce, J.A. Restrepo, K.S. Reisinger, J. Mehlsen, A. Chatterjee, O.-E. Iversen, H.L. Sings, Q. Shou, T.A. Sausser, and A. Saah. (2014) "Long-term study of a quadrivalent human papillomavirus vaccine." *Pediatrics* 134 (3), e657–e665.

Fischer, K.R., A. Durrans, S. Lee, J. Sheng, F. Li, S.T.C. Wong, H. Choi, T. El Rayes, S. Ryu, J. Troeger, R.F. Schwabe, L.T. Vahdat, N.K. Altorki, V. Mittal, and D. Gao. (2015) "Epithelial-to-mesenchymal transition is not required for lung metastasis but contributes to chemoresistance," *Nature* 527 (7579), pp. 472–476.

Flannery, B., J. Clippard, R.K. Zimmerman, M.P. Nowalk, M.L. Jackson, L.A. Jackson, A.S. Monto, J.G. Petrie, H.Q. McLean, E.A. Belongia, M. Gaglani, L. Berman, A. Foust, W. Sessions, S.N. Thaker, S. Spencer, and A.M. Fry. (2015) "Early estimates of seasonal influenza vaccine effectiveness—United States, January 2015," *Morbidity and Mortality Weekly Report (MMWR)* 64 (1), pp. 10–15.

Fojo, T., S. Mailankody, and A. Lo (2014) "Unintended consequences of expensive cancer therapeutics—The pursuit of marginal indications and a me-too mentality that stifles innovation and creativity: The John Conley Lecture," *JAMA Otolaryngology—Head & Neck Surgery* 140 (12), pp. 1–12.

Fothergill, J.L., M.J. Walshaw, and C. Winstanley (2012) "Transmissible strains of *Pseudomonas aeruginosa* in cystic fibrosis lung infections," *The European Respiratory Journal* 40 (1), pp. 227–238.

Foxman, B., and L. Riley (2001) "Molecular epidemiology: Focus on infection," *American Journal of Epidemiology* 153 (12), pp. 1135–1141.

Gatenby, R. A. (2009) "A change of strategy in the war on cancer," *Nature* 459, pp. 508–509.

Gatenby, R.A., J. Brown, and T. Vincent (2009) "Lessons from applied ecology: Cancer control using an evolutionary double bind," *Cancer Research* 69 (19), pp. 7499–7502.

Greaves, M., and C.C. Maley (2012) "Clonal evolution in cancer," *Nature* 481 (7381), pp. 306–313.

Greig, D., and M. Travisano (2004) "The prisoner's dilemma and polymorphism in yeast SUC genes," *Proceedings of the Royal Society B: Biological Sciences* 271 (Suppl_3), S25–S26.

— (2008) "Density-dependent effects on allelopathic interactions in yeast," *Evolution* 62 (3), pp. 521–527.

Hari, D.M., A.M. Leung, J.-H. Lee, M.-S. Sim, B. Vuong, C.G. Chiu, and A.J. Bilchik (2013) "AJCC Cancer Staging Manual 7th edition criteria for colon cancer: Do the complex modifications improve prognostic assessment?" *Journal of the American College of Surgeons* 217 (2), pp. 181–190.

Hoshino, A., B. Costa-Silva, T.-L. Shen, G. Rodrigues, A. Hashimoto, M.T. Mark, H. Molina, S. Kohsaka, A. Di Giannatale, S. Ceder, S. Singh, C. Williams, N. Soplop, K. Uryu, L. Pharmer, T. King, L. Bojmar, A.E. Davies, Y. Ararso, T. Zhang, H. Zhang, J. Hernandez, J.M. Weiss, V.D. Dumont-Cole, K. Kramer, L.H. Wexler, A. Narendran, G.K. Schwartz, J.H. Healey, P. Sandstrom, K.J. Labori, E.H. Kure, P.M. Grandgenett, M.A. Hollingsworth, M. de Sousa, S. Kaur, M. Jain, K. Mallya, S.K. Batra, W.R. Jarnagin, M.S. Brady, O. Fodstad, V. Muller, K. Pantel, A.J. Minn, M.J. Bissell, B.A. Garcia, Y. Kang, V.K. Rajasekhar, C.M. Ghajar, I. Matei, H. Peinado, J. Bromberg, and D. Lyden. (2015) "Tumour exosome integrins determine organotropic metastasis," *Nature* 527 (7578), pp. 329–335.

Ignatiadis, M., C. Desmedt, C. Sotiriou, E. de Azambuja, and M. Piccart (2009) "HER-2 as a target for breast cancer therapy," *Clinical Cancer Research* 15 (6), pp. 1848–1852.

Jones, M.R. (2005) "Idealization and abstraction: A framework." In *Idealization XII: Correcting the Model: Idealization and Abstraction in the Sciences*, edited by M.R. Jones and N. Cartwright, pp. 173–217. Amsterdam/New York: Rodopi.

Kalmanti, L., S. Saussele, M. Lauseker, M.C. Müller, C.T. Dietz, L. Heinrich, B. Hanfstein, U. Proetel, A. Fabarius, S.W. Krause, S. Rinaldetti, J. Dengler, C. Falge, E. Oppliger-Leibundgut, A. Burchert, A. Neubauer, L. Kanz, F. Stegelmann, M. Pfreundschuh, K. Spiekermann, C. Scheid, M. Pfirrmann, A. Hochhaus, J. Hasford, and R. Hehlmann. (2015) "Safety and efficacy of Imatinib in CML over a period of 10 years: Data from the randomized CML-Study IV," *Leukemia* 29 (5), pp. 1123–1132.

Kam, Y., K.A. Rejniak, and A.R.A. Anderson (2012) "Cellular modeling of cancer invasion: integration of in silico and in vitro approaches," *Journal of Cellular Physiology* 227 (2), pp. 431–438.

Kareva, I. (2015) "Cancer ecology: Niche construction, keystone species, ecological succession, and ergodic theory," *Biological Theory* 10 (4), pp. 283–288.

Kaznatcheev, A. (2014) "Ecology of cancer: Mimicry, Eco-engineers, morphostats, and nutrition," *Theory, Evolution, and Games Group.* https://egtheory.wordpress.com/2014/10/09/ecology-of-cancer/.

Kerem, E., S. Conway, S. Elborn, and H. Heijerman. (2005) "Standards of care for patients with cystic fibrosis: A european consensus," *Journal of Cystic Fibrosis* 4 (1), pp. 7–26.

Loh, R., J. Bergfeld, D. Hayes, A. O'Hara, S. Pyecroft, S. Raidal, and R. Sharpe (2006) "The immunohistochemical characterization of Devil Facial Tumor Disease (DFTD) in the Tasmanian Devil *(Sarcophilus harrisii),*" *Veterinary Pathology* 43 (6), pp. 890–895.

Love, A.C. (2010) "Idealization in evolutionary developmental investigation: A tension between phenotypic plasticity and normal stages," *Philosophical Transactions of the Royal Society B: Biological Sciences* 365 (1540), pp. 679–690.

Lugo, T.G., A.-M. Pendergast, A.J. Muller, and O.N. Witte (1990) "Tyrosine kinase activity and transformation potency of Bcr-Abl oncogene products," *Science* 247 (4946), pp. 1079–1082.

Lyczak, J.B., C.L. Cannon, and G.B. Pier (2002) "Lung infections associated with cystic fibrosis" *Clinical Microbiology Reviews* 15 (2), pp. 194–222.

Maley, C.C., P.C. Galipeau, J.C. Finley, V.J. Wongsurawat, X. Li, C.A. Sanchez, T.G. Paulson, P.L. Blount, R.-A. Risques, P.S. Rabinovitch, and B.J. Reid. (2006) "Genetic clonal diversity predicts progression to esophageal adenocarcinoma," *Nature Genetics* 38 (4), pp. 468–473.

Markowitz, L.E., S. Hariri, C. Lin, E.F. Dunne, M. Steinau, G. McQuillan, and E.R. Unger. (2013) "Reduction in human papillomavirus (HPV) prevalence among young women following HPV vaccine introduction in the United States, National Health and Nutrition Examination Surveys, 2003–2010," *Journal of Infectious Diseases* 208 (3), pp. 385–393.

Massagué, J., and A.C. Obenauf (2016) "Metastatic colonization by circulating tumor cells," *Nature* 529 (7586), pp. 298–306.

Merlo, L.M.F., J.W. Pepper, B.J. Reid, and C.C. Maley (2006) "Cancer as an evolution-ary and ecological process," *Nature Reviews. Cancer* 6 (12), pp. 924–935.

Metzger, M.J., C. Reinisch, J. Sherry, S.P. Goff, M.J. Metzger, C. Reinisch, J. Sherry, and S.P. Goff (2015) "Horizontal transmission of clonal cancer cells causes leukemia in soft-shell clams" *Cell* 161 (2), pp. 255–263.

Miller, A.B., C. Wall, C.J. Baines, P. Sun, T. To, and S.A. Narod (2014) "Twenty five year follow-up for breast cancer incidence and mortality of the Canadian National Breast Screening Study: Randomised screening trial," *BMJ* 348, g366.

Moja, L., L. Tagliabue, S. Balduzzi, E. Parmelli, V. Pistotti, V. Guarneri, and R. D'Amico. (2012) "Trastuzumab containing regimens for metastatic breast cancer," *Cochrane Database of Systematic Reviews* (4), Art. No.: CD006243, doi:10.1002/14651858.CD006243.pub2.

Murchison, E.P. (2009) "Clonally transmissible cancers in dogs and Tasmanian Devils," *Oncogene* 27, S19–S30.

Murchison, E.P., C. Tovar, A. Hsu, H.S. Bender, P. Kheradpour, C.A. Rebbeck, D. Obendorf, C. Conlan, M. Bahlo, C.A. Blizzard, S. Pyecroft, A. Kreiss, M. Kellis, A. Stark, T.T. Harkins, J.A. Marshall Graves, G.M. Woods, G.J. Hannon, and A.T. Papenfuss. (2010) "The Tasmanian Devil transcriptome reveals Schwann cell origins of a clonally transmissible cancer," *Science* 327, pp. 84–87.

Narod, S.A., J. Iqbal, V. Giannakeas, V. Sopik, and P. Sun (2015) "Breast cancer mortality after a diagnosis of ductal carcinoma in situ," *JAMA Oncology* 1 (7), pp. 888–896.

National Cancer Institute (2015) "What is cancer?" Accessed January 30th, 2016. www. cancer.gov/about-cancer/what-is-cancer.

National Cancer Institute (2016a) "Home—The cancer genome atlas." Accessed January 30th, 2016. www.cancergenome.nih.gov.

National Cancer Institute (2016b) "Translational research program." Accessed January 30th, 2016. www.trp.cancer.gov.

Naud, P.S., C.M. Roteli-Martins, N.S. De Carvalho, J.C. Teixeira, P.C. de Borba, N. Sanchez, T. Zahaf, G. Catteau, B. Geeraerts, and D. Descamps (2014) "Sustained efficacy, immunogenicity, and safety of the HPV-16/18 AS04-adjuvanted vaccine," *Human Vaccines & Immunotherapeutics* 10 (8), pp. 2147–2162.

Nigro, J.M., S.J. Baker, A.C. Preisinger, J.M. Jessup, R. Hosteller, K. Cleary, S.H. Signer, N. Davidson, S. Baylin, P. Devilee, T. Glover, F.S. Collins, A. Weslon, R. Modali, C.C. Harris, and B. Vogelstein. (1989) "Mutations in the p53 gene occur in diverse human tumour types," *Nature* 342 (6250), pp. 705–708.

Nowell, P.C. (1976) "The clonal evolution of tumor cell populations," *Science* 194 (4260), pp. 23–28.

O'Sullivan, B.P., and S.D. Freedman (2009) "Cystic fibrosis," *The Lancet* 373 (9678), pp. 1891–1904.

O'Sullivan, C.C., I. Bradbury, C. Campbell, M. Spielmann, E.A. Perez, H. Joensuu, J.P. Costantino, S. Delaloge, P. Rastogi, D. Zardavas, K.V. Ballman, E. Holmes, E. de Azambuja, M. Piccart-Gebhart, J.A. Zujewski, and R.D. Gelber. (2015) "Efficacy of adjuvant trastuzumab for patients with human epidermal growth factor receptor 2-positive early breast cancer and tumors ≤= 2 cm: A meta-analysis of the randomized trastuzumab trials," *Journal of Clinical Oncology* 33 (24), pp. 2600–2608.

Ogden, L.E. (2015) "Cross-disciplinary approaches in cancer research," *BioScience* 65 (8), pp. 750–756.

Ohmit, S.E., M.G. Thompson, J.G. Petrie, S.N. Thaker, M.L. Jackson, E.A. Belongia, R.K. Zimmerman, M. Gaglani, L. Lamerato, S.M. Spencer, L. Jackson, J.K. Meece, M.P. Nowalk, J. Song, M. Zervos, P.-Y. Cheng, C.R. Rinaldo, L. Clipper, D.K. Shay, P. Piedra, and A.S. Monto. (2014) "Influenza vaccine effectiveness in the 2011–2012 season: Protection against each circulating virus and the effect of prior vaccination on estimates," *Clinical Infectious Diseases* 58 (3), pp. 319–327.

Pace, L.E., and N.L. Keating (2014) "A systematic assessment of benefits and risks to guide breast cancer screening decisions." *JAMA* 311 (13), pp. 1327–1335.

Park, S.Y., M. Gonen, H.J. Kim, F. Michor, and K. Polyak (2010) "Cellular and genetic diversity in the progression of in situ human breast carcinomas to an invasive pheno-type," *The Journal of Clinical Investigation* 120 (2), pp. 636–644.

Parmelee, A.H. (1935) "The pathology of steatorrhea," *American Journal of Diseases* 50 (6), pp. 1412–1428.

Pebody, R.G., N. Andrews, J. McMenamin, H. Durnall, J. Ellis, C.I. Thompson, C. Robertson, et al. (2013) "Vaccine effectiveness of 2011/12 trivalent seasonal influenza vaccine in pre-venting laboratory-confirmed influenza in primary care in the United Kingdom: Evidence of waning intra-seasonal protection," *Eurosurveillance* 18 (5), pp.1–8.

Perez, E.A., E.H. Romond, V.J. Suman, J.-H. Jeong, G. Sledge, C.E. Geyer, S. Martino, P. Rastogi, J. Gralow, S.M. Swain, E.P. Winer, G. Colon-Otero, N.E. Davidson, E. Mamounas, J.A. Zujeski, and N. Wolmark. (2014) "Trastuzumab plus adjuvant chemotherapy for human epidermal growth factor receptor 2-positive breast cancer:

Planned joint analysis of overall survival from NSABP B-31 and NCCTG N9831," *Journal of Clinical Oncology* 32 (33), pp. 3744–52.

The Polio Eradication Initiative (2016) "Polio this week." Accessed January 30th, 2016. www.polioeradication.org/Dataandmonitoring/Poliothisweek.aspx.

Prasad, V., J. Lenzer, and D.H. Newman (2015) "Why cancer screening has never been shown to 'save lives'— and what we can do about it," *BMJ* 352, h6080.

Restifo, N.P., M.E. Dudley, and S.A. Rosenberg (2012) "Adoptive immunotherapy for cancer: Harnessing the T cell response," *Nature Reviews Immunology* 12 (4), pp. 269–281.

Sawyers, C.L. (1999) "Chronic myeloid leukemia," *The New England Journal of Medicine* 340 (17), pp. 1330–1340.

Scudellari, M. (2015) "Myths that will not die," *Nature* 528 (7582), pp. 322–325.

Sibley, C.D., H. Rabin, and M.G. Surette (2006) "Cystic fibrosis: A polymicrobial infectious disease," *Future Microbiology* 1 (1), pp. 53–61.

Siddle, H.V., A. Kreiss, M.D.B. Eldridge, E. Noonan, C.J. Clarke, S. Pyecroft, G.M. Woods, and K. Belov (2007) "Transmission of a fatal clonal tumor by biting occurs due to depleted MHC diversity in a threatened carnivorous marsupial," *Proceedings of the National Academy of Sciences of the United States of America* 104 (41), pp. 16221–16226.

Siegel, R.L., K.D. Miller, and A. Jemal (2015) "Cancer statistics, 2015," *CA: A Cancer Journal for Clinicians* 65 (1), pp. 5–29.

Sun, L., R.-Z. Jiang, S. Steinbach, A. Holmes, C. Campanelli, J. Forstner, U. Sajjan, T. Tan, M. Riley, and R. Goldstein (1995) "The emergence of a highly transmissible lineage of Cbl+ *Pseudomonas (Burkholderia) cepacia* causing CF centre epidemics in North America and Britain," *Nature Medicine* 1 (7), pp. 661–666.

Travisano, M., and G.J. Velicer (2004) "Strategies of microbial cheater control," *Trends in Microbiology* 12 (2), pp. 72–78.

van Heyningen, V., and P.L. Yeyati (2004) "Mechanisms of non-Mendelian inheritance in genetic disease," *Human Molecular Genetics* 13 (REV. ISS. 2), pp. 225–233.

Wadman, M. (2011) "NIH revamp rushes ahead," *Nature* 471 (7336), pp. 15–16.

Walboomers, J.M. M., M.V. Jacobs, M.M. Manos, F.X. Bosch, J.A. Kummer, K.V. Shah, P.J.F. Snijders, J. Peto, C.J.L.M. Meijer, and N. Muñoz (1999) "Human papillomavirus is a necessary cause of invasive cervical cancer worldwide," *Journal of Pathology* 189 (1), pp. 12–19.

Walker, F.O. (2007) "Huntington's disease," *The Lancet* 369 (9557), pp. 218–228.

Webster, R.G., W.J. Bean, O.T. Gorman, T.M. Chambers, and Y. Kawaoka (1992) "Evolution and ecology of influenza A viruses," *Microbiological Reviews* 56 (1), pp. 152–179.

Weisberg, M. (2007) "Three kinds of idealization," *The Journal of Philosophy* 104 (12), pp. 639–659.

Wiemann, B., and C.O. Starnes (1994) "Coley's toxins, tumor necrosis factor and cancer research: A historical perspective," *Pharmacology & Therapeutics* 64 (94), pp. 529–564.

Wilson, M. (2013) "What is 'classical mechanics' anyway?" In *The Oxford Companion to the Philosophy of Physics*, edited by Robert Batterman, pp. 43–106. New York: Oxford University Press.

Zerhouni, E. (2003) "The NIH roadmap," *Science* 302 (5642), pp. 63–64, 72.

Zerhouni, E.A. (2005) "Translational and clinical science—time for a new vision," *New England Journal of Medicine* 353 (15), pp. 1621–1623.

6 Evolutionary Perspectives on Molecular Medicine
Cancer from an Evolutionary Perspective

Anya Plutynski

Abstract

There is an active research program currently underway that treats cancer progression as an evolutionary process. This chapter investigates the ways that cancer progression is like and unlike evolution in other contexts. The aim is to take a multilevel perspective on cancer, investigating the levels at which selection may be acting, the unit or target of selection, the relative roles of selection and drift, and the idea that cancer progression may be a by-product of selection at other levels of organization. The chapter integrates data and theory from molecular biology and in situ studies of cancer progression, as well as dynamical models of cancer that represent progression as a multistage process.

1 Introduction: Nothing in Cancer Makes Sense Except in Light of Evolution?

Dobzhansky (1973) wrote that "nothing in biology makes sense except in light of evolution." How can evolutionary thinking shed light on cancer? After all, cancer is ordinarily understood to be a case of the failure of otherwise functional controls on cell birth and death. At first pass, this view seems fundamentally at odds with taking an evolutionary perspective on cancer. For how can something that is an exemplary case of "dysfunction" count as an evolutionary process or product of adaptive evolution?

There are many ways of investigating cancer; this is in part because cancer is a product of many causes, both remote and proximate, acting at a variety of temporal and spatial scales. Different disciplines frame different questions about cancer initiation and progression, scaling up from proximate mechanisms to remote etiology. Molecular biologists identify molecular and genetic mechanisms that are associated with cancer, using either experimental work on cells in culture, gene knockout experiments, genome-wide association studies (GWASs), proteomics, epigenomics, or transcriptomics. Geneticists and developmental and systems biologists develop models of genetic regulatory networks to better understand how genes and their products act and interact in either preventing or advancing cancer progression. Multilevel models of interactive causal factors in cancer draw upon a variety of evidence from epidemiology, genetics, and developmental biology. The study of familial patterns of cancer incidence using classic Mendelian models can be used in concert with evidence from genetic and genomic investigations to identify genes or gene families associated with familial cancer syndromes such as Li Fraumeni syndrome. Epidemiological investigations into environmental risk factors in cancer, such as smoking, radiation exposure, or endocrine disruptors, use case-control, cohort, and ecological studies to tack risk exposure and find correlative body burdens of toxins as well as rates of incidence of disease. Studies of the roles in cancer of the immune system and tissue microenvironment complement this research and have immense potential for therapy. Evolutionary perspectives are merely one additional approach, drawing upon evolutionary models or the study of evolutionary history to better understand this complex and heterogeneous causal process.

However, Dobzhansky's dictum suggests a stronger thesis: that literally "nothing" in cancer makes sense in the absence of an evolutionary perspective. To be sure, this claim may seem unduly strong. However, here are two suggestions of what someone defending such a view might have in mind. First, one could argue that what makes some organisms distinctively vulnerable to cancer is their belonging to one or another lineage in the tree of life. For instance, all and only multicellular organisms may get cancer. To "make sense" of cancer in a particular species, moreover, one needs to know how and why that lineage is distinctively vulnerable to cancer. Understanding a species' evolutionary history—e.g., the selective trade-offs they face in development and life history—can inform our understanding of how and why they are more or less vulnerable to cancer. For example, elephants are less vulnerable to cancer than they would be if they did not have multiple copies of *p53*, likely an adaptation to the large number of replications of somatic cells in development (Abegglen et al. 2015).

Moreover, to understand cancer, one needs to understand that the emergence of multicellularity involved a compromise in fitness of parts in service of collectives; cells had to come together and cooperate in service of the survival and reproductive success of the collective. Any collective of cells, especially collectives whose survival and reproductive success depends on functional organization or the "division of labor" among cells are at least potentially vulnerable to breakdown in cooperative organization. Somatic cells divide and acquire mutations during our lifetimes; some of these mutations involve failures in regulatory pathways that ordinarily enforce functional organization and thus cooperation. In this way, an evolutionary perspective—understanding how the evolution of multicellularity required the emergence of cooperative organization, and understanding how and why multicellular organisms are thus distinctively vulnerable to failure—is essential to understanding cancer.

Second, advanced carcinomas in complex metazoans coopt signaling pathways that are ordinarily adaptive at the organismic level; this is a classic example of a cross-level by-product. This process can only "make sense," in other words, given a multilevel perspective on evolution. On the multilevel perspective, any entity in the biological hierarchy with heritable variation that makes a difference to survival and/or reproductive success may be subject to selection. But multilevel processes are subject to cooption, and cross-level by-products abound. Traits that are advantageous at one level or with respect to units of selection that are component parts of some higher level can compromise (or enhance) fitness at other levels; classic examples are "jumping genes" or meiotic drive. Traits at the level of cells that ordinarily enhance fitness at the level of the organism include *apoptosis* or programmed cell death. Some of the capacities that invasive cancer cells acquire (the capacity to invade and metastasize) are in fact unique features of metazoans that ordinarily enhance fitness (e.g., the capacity for cells to undergo a change in phenotype from epithelial to mesenchymal cells, or the *epithelial–mesenchymal transition*). So understanding how metastasis is possible requires understanding the distinctive adaptive features of metazoans. We will now develop some of these considerations further.

There are at least three ways in which evolutionary perspectives shed light on cancer causation and potentially also unify diverse lines of inquiry in the biomedical sciences: comparative biology of cancer, evolutionary medicine, and evolutionary dynamics of cancer.

- First, we can compare different species and higher taxa in order to understand what makes some lineages particularly vulnerable to cancer. In other words, comparative biology may help us identify mechanisms associated with cancer vulnerability, onset, and progression; when they arose; and where and why they are shared, as well as how they have diverged (Aktipis et al. 2015).[1]
- Second, we can look to unique features of our own evolutionary history in order to explain patterns of cancer incidence in humans or identify distinctive causes of vulnerability to cancer in different human populations. For instance, why do women with lower parity have higher rates of cancer? Or why are men more vulnerable to cancer than women? What aspects of our evolutionary history might explain these differences? (Gluckman et al. 2009; Stearns and Koella 2008; Sun et al. 2014)
- Third and last, we can consider cancer progression itself as an evolutionary process, with corresponding evolutionary dynamics. We might develop theoretical models of this process and link these models with empirical data (Frank 2007; Wodarz and Komarova 2015).

Each approach yields important insights. For instance, work on the comparative biology of cancer identifies common mechanisms associated with the prevention of cancer: mechanisms involved in the inhibiting of cell proliferation, regulation of cell death, division of labor, resource transport, and creation and maintenance of the extracellular environment (Aktipis et al. 2015). Comparing and contrasting how tissue architecture and development and other mechanisms of the suppression of cancer-like growth across species can help cancer researchers identify targets of treatment or prevention of cancer. For instance, the relative absence of invasive cancer in the naked mole rat and blind mole rat may be due to a variety of mechanisms that enhance multicellular cooperation and suppress dysplastic growth, as well as unique features of the extracellular matrix and relatively low metabolic rates. Understanding the operation of each factor's role in preventing disease in mole rats might help identify targets for intervention or tools for thinking about how best to prevent the advance of the disease in humans. Evolutionary thinking shapes this kind of research in (at least) two ways: first, in helping us understand the selective context, trade-offs, and thus origins of such traits; and second, in helping uncover their shared and disparate mechanistic bases. Knowing how a trait is realized (or can be decomposed into parts, processes, etc.) in one organism can give us insight into how a trait can be decomposed or mechanistically realized in closely related organisms. Shared ancestry is (at least sometimes) good reason to suspect shared genetics and shared mechanistic and developmental bases for many traits (for a discussion of the logic behind this inference, see, e.g., Sober 1991, and more recently, 2008).

Each approach also faces various challenges. Like comparative biology, evolutionary medicine is concerned with how differential vulnerability to cancer evolved. Evolutionary medicine's focus is largely on *human* vulnerability to disease and how features of our selective environment or selective trade-offs yielded these vulnerabilities. Testing hypotheses about the evolutionary past is difficult, to say the least. It requires a diverse array of evidence, and of course, often at best we can say that one or another hypothesis is most consistent with the widest array of evidence and theoretical considerations. Claims about how evolution shaped our vulnerability to disease have been contentious. Some have argued that such claims make "adaptationist" assumptions, that is, assumptions that a given trait is adaptive or selectively advantageous, founded on at best "just so" stories (Valles 2011; see also Murphy 2006).

However, there are better and worse such arguments; the best arguments consider not only the widest array of evidence but also trade-offs in fitness, as well as the role of constraints arising out of development and life history. Many arguments from evolutionary medicine concern trade-offs in fitness. For instance, traits adaptive early in life may yield fitness costs later in life. A vivid example is androgenic hormones; male hormones predispose men to higher prostate cancer risk, but they may also yield an advantage early in life in terms of increasing sperm production, relative growth and size at sexual maturity, and thus (potentially) access to mates and resources. Of course, large size may be less of a fitness advantage in current environments; this may be a case of a "mismatch" between our ancestral and current contexts (see, e.g., Summers et al. 2008 for discussion). "Mismatch" hypotheses suggest that traits that may have been adaptive in the past leave us vulnerable to disease in our current environment. For instance, many advocates of evolutionary medicine have argued that women in the modern world are at higher risk of breast cancer because they delay or prevent pregnancy. This exposes them to more cycles of estrogen, which increases breast cancer risk. Presumably, higher rates of pregnancy reduced estrogen exposure in our evolutionary past. Of course, such hypotheses are contentious; there is always the potential for confounding causes (in this case, of increased cancer risk due to a variety of risk factors at work in modern society) (Greaves, 2000). Claims about human psychology, behavior, and social conditions at work in our evolutionary past are particularly contentious (see, e.g., Adriens and DeBlock 2011), because claims about our ancestral social environment are so difficult to substantiate. Nonetheless, evolutionary medicine can help us better understand patterns of incidence of disease in different environments, while it has the potential to inform practical applications.

The last of the three approaches has already been applied in contexts of cancer treatment and prevention. For instance, the evolution of multidrug resistance is one of the major causes of cancer mortality. This is the case not only for standard chemotherapy but also for targeted or "precision" drugs; such drugs can be more or less effective in different patients and lose their effectiveness over time. With some caveats (see, e.g., Pisco et al. 2012), an evolutionary perspective on cancer may shed light on how drug resistance comes about.

In this paper, the focus will be primarily on the third approach. I will provide a brief summary of the basic presuppositions about cancer progression that one must accept in order to defend this view, as well as how this perspective fits into a multilevel perspective on evolution (Section 2); a brief discussion of several mathematical models that have been developed to characterize this process (Section 3); a discussion of the variety of empirical data that has been brought to bear on the theory (Section 4); and a discussion of the broader implications of taking this perspective for cancer diagnosis and treatment (Section 5).

2 The Evolutionary Perspective on Cancer Progression: A General Introduction

According to an ongoing research program, cancer arises from a Darwinian[2] process of mutation and selection among somatic cells (Greaves 2000, 2007; Frank and Nowak, 2004; Merlo et al. 2006; Greaves and Maley 2010). Cancer cells are cells that have acquired a series of somatic mutations and epigenetic alterations that allow them to escape regulation of cell birth and death, leading to disorderly growth, invasion, and metastasis. This is a long process that can start as early as the womb (Mori et al. 2002). Over the course of the average human's lifetime, there are many millions of cell divisions in the body. Thus, by chance alone, mutations and epigenetic changes occur. Some such mutations are associated with cancer; they may lead to chromosomal instability, failures of DNA repair, or failures in regulation of cell birth and death.

Estimations of mutation rates per gene per cell division are about 10^{-7}. One question that an evolutionary or "dynamic" perspective on cancer can shed light on is whether the somatic mutation rate (the rate at which mutations are acquired in somatic cells) is high enough to eventuate in cancer during the lifetime of the average individual. There is some disagreement about this in the literature; some argue that it is sufficient; indeed, some contend that we should be surprised that cancer does not occur more often (Tomlinson et al. 1996; Sieber et al. 2003). Others disagree (Loeb 1991, 2011) and argue that a "mutator phenotype" needs to come on the scene first, accelerating cancer development. Settling this debate is not easy; one needs to know not only how many mutations are necessary or typical for a cancer cell to eventuate and the typical rate of mutations but also have some sense of how effective the immune system and interactions between cancer-precursor cells and the tissue microenvironment may be in halting cancer progression. In addition, tissue architecture, or the hierarchical subdivision of cells into stem and differentiated cells, could also play an important role in preventing the advance of disease. That is, most cells with mutations that may otherwise have yielded cancer are prevented from doing so for a variety of reasons. How many such incipient cancer cells are there?

According to one recent study, 18–32 percent of normal skin cells in the average sun-exposed adult have clonal populations of cells with 2–3 "driver" mutations (at a density of ~140 driver mutations per square centimeter) (see Martincorena et al. 2015). There is, in other words, a "vast reservoir" of mutations in healthy normal cells, but most such cells do not eventuate in cancer. This naturally leads to a question: Why don't we get cancer more often than we do? When we do, what are the main reasons why? What can cause some such clonal populations of cells to become cancer? Are the causes primarily cell-intrinsic? Are most precancerous cells so unstable that they eventually die on their own? Does their relative success have to do with competitive interactions between cells? Or, are there population-level interactions between clonal populations or subpopulations or perhaps between whole tumors for multifocal lesions? In other words, at what "level," or between what entities, might such interactions take place? Could natural selection be acting at multiple "levels"?

In normal tissue, contact inhibition, tissue architecture, and various other mechanisms control cellular growth and prevent overgrowth of cells. How and when cells are born and die—i.e., the particular mode of regulation of growth—is very specific to the type of tissue. Epithelial cells in the skin or colon, for instance, regularly slough off and die over the course of a lifetime; in contrast, bone growth and renewal is relatively slow post-adolescence. Cancer occurs when the tissue-specific signals that regulate cell birth and death fail. What enables cancers to escape these regulatory signals? What most cancer researchers will say is that the primary causes are chromosomal alterations or the acquisition of mutations or epigenetic changes to cells that give them distinctive capacities (or, perhaps better, *incapacities*), such as the capacity to resist apoptosis, attract a blood supply, or continue to divide.

But while this explanation describes properties of cancer cells, cancer itself is not simply a disease of cells. That is, whether populations of such cells eventuate in cancer has to do with interactions between cells and the tissue microenvironment. What evolutionary approaches to cancer investigate is whether differences in cells and populations of cells descended from common ancestors are more or less successful at persisting and eventuating in disease. This has to do not only with cell-intrinsic features but also population-level features: population size, mutation rates, death rates, genetic heterogeneity, and so on. These differences in populations of cells may make some populations more or less "evolvable." Population level features of certain tissue types or tissue architectures may be more vulnerable to cancer. For instance, populations with stem-cell hierarchies may be more (or less) vulnerable to cancer, depending upon the features of the stem cells (whether they divide in one way or another), as well as the relative number of stem cells. That is, an evolutionary approach to cancer is not simply a matter of understanding properties of cancer cells but also the dynamic properties of populations of such cells in interaction with the "ecology" of the tissue microenvironment.[3]

Once a population of cells has become invasive, evolutionary approaches to cancer progression explore how competition among cancer cells or cell lineages within a tumor for space and resources affects cancer progression. Or they might investigate the possibility of cooperative interactions between cell lineages and between the tumor and the tissue microenvironment, and how these cooperative interactions could be selected for. A variety of different theoretical models, as we will see, have been used to investigate these questions.

What presuppositions about cancer does this approach make? All start with the assumption that multicellular organisms are a product of a long history of evolution, from solitary replicators to networks of replicators enclosed within compartments, from genes to chromosomes, from prokaryotic cells to eukaryotic cells containing organelles, from unicellular to multicellular organisms, and from solitary organisms to colonies. These are sometimes called the "major transitions" of evolution (Smith and Szathmary 1995). In each of these transitions, entities capable of surviving and reproducing autonomously aggregated into a single, larger unit, and a new level of biological organization. In order for this transition to happen, it was necessary for individual units to benefit in some way from participation in collectives. Cancer, in this view, illustrates that cooperative organization in biology is always an unstable compromise. That is, the mechanisms that reinforce cooperation in collectives are subject to breakdown. As long as there are entities with heritable variation in fitness within any collective, the collective is vulnerable to "defection" from within. Cancer is a vivid example of breakdown in cooperative organization of cells and tissues due to the acquisition of a series of mutations and epigenetic and genomic changes—changes that allow cancer cells to become relatively "autonomous." According to the evolutionary perspective, then, cancer is both a by-product and a process of evolution: cancer cells are populations of cells that have acquired one or more mutations. Such populations may eventually yield the cancer phenotype: self-sufficiency of growth, failure to respond to apoptotic signals, acquisition of a blood supply, and so on. Insofar as these variations are heritable and make a difference to the relative fitness of cells, cancer progression is driven (in part) by natural selection on heritable variations in these precancerous cells, cancer cells, and cell lineages. These interactions take place within the "ecology" of the tissue microenvironment. This ecology is shaped both by the circulating molecules and by tissue architecture.

Implicit in this approach is a "multilevel" perspective: this is the idea that selection can, in principle, operate at more than one level in the biological hierarchy, provided that each level of analysis consists of populations of entities with heritable variation and such variation makes a difference to relative survival or reproductive success of the selected entities. This idea is not controversial, nor is it new; indeed, Darwin himself imagined that selection could occur both at the individual and the group level (cooperative social groups, he thought, might have a fitness advantage over collectives wherein every man sought his own advantage). A corollary of this view is that traits adaptive at one level of organization can be coopted at another level. Put more generally, evolutionary processes at one level of analysis can affect evolutionary processes at other levels of organization, either by constraining available trajectories, or providing traits that can be "coopted" toward new ends. Cancer is a vivid example: traits that are otherwise advantageous at the organismic level can be coopted by cancer. To be clear, this is not the claim that a single "event" is caused by selection acting (simultaneously) at two levels, but that the same process can be described as a process of selection at one level (that of cells) and as a by-product of selective processes (in the past) at another level (that of the organism as a whole). (For discussion of multilevel selection and, in particular, the possibility of cooption and cross-level by-products, see, e.g., Okasha 2006.)

Multilevel selection theory has been used to explore how cooperation or multicellular interactions evolve. For instance, it is generally believed that altruistic or cooperative behaviors are unlikely to evolve because individual level selection will override "group" selection. The higher rates of turnover of individuals as opposed to groups means that a "free rider" can always invade a group with a high level of cooperation, thus preventing the overall increase in cooperation. Multilevel selection theory has been deployed in order to explain how "major transitions" in evolution could have come about (Michod 1997, Michod and Herron 2006; Smith and Szathmary 1995; Frank 1998; Okasha 2006). Cooperative behaviors leading to the emergence of such collective benefits can be found today, even among organisms that are ordinarily understood to be solitary and self-interested. For instance, bacteria join together in service of production of what is sometimes called a "public good," such as access to nutrients or escape from predation (West et al. 2007). Such collectives are always subject to "free riding" and "defection" from within; it is always possible that an individual could take advantage of a collective resource and either fail to cooperate or use that resource to benefit itself at the expense of its cohort. Cancer is yet one more example.

This picture of cancer represents cancer progression as the process of selection within a population of cells in an individual organism, or in a tumor, where competition might occur for space, resources, or the simple capacity to divide at relatively high rates. Cancer progression could, however, involve selective processes operating at a variety of levels of organization. These are not mutually exclusive options. For instance, there may be competition between clonal populations of precancerous cells in somatic tissue. Some such populations may acquire an advantage relatively early or late that enable it to outcompete neighboring tissue and acquire a cancerous phenotype earlier. Or competition could occur between subpopulations of cells within a tumor for access to space or resources. Tumors are often composed of relatively heterogeneous subpopulations of cells, some of which have acquired mutations that enable them to survive in unoccupied niches. For instance, hypoxia is a capacity to survive without access to oxygen; some cancer cell lineages acquire mutations that enable them to survive in relatively oxygen-poor environments. Such lineages might outcompete other lineages in certain environments. Moreover, whole tumors might perturb the environment in ways that are optimal for their survival by producing products that are toxic to the normal cells with which they compete for space and resources. Alternatively, they might make it difficult for neighboring early-stage tumors to grow and survive.

What are some of the more contentious assumptions of this model? First, selection requires heritable variation in fitness, and, cancer cells have far less than perfect heritability. Many cancer cells are characteristically "CIN-ful," that is, they have "chromosomal instability." This is when replication is imperfect because the mechanisms that control cell mitosis have broken down. As a result, many cancer cells have huge chromosomal duplications and inversions. Selection cannot act as effectively on entities with low heritability. Second, cancer cell populations are characteristically short-lived, because they are either quickly subject to attack from the immune system or "drift" to extinction because of relatively small population sizes and low variation. Moreover, cancer cells' "adaptations" are not very sophisticated but may involve a single phenotypic change, such as failure to respond to a specific apoptotic signal (Germain 2012). This is very far from paradigmatic cases of complex adaptation, such as the compound eye or vertebrate limb. Most populations of protocancer cells have relatively low "evolvability"; they simply die off because of intrinsic failures due to excessive chromosomal instability. By and large, the evolution of complex adaptations requires ample heritable variation in fitness, high levels of heritability, complex selective environments with distinctive ecological features or local adaptive niches, and time. All of these are lacking in cancer. Even "successful" cancers (eventually) kill their host. Only rarely have cancers that survive the death of the host arisen in nature; canine viral cancers, for instance, appear to be an example. Thus, if cancer is an evolutionary process, it is a particularly short-lived one.

These are not necessarily devastating objections to the evolutionary perspective on cancer, however. As Godfrey-Smith (2009) has argued, the extent to which a population of entities approaches the "paradigmatic" case of Darwinian evolution is a matter of degree. A paradigmatic case has ample heritable variation, high heritability, and significant differences in fitness due to intrinsic rather than extrinsically varying, "contingent" circumstances. Only in such cases can selection make a more significant difference to the distribution of variation in a population than drift. Cancer cells have—admittedly—low heritability, but the heritable variation that they do have can (and does) make a difference to survival and reproductive success. This is evident in the emergence of chemotherapy resistance. And though evolution in cancer-cell populations is short term, short-term evolutionary change is—arguably—still evolution. After all, most species in the history of life have become extinct. So also, most precancerous lesions may be a particularly vivid example of a short-lived evolutionary process. While there can be multiple "showers" of metastases, which are more or less successful in variable environments, by and large, even successful metastatic cancers are short-lived—at least from a geological perspective. So cancer is a particularly short-lived evolutionary process, but so is most evolution in "solution"—the typical processes of cell division and growth that occur in culture in most laboratories in experimental evolution. Nonetheless, even in a single short-lived cycle of selection on dividing cells in culture, adaptations can be acquired and take over relatively quickly (see, e.g., Lenski et al. 1991).

3 Mathematical Models of Tumorigenesis

Mathematical models of cancer's dynamics vary in terms of their explanatory targets—or the questions they are used to investigate. Their assumptions thus vary, and, so also, their "realism," or representativeness of various aspects of the complex process of tumorigenesis. Mathematical models are deliberate simplifications (though of course all models are, to a greater or lesser extent). Indeed, when building a model, mathematical biologists often start with the simplest possible scenario and gradually add in complications as relevant to the case at hand. Detailed information about the variety of initiating conditions, constraints, and mechanistic bases of cancer is not necessarily of relevance to a dynamical model of cancer progression. A model may be very effective at representing one aspect of a dynamic process or addressing one very specific question. Of course, cancer is a massively heterogeneous, complex causal process. Thus, any simple model will be in some sense a deliberate fiction.

However, as Wimsatt (1987) has argued, "false models" may lead to "true theories." That is, deliberate simplifications of complex processes can yield important insights about what outcomes can be expected under what initial conditions. While key assumptions of one's model should be at least consistent with our best scientific understanding of cancer causation and progression, it is permissible in model building to deliberately represent the system falsely (cf. Love and Nathan 2015). That is, provided that the falsehoods in question are irrelevant or instrumentally useful, many models contain overt and explicit falsehoods. For instance, one can define equilibrium conditions (when it is known that in nature, such conditions never or very rarely obtain), or one can identify what is expected for extreme cases on a continuum or under highly idealized circumstances. Such models provide a baseline or "null" case, so that departures from the null can be explained as expected due to the presence of X, Y, or Z conditions. Whether a model is successful ultimately depends upon one's purposes. These vary from merely theoretical to practical. What matters is whether the model can answer the question we care about, or, alternatively, whether it provokes new lines of investigation or suggests hypotheses worth exploring further. No model is intended to cover or represent all possible cancers. The aim is often conditional generalizations. That is, the aim is to find out, under the following conditions, what might we expect for systems of this type?

There are deterministic and stochastic models, individual-based or "cellular automaton" models, optimality models, models of competition, spatial dynamics, and hierarchical dynamics. One can use a model to explore different questions about different kinds of populations of cancer cells that have different hierarchical structures due to the presence of stem-like cells. Such models are used to pose different kinds of questions, test various hypotheses, and simulate different aspects of cancer progression. Over the course of the latter half of the twentieth century, as new information about cancer incidence, initiation, progression, and molecular biology became available, new models with more sophisticated targets were developed. The first mathematical (or formal) models of cancer's dynamics were developed long before the molecular mechanisms or genetic mutations associated with cancer were well understood. Starting in the 1950s, these models represented cancer progression as a rate-limited multistage process, drawing upon epidemiological data—specifically, patterns of cancer incidence. Thus, for instance, Armitage and Doll (1954) developed dynamical models that predicted patterns of cancer incidence; essentially, they argued that since cancer incidence increases as a power of age, cancer progression is a product of a multistage, rate-limited process of acquisition of mutations. Today, models of cancer progression are more sophisticated and involve the use of different kinds of mathematical tools—ranging from ordinary differential equations to agent-based approaches to elaborate simulations of spatial growth dynamics. We will consider two examples from the recent literature: a simple model of competition between two cell populations, and a more complex model of the evolution of chemotherapy resistance.

The former model is applicable to early stages of development of two "proto-cancer" cell populations. As we saw in Section 1, most such populations do not progress to invasive disease. This model explores what features enable some cell populations to succeed in progressing. The second model examines how evolution of chemotherapy resistance is dependent on a variety of factors: treatment with one or more chemotherapeutic drugs, population size, birth and death rates of cells, and mutation rates. In the following section, we will see how various models of cancer have been applied to specific observations or how data has been brought to bear on the models.

The simplest possible model of cancer competition dynamics represents two populations of incipient cancer cells; these populations might vary in a number of ways, but we'll consider the case of "stable" versus "unstable" populations. Stable cells have wild-type or relatively normal somatic mutation rates. Unstable cells have the "mutator" phenotype—or, they are characterized by a lack of appropriate DNA repair mechanisms and so have elevated rates of mutation. They accumulate mutations faster than normal somatic cells. Under what circumstances will such unstable cells come to dominate stable cells? In order to answer this question, we need to know some of the fitness trade-offs associated with stability versus instability. We might predict that stable cells with intact repair systems face a cost; DNA repair takes time. Cell-cycle arrest and repair results in an overall slower rate of growth in stable populations than unstable populations. On the other hand, unstable cells suffer from high levels of DNA damage—they bear a larger proportion of mutations, many of which would be expected to be deleterious. Using such a set of differential growth equations, one can represent competitive interactions between such populations as a function of their intrinsic replication rate, mutation rate, and rate of DNA repair. Such models yield a variety of interesting results. As Wodarz and Komarova explain, "If the intrinsic replication rate of the mutator (M) is higher than that of the stable cells (S), then a high DNA hit rate can select for stable cells (S)," but a low DNA hit rate selects for genetic instability (M). However, the reverse is the case if the intrinsic growth rate of stable cells is higher than unstable cells. In other words, we can use such models to represent the relative fitness of cell types given the costs and benefits of high rates of mutation versus DNA repair.

Of course, this is a very simple model, and we also might wish to consider a variety of further complications appropriate to different contexts or stages of cancer progression. In early stages of cancer progression—e.g., before an incipient population of cells becomes a tumor—genetic instability may be relatively more advantageous than in later stages. This is because "mutators" might be more likely to acquire mutations that enable escape of the variety of controls on cell division in the tumor microenvironment—for example, apoptotic signals. On the other hand, too many mutations will result in a highly unstable cell and perhaps eventually in high rates of cell death.

Or we can model the effects of chemotherapy on stable and unstable populations of cells, given different assumptions about apoptotic response in such populations. Wodarz and Komarova (2014) show that if we assume that apoptotic response is intact in both such populations (i.e., if cells shut down provided a sufficient number of mutations is acquired) at high DNA hit rates, the mutator cells with the faster instrinsic rate of growth will overcome stable cells. On the other hand, with a low DNA hit rate, stable cells will overcome mutators. The reverse is true when apoptosis is impaired. This has important implications for chemotherapy. Chemotherapy can induce a high DNA hit rate by impairing DNA repair mechanisms. So in populations of healthy cells, we can expect chemotherapy to select for genetic instability and induce tumors. On the other hand, where apoptotic response is weak, chemotherapy may reverse progression and increase the fitness of stable cells.

Competition models such as these represent the two populations as coevolving, with two differential equations representing their intrinsic replication rate, mutation rate (assuming all mutations are deleterious), and chance of DNA repair—or rate of arrest of replication and repair. This results in a set of coupled equations where competition between two such populations is a function of rate or replication, which can predict and explain the behavior of populations of coevolving cells both before and during chemotherapy.

More sophisticated models can represent the emergence of drug resistance in a tumor or in leukemia for one or more drugs. Such models might start with a simple set of assumptions: for example, we can imagine that the population size of a tumor or leukemia cells in the body as N, the growth rate of cells as L, and the death rate as D. L > D corresponds to clonal expansion. Mutations can lead to the generation of cell types that are resistant. Let the mutation rate be u; assume resistant cells proliferate and the rate of death of the wild-type (nonresistant) to be H. We can use this simple mathematical model to predict when and how quickly resistance will evolve in one specific type of leukemia (CML), depending on the stage of growth or given treatment with one or more drugs. This can help us to predict when treatment failure is likely, and so which kinds of combination therapy will be most effective and at what stages. Chronic myeloid leukemia has three stages: a chronic phase (which is asymptomatic), the accelerated phase, and the blast crisis. The latter stages are associated with a very rapid rate of increase in undifferentiated cells.

So, on the basis of these models, Wodarz and Komarova (2005, 2014) have shown that depending upon initial population size of cells in a cancer, treatment failure is expected to occur when turnover rates of cells is high and mutation rates are high. Larger tumors evolve resistance more quickly, even with a relatively low rate of turnover. On the other hand, resistance can arise at lower tumor sizes if rate of turnover of cells (D) is high. They predict that combination therapy will prevent treatment failure except when a cancer has a high turnover rate, or resistance mutations can be generated at rates several orders of magnitude higher than the physiological mutation rate. They extend the model to cases where there is complex tumor architecture (tumor stem cells), as well as cross resistance to multiple chemotherapies. In fact, the model was predictive: they offered treatment schedules that maximize the chances of successful therapy, given different sizes of tumor or stages of CML. The model could be extended to consider further complications, such as tumor stem cells or hierarchical structure of cell lineages in a tumor.

What do these models illustrate about (a) the value of approaching cancer from an evolutionary perspective, and (b) the nature of modeling in science? There are (at least) three general conclusions we can draw from such cases. First, as mentioned above, cancer can and should be understood as a dynamic process of change in populations of cells undergoing complex interactions with their surrounding environment. This is—broadly understood—a "Darwinian" process of natural selection, one involving changes in proportions of distinct phenotypes and genotypes, over time. At the same time, we know that this process is far more complex and can be disrupted or influenced by a variety of factors both remote and proximate; for example, "drift" or chance factors to do with proximity to a blood supply or features of the tissue architecture could "accidentally" cause a population of otherwise highly "fit" cancer cells to die off. Or by-products of selection at other levels of organization can either be coopted in cancer or (as is far more often the case) halt progression. Second, of course, the evolutionary perspective is a kind of idealization. Cancer is far from the "paradigmatic" case of adaptive evolution. Nonetheless, thinking about cancer from an evolutionary perspective can help us discover new hypotheses worth testing, both about our distinctive vulnerability to cancer and the dynamics of cancer progression. Third, the case illustrates how modeling is a process of picking out what we take to be causally significant factors given some target or phenomenon of interest. Whether it is chemotherapy resistance or the emergence of hypoxic cells, we can use evolutionary modeling to represent the dynamics of the emergence of these traits in cancer. Of course, each cancer is genetic and epigenetically unique, and each cancer's environment is likewise unique. So while some of the conclusions of these models can be generalized across distinct cases, this is of course "all else being equal."

What kind of knowledge do we gain from using evolutionary modeling to represent the dynamics of the emergence of these traits in cancer? First, at minimum, this is predictive knowledge. That is, we can use evolutionary models to predict and describe the dynamics of populations of cancer cells and the emergence of complex tumors as in part a product of dynamic interactions between cells, cell lineages, and the tumor microenvironment; and in part a by-product of organismic adaptations. In other words, these models help predict when and why cancer progresses or fails to progress. Second, if one takes selection and drift to be causes of changes in populations over time, these models do not merely describe but also explain. That is, to the extent that these models are accurate representations of cancer, they are not merely providing predictive but also causal knowledge.[4]

Does adopting the evolutionary perspective on cancer progression require that we jettison other perspectives? That is, does suggesting that cancer can be understood in this way require setting aside causal explanations from epidemiology, developmental biology, or genetics? In my view, it does not. All of these perspectives are important for understanding the complex causes of cancer. Cancer is a case study in complex causation; there are both remote and proximate causes of cancer, operating at very distinct temporal and spatial scales. In my view, it is a mistake to exclusively take one approach to cancer causation as "correct." Indeed, as several others have argued, the history of cancer research has been a history of a search for the "magic bullet"—one "essential" or most important cause (Mukherjee, 2011). But the closer we look, the more heterogeneous the causes of cancer. So while a search for one unified causal basis of cancer may have been a fruitful research strategy, it is a mistake to infer from the success of this strategy that causes acting at one temporal or spatial scale, or one disciplinary approach, can or should supplant all others.

4 Empirical Data

Formal models of cancer progression and evolution have been developed for decades, and these models were revised or updated as new epidemiological, molecular, and clinical data became available. The simplest models of cancer as a rate-limited process of acquisition of somatic mutations, for instance, were developed more or less based on epidemiological patterns of cancer incidence, which rises by and large as a power of age (Armitage and Doll, 1954). More recently, a wide array of data from molecular biology, immunology, genomics, and clinical medicine, has been used to either provide more precise parameters, or test applications of mathematical models of cancer's evolutionary dynamics.

The most vivid examples of these are efforts as sequencing tumors and metastases to better understand the dynamics of progression, either within a single patient or across classes of patients with a common cancer type or subtype. For instance, Campbell et al. (2010) sequenced the genomes of 13 patients' pancreatic adenocarcinomas to identify somatically acquired genomic rearrangements and explore clonal relationships among metastases. Navin et al. (2011) used heterogeneous breast tumors to study human tumor progression, arguing that they "still contain evidence of early and intermediate subpopulations in the form of the phylogenetic relationships." Navin et al. developed algorithms to compare the genomes of tumor subpopulations within patients to assess their divergence and to identify genetic elements that may be involved in tumor progression. Genomic tools such as expression profiling, array-based copy number analysis, high-throughput DNA sequencing, and DNA methylation analysis have accelerated the accumulation of data about cancers and how they evolve. A variety of generalizations have arisen out of this research, generalizations that may in turn be useful in developing more detailed models of cancer progression and in developing more targeted methods of prevention and treatment.

- For example, while it has long been known that cancers can be enormously heterogeneous, it is possible now to not only observe pathological heterogeneity but also to measure the extent of genetic heterogeneity in a tumor and even track the historical progression of multiple different subpopulations of cells. However, histological heterogeneity does not imply genetic heterogeneity or vice versa (Navin et al. 2011).
- Some tumors are relatively *monogenomic* (consisting of an apparently homogeneous population of tumor cells with highly similar genome profiles

throughout the tumor mass), whereas others are *polygenomic* (containing multiple tumor subpopulations that can be distinguished and grouped by similar genome structure) (cf. Navin et al. 2011).

- This heterogeneity suggests that GWASs derived from multiple samples of single regions of a given tumor type or subtype (e.g., in the typical tumor biopsy) may not be representative of the entire tumor when subpopulations are anatomically segregated. This suggests that more comprehensive genomics of cancer should be based on multiple rather than single samples from a tumor.
- Campbell et al. (2010) and Navin et al. (2011) both show that major chromosomal events and the amplification of cancer genes occur predominantly in early cancer development rather than the later stages of the disease.
- Campbell et al. (2010) also suggest that there is ongoing, parallel, and even convergent evolution among different metastases and also organ-specific branches, suggesting site-specific metastatic evolution.
- In all such studies, there were also patterns of change and distinctive types of mutations and chromosomal alterations typical of distinct cancers arising in distinct tissue types.

In sum, these studies suggest a variety of insights of relevance to both the theoretical understanding of cancer and to cancer diagnosis and treatment. A careful study of the distribution of variation in cancers of distinct types may assist in predicting how and whether cancers are likely to progress. Data on subpopulations in a tumor and their spatial organization can be used to refine and explore theories of cancer progression, patterns of growth, migration, and metastasis. Such patterns also suggest interesting questions worth exploring further. For instance, are more heterogeneous tumors more likely to generate more or more successful metastases? Which particular mutations in which tumors are likely to yield metastases in which remote sites? How do subpopulations in a tumor evolve or coevolve? How does clonal architecture shape tumor progression? In what ways are metastases preadapted to distinctive remote locales? How do metastatic cells evolve and coevolve in novel environments?

There have also been interesting studies of how the tissue microenvironment prevents cancer and which kinds of alterations to the tissue microenvironment promote cancer. For instance, inflammatory environments are by and large tumorigenic. Experimental work on radiation and various other insults (wounding) to the tissue microenvironment can, in fact, initiate cancer. On the other hand, the appropriate tissue architecture or the normalization of signal transduction pathways appears to suppress cancer (Bissell and Hines 2011). Cancer has been called the "wound that never heals" because the cancer cells initiate an inflammatory response and attract cells from the immune system that in part assist the cancer in attracting a blood supply and building up stromal tissue. That is, inflammatory cells are mobilized in response to signals emanating from the tumor microenvironment. Modeling this dynamic process has yielded some important predictions and even novel treatments; for instance, the recognition that the immune system has a very specific response to cancer has led to the development of checkpoint therapies that take advantage of the body's own immune system. For a review and discussion of the role of modeling in this process, see Merlo et al. (2006).

Other work on the role of tissue architecture in preventing cancer has lent support to models of differential cancer risk in different tissue types. Recently, Tomasetti and Vogelstein argued in a rather provocative paper (2015) that "only a third of the variation in cancer risk *among tissues* is attributable to environmental factors or inherited predispositions. The majority is due to 'bad luck,' that is, random mutations arising during DNA replication in normal, noncancerous stem cells." (This assumes a somewhat artificial partitioning of causes.) Their argument, it turns out, was a very simple mathematical argument, drawing upon data showing that the frequency of a given cancer type appears to correlate with the tissue architecture of the tissue of origin in which that cancer occurs. In other words, we know that cells divide and mutations happen, but some tissues have more stem-cell divisions than others. They conclude that cells that divide more often have more mutations. This is consistent with the observation that tissues with more stem-cell divisions and more mutations are more frequently subject to cancers (see Figure 6.1).

This is a very simple linear regression model, however, and some have argued that there are exceptions to the general rule.

Indeed, exceptions to these general law-like claims about cancer progression are to be expected. Cancer is a massively heterogeneous disease, due to the fact that where, when, and how cancer arises in different tissues and due to different remote and proximate causes yields a widely disparate pattern of incidence and progression. Some cancers progress relatively quickly due to massive chromosomal aberrations that occur early on in cancer progression, what some scientists call "chromothripsis" (Jones et al. 2012). However, this observation does not (*per se*) counter the idea that cancer progression is an evolutionary process; some evolutionary processes occur more quickly than others.

Some have argued that tumor stem cells complicate the evolutionary picture of cancer. However, there is no reason (in principle) why tissue architecture in a tumor could not be modeled using an evolutionary dynamic perspective. The only difference is that instead of using a model of growth such as exponential or logistic growth of clonal populations, one might represent a tumor composed of multiple cell lineages, some dividing and differentiating and others dividing indefinitely. Different tumors have different pathways to progression, and representing this complexity requires looking to different kinds of modeling strategies. Thus, instead of modeling "gain of function" mutations (oncogenes) or "loss of function" mutations (tumor suppressor genes) in relatively uniform populations, one might model interactions between different subpopulations. For instance, in stem cell–driven tumors, stem cells need to acquire

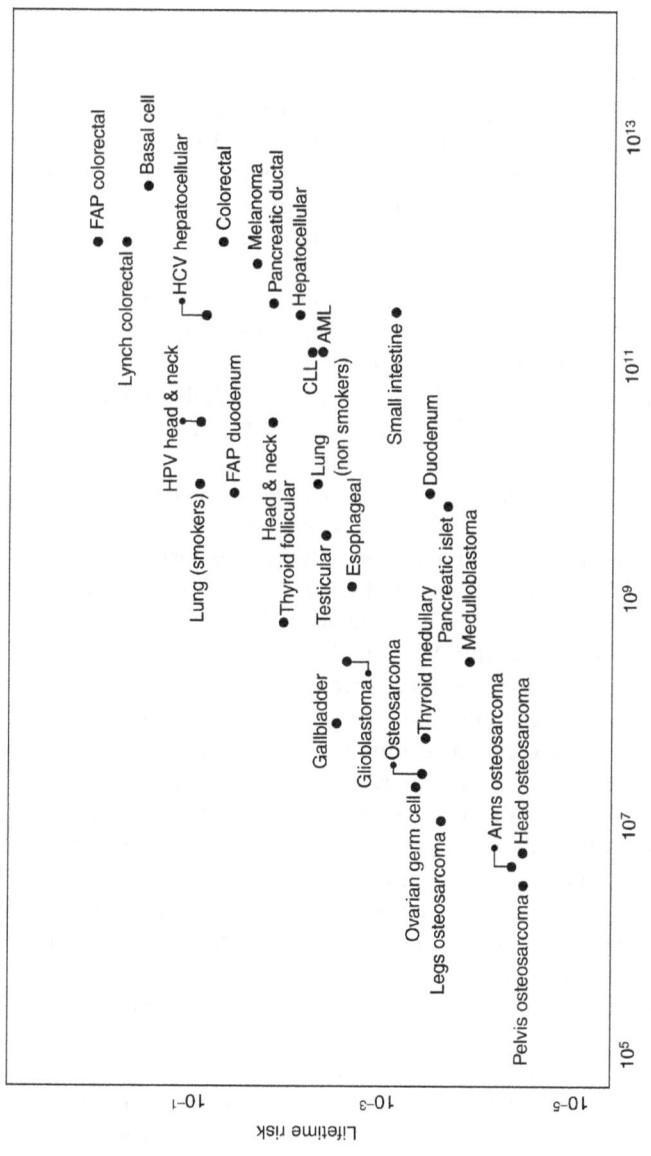

Figure 6.1 The relationship between the number of stem cell divisions in the lifetime of a given tissue and the lifetime risk of cancer in that tissue.

FAP = Familial Adenomatous Polyposis ◇ HCV = Hepatits C virus ◇ HPC = Human papillomavirus ◇ CLL = Chronic lymphocytic leukemai ◇ AML = Acute myeloid leukemai

Source: Values are from Table S1, the derivation of which is discussed in the supplementary materials.

the capacity to overcome inhibition of cell division by differentiated cells. One can represent the dynamics of escape of stem cells from this inhibition using various feedback models with coupled differential equations for stem cells and differentiated cells.

5 Clinical Implications

The above modeling strategies have a variety of implications for the clinic:

- First, as we saw above, knowing the population dynamics of the emergence of chemotherapy resistance is enormously valuable for designing a treatment schedule. Thus, assessing the size, mutation rate, and particular mutations of a tumor or leukemia can assist in decisions as to when and how to treat, with one or more drugs. Or one can determine when and how radiation therapy is more or less likely to eventuate in secondary cancers.
- Second, using such models, one can, in principle, act to prevent cancer by intervening with the conditions under which cancers progress.
- Third, one might use evolutionary models to predict the likely progression of a tumor of a given size with given features and thus provide more precise prognoses, as well as better-timed and targeted therapies.
- Indeed, one might better predict which cancers are unlikely to progress and so when it is better to engage in less aggressive treatment.

More generally, thinking of cancer as an evolving and coevolving population of cells reinforces a fundamental insight about cancer and our vulnerability to cancer. First, cancer is a process, not a state of affairs. This insight is essential to understanding how better to prevent and treat cancer. Knowing that cancer is a process with variable natural histories suggests that when we screen, we need to be sensitive to the possibility that not all cancers progress uniformly to metastasis. This means that some cancers or precancerous lesions should not be intervened in; less is sometimes more, and viewing cancer from an evolutionary perspective shows us why. Second, the evolutionary perspective on cancer reinforces the idea that ecology or tissue microenvironment of a cancer is enormously important; we need to devote more time and attention to why cancer does not progress and which environments are less tumorigenic. This might lead to better preventive measures and less aggressive intervention when it is far too late. Last but not least, the evolutionary perspective can shed light on why some of us are more or less vulnerable to cancer, given our different trade-offs in life-history traits and different mechanisms for regulation.

These considerations have larger import for both philosophy, and the clinic. First, as I hope to have shown, they illustrate the important role of modeling in biology and in the biomedical sciences in particular. Models (however idealized) help us identify and isolate causally relevant factors in cancer progression. Moreover, we've seen that these models often have very specific targets; they are not intended to be "complete" explanations of cancer but only to pick out factors relevant to specific outcomes of interest. Second, such outcomes have broader relevance to clinical practice. The better we can get at identifying which cancers are more likely to progress more rapidly or evolve resistance to chemotherapy, the better position we are in to intervene earlier, provide patients with information, and (ideally) develop targeted, effective interventions.

Notes

1   To be sure, comparative evolutionary biology is only one of several approaches to comparative biology; comparative molecular and developmental biology are also fruitful. But knowing the shared history of two organisms can help direct one to better identify and uncover shared molecular and developmental structures and mechanistic bases.

2   I am here following Godfrey-Smith's (2009) sense of a "minimal" Darwinian population; this is any population wherein there are heritable variations in fitness in a population (see also e.g., Lewontin 1970). A population that is undergoing changes due to such heritable variations in fitness is undergoing a "Darwinian" process. Godfrey-Smith argues that there are more or less "paradigmatic" cases of Darwinian populations. For instance, higher heritability is one condition that is optimal for selection; so also is greater availability of variation for selection to act, and so on. This idea is (roughly) similar to a minimal sense of what makes a population "evolvable" (see, e.g., Pigliucci 2008).

3   I prefer not to take a dogmatic stance regarding what cancer is "ultimately" a disease "of." Cancer is a disease of cells, but it is also (at least in some cases) a heritable disorder—a disease of genes, the immune system, tissue architecture, stem-like cells, development, tissue disorganization, viral infection, environmental toxins, and much else besides. All of these factors are causally relevant to explaining cancer in its diversity. Cancer is an enormously heterogeneous set of disease processes with very different causes, acting at a variety of temporal and spatial scales. Philosophers (and scientists) err when they attempt to reduce all these causal explanations to a single scale or single theoretical framework (see, e.g., Wimsatt 1972, reprinted in 2007). Thanks are due to reviewers for challenging me to address this concern. See, for example, Nathan (2014) for a fascinating discussion of how (and how not) to think about the molecular environment as an "ecosystem."

4   In my view, Stephens (2004), Millstein (2006), and Forber and Riesman (2007) (among others) have argued persuasively that selection and drift may be viewed as causes of evolution. Such a view is consistent with a variety of theories of causation. Addressing this debate at any length, however, would go well beyond the scope of this paper.

References

Abegglen, L.M., A.F. Caulin, A. Chan, K. Lee, R. Robinson, M.S. Campbell, W.K. Kiso et al. (2015) "Potential mechanisms for cancer resistance in elephants and comparative cellular response to DNA damage in humans," *JAMA*. Published online October 8th, 2015. doi:10.1001/jama.2015.13134.

Adriaens, P.R., and A. De Block (2011) *Maladapting Minds: Philosophy, Psychiatry, and Evolutionary Theory*. Oxford University Press.

Aktipis, C.A., A.M. Boddy, G. Jansen, U. Hibner, M.E. Hochberg, C.C. Maley, and G.S. Wilkinson (2015) "Cancer across the tree of life: cooperation and cheating in multicellularity," *Phil. Trans. R. Soc. B*, 370(1673), 20140219.

Armitage, P., and R. Doll (1954) "The age distribution of cancer and a multi-stage theory of carcinogenesis," *British Journal of Cancer*, 8(1), p. 1.

Bissell, M.J., and W.C. Hines (2011) "Why don't we get more cancer? A proposed role of the microenvironment in restraining cancer progression," *Nature Medicine*, 17(3), pp. 320–329.

Campbell, P.J., S. Yachida, L.J. Mudie, P.J. Stephens, E.D. Pleasance, L.A. Stebbings, L.A. Morsberger, C. Latimer, S. McLaren, M.-L. Lin, D.J. McBride, I. Varela, S.A. Nik-Zainal, C. Leroy, M. Jia, A. Menzies, A.P. Butler, J.W. Teague, C.A. Griffin, J. Burton, H. Swerdlow, M.A. Quail, M.R. Stratton, C. Iacobuzio-Donahue, and P.A. Futreal (2010) "Patterns and dynamics of genomic instability in metastatic pancreatic cancer," *Nature*, 467, pp. 1109–1113.

Dobzhansky, T. (1973) "Nothing in biology makes sense except in the light of evolution," *The American Biology Teacher*, 35, pp. 125–129.

Forber, P., and K. Reisman (2007) "Can there be stochastic evolutionary causes?" *Philosophy of Science*, 74(5), pp. 616–627.

Frank, S.A. (1998) *Foundations of Social Evolution*. Princeton University Press.

Frank, S.A. (2007) *Dynamics of Cancer: Incidence, Inheritance, and Evolution*. Princeton University Press.

Frank, S.A., and M.A. Nowak (2004) "Problems of somatic mutation and cancer," *Bioessays*, 26(3), pp. 291–299.

Germain, P.-L. (2012) "Cancer cells and adaptive explanations," *Biology and Philosophy*, 27(6), pp. 785–810.

Gluckman, P.D., A. Beedle, and M.A. Hanson (2009) *Principles of Evolutionary Medicine*. Oxford: Oxford University Press.

Godfrey-Smith, P. (2009) *Darwinian Populations and Natural Selection*. Oxford University Press.

Greaves, M. (2000) *Cancer: The Evolutionary Legacy*. New York: Oxford University Press.

— (2007) "Darwinian medicine: A case for cancer." *Nature Reviews Cancer*, 7(3), pp. 213–221.

Greaves, M., and C.C. Maley (2012) "Clonal evolution in cancer," *Nature*, 481(7381), pp. 306–313.

Jones, M.J.K., and P.V. Jallepalli (2012) "Chromothripsis: Chromosomes in crisis," *Developmental Cell*, 23(5), pp. 908–917.

Komarova, N.L., and D. Wodarz (2005) "Drug resistance in cancer: Principles of emergence and prevention," *Proceedings of the National Academy of Sciences*, 102(7), pp. 9714–9719.

Lauren, M.F., J.W. Pepper, B.J. Reid, and C.C. Maley (2006) "Cancer as an evolutionary ecological process," *Nature Reviews Cancer*, 6(12), pp. 924–935.

Lenski, R.E., M.R. Rose, S.C. Simpson, and S.C. Tadler (1991) "Long-term experimental evolution in Escherichia coli. I. Adaptation and divergence during 2,000 generations," *American Naturalist*, 138(6), pp. 1315–1341.

Lewontin, R.C. (1970) "The units of selection," *Annual Review of Ecology and Systematics*, pp. 1–18.

Loeb, L.A. (1991) "Mutator phenotype may be required for multistage carcinogenesis," *Cancer Research*, 51(12).

Loeb, L.A. (2011) "Human cancers express mutator phenotypes: Origin, consequences and targeting," *Nature Reviews Cancer*, 11(6), pp. 450–457.

Love, A.C., and M.J. Nathan (2015) "The idealization of causation in mechanistic explanation," *Philosophy of Science*, 82, pp. 761–774.

Martincorena, I., A. Roshan, M. Gerstung, P. Ellis, P. Van Loo, S. McLaren, D.C. Wedge, A. Fullam, L.B. Alexandrov, J.M. Tubio, L. Stebbings, A. Menzies, S. Widaa, M.R. Stratton, P.H. Jones, and P.J. Campbell (2015) "High burden and pervasive positive selection of somatic mutations in normal human skin," *Science*, 348(6237), pp. 880–886.

Merlo, L.M.F., J.W. Pepper, B.J. Reid, and C.C. Maley (2006) "Cancer as an evolutionary and ecological process," *Nature Reviews Cancer*, 6(12), pp. 924–935.

Michod, R.E. (1997) "Evolution of the individual," *The American Naturalist*, 150(S1), S5–S21.

Michod R.E., and M.D. Herron (2006) "Cooperation and conflict during evolutionary transitions in individuality," *Journal of Evolutionary Biology*, 19, pp. 1406–1409.

Millstein, R.L. (2006) "Natural selection as a population-level causal process," *The British Journal for the Philosophy of Science*, 57(4), pp. 627–653.

Mori, Hiroshi, S.M. Colman, Z. Xiao, A. M. Ford, L. E. Healy, C. Donaldson, J. M. Hows, C. Navarrete, and M. Greaves (2002) "Chromosome translocations and covert leukemic clones are generated during normal fetal development," *Proceedings of the National Academy of Sciences*, 99(12), pp. 8242–8247.

Mukherjee, S. (2011) *The Emperor of All Maladies: A Biography of Cancer*. Simon and Schuster.

Murphy, D. (2006) *Psychiatry in the Scientific Image*. Cambridge, MA: MIT Press.

Nathan, M.J. (2014) "Molecular ecosystems," *Biology and Philosophy*, 29(1), pp. 101–122.

Navin, N., J. Kendall, J. Troge, P. Andrews, L. Rodgers, J. McIndoo, K. Cook, A. Stepansky, D. Levy, D. Esposito, L. Muthuswamy, A. Krasnitz, W.R. McCombie, J. Hicks, and M. Wigler (2011) "Tumour evolution inferred by single-cell sequencing," *Nature*, 472(7341), pp. 90–94.

Okasha, S. (2006) *Evolution and the Levels of Selection*. Oxford University Press, Oxford.

Pigliucci, M. (2008) "Is evolvability evolvable?" *Nature Reviews Genetics*, 9, pp. 75–82.

Pisco, A.O., A. Brock, J. Zhou, A. Moor, M. Mojtahedi, D. Jackson, and S. Huang (2013) "Non-Darwinian dynamics in therapy-induced cancer drug resistance," *Nature Communications*, 4.

Sieber, O.M., K. Heinimann, and I.P.M. Tomlinson (2003) "Genomic instability—the engine of tumorigenesis?" *Nature Reviews Cancer*, 3(9), pp. 701–708.

Smith, J.M., and E. Szathmary (1995) *The Major Transitions in Evolution*. Oxford University Press.

Sober, E. (1991) *Reconstructing the Past: Parsimony, Evolution and Inference*. Bradford Books.

— (2008) *Evidence and Evolution: The Logic Behind the Science*. Cambridge University Press.

Sprouffske, K., L.M.F. Merlo, P.J. Gerrish, C.C. Maley, and P.D. Sniegowski (2012) "Cancer in light of experimental evolution," *Current Biology*, 22(17), pp. R762–R771.

Stearns, S.C., and J.C. Koella, eds. (2007) *Evolution in Health and Disease*, 2nd edition, Oxford University Press.

Stephens, C. (2004) "Selection, drift, and the 'forces' of evolution," *Philosophy of Science*, 71(4), pp. 550–570.

Summers, K., and B. Crespi (2008) "The androgen receptor and prostate cancer: A role for sexual selection and sexual conflict?" *Medical Hypotheses*, 70(2), pp. 435–443.

Sun, T., A. Plutynski, S.Ward, and J. Rubin (2015) "An integrative view on sex differences in brain tumors," *Cell and Molecular Life Sciences*, 72(17), pp. 3323–3342.

Tomasetti, C., and B. Vogelstein (2015) "Variation in cancer risk among tissues can be explained by the number of stem cell divisions," *Science*, 347(6217), pp. 78–81.

Tomlinson, I.P.M., M.R. Novelli, and W.F. Bodmer (1996) "The mutation rate and cancer," *Proceedings of the National Academy of Sciences*, 93(25), pp. 14800–14803.

Valles, S.A. (2012) "Evolutionary medicine at twenty: Rethinking adaptationism and disease," *Biology and Philosophy*, 27(2), pp. 241–261.

West, S.A., P. Stephen, A. B. Diggle, A. Gardner, and A.S. Griffin (2007) "The social lives of microbes," *Annual Review of Ecology, Evolution, and Systematics*, 38, pp. 53–77.

Wimsatt, W.C. (1972) "Complexity and organization," reprinted in Wimsatt, 2007.

Wodarz, D., and N.L. Komarova (2014) *Dynamics of Cancer: Mathematical Foundations of Oncology*. Singapore: World Scientific Publishing Co.

— (1987) "False models as means to truer theories," *Neutral Models in Biology*, 23–55.

— (2007) *Reengineering Philosophy for Limited Beings: Piecewise Approximations to Reality*. Harvard University Press.

Yachida, S., S. Jones, I. Bozic, T. Antal, R. Leary, B. Fu, M. Kamiyama, R.H. Hruban, J.R. Eshleman, M.A. Nowak, V.E. Velculescu, K.W. Kinzler, B. Vogelstein, and C.A. Iacobuzio-Donahue (2010) "Distant metastasis occurs late during the genetic evolution of pancreatic cancer," *Nature*, 467, p. 1114.

Part III
Representation and Modeling

7 Toward a Notion of Intervention in Big-Data Biology and Molecular Medicine

Federico Boem and Emanuele Ratti

Abstract

We claim that in contemporary studies in molecular biology and biomedicine, the nature of "manipulation" and "intervention" has changed. Traditionally, molecular biology and molecular studies in medicine are considered experimental sciences, whereas experiments take the form of material manipulation and intervention. On the contrary "big science" projects in biology focus on the practice of data mining of biological databases. We argue that the practice of data mining is a form of intervention although it does not require material manipulation. We also suggest that material manipulation, although still present in the practice of data mining, fulfills a different epistemic role.

1 Introduction

In this chapter, we argue that contemporary molecular biology has changed the nature of "manipulation" and "intervention." While molecular biology and molecular studies in medicine have been traditionally associated with experimentation in the form of *material manipulation* and *intervention*, "Big Science" projects in biology focus on the practice of data mining of biological databases. We claim that data mining is a kind of intervention, though formal and not material. This, of course, does not mean that contemporary biologists cast off material manipulation and intervention. However, while material manipulation has still a role in the practice of data mining, that role is merely confirmatory. In other words, we argue/defend two claims: First, data mining of biological databases can achieve similar goals to those of traditional interventionist and experimental practices but without material manipulation. Second, such a practice is indeed a form of intervention, but it must be distinguished from the traditional account.

The structure of the chapter is as follows. In Section 2, we recall Hacking's idea of intervention and apply it to the received view of molecular biology and in particular to molecular studies of biomedicine. In Section 3, we illustrate in biomedicine the practice of data mining or, as we call it, *ordering data*. In Section 4, we argue that ordering data is a kind of intervention, though different from material intervention. Finally, in Section 5, we argue that such an approach does not eliminate the material manipulation part but rather changes the priority of material manipulation. Contrary to Hacking's famous dictum, experiments do not have a life of their own; especially in molecular studies of biomedicine, experiments are confirmatory tools of hypotheses generated through bioinformatics analyses.

2 Intervention as Manipulation

In his *Representing and Intervening* (1983), Ian Hacking famously challenged the received view of the neopositivist tradition in the philosophy of science on the role of theory in experimental research (see Boniolo's contribution in this volume for previous accounts of experimental practice in philosophy of science). According to Hacking, philosophers of science have generally focused on theories, often neglecting the experimental side of science. In his view, scientific research is not just *theorizing* but rather—and more often—*doing*. Nature cannot just be observed untouched; it must be also actively interrogated. Experimenting is indeed questioning nature (Bacon 1620/1898).

Moreover, Hacking claims that experimental work is not subordinated to theory. This means that, contrary to classic accounts of the logic of scientific discovery, scientists do not make experiments just to confirm or refute a particular theory concerning the phenomena under investigation. As he puts it, "[i]n schools and colleges experiments are repeated *ad nauseam*. The point of those classroom exercises is never to test or elaborate the theory. The point is to teach people how to become experimenters" (Hacking 1983, p. 231).

Indeed, scientists are often questioning nature by changing it and observing how the natural world responds. *Manipulation* is a genuine way to know about the world. However, Hacking goes further. He thinks that scientific phenomena are, somehow, *created*. Again, such a claim is prone to misinterpretation. Hacking does not argue that natural phenomena are either constructed or *solipsistically* generated. On the contrary, "in nature there is just complexity, which we are remarkably able to analyse. We do so [...] by presenting, in the laboratory, pure, isolated phenomena" (Hacking 1983, p. 226). Indeed, in order to work, experiments must be performed under certain particular and controlled conditions. The work of a scientist consists in tinkering with the experimental system in order to let phenomena "emerge" from the chaos of everyday experience. It is in this sense that experiments "create" the phenomena. This also explains why and how Hacking sees the nature of scientific knowledge as an activity rather than as speculation. "Intervening" represents a key feature of the experimental culture. By that we mean that "experimenting" as a privileged way of knowing nature (at least since the so-called scientific revolution) contrasts the traditional account of knowledge, which was based on observation and speculation.

2.1 Experimental Culture and Different Types of Experiments

The experimental culture tends to see "direct intervention" as "material manipulation." Manipulation is a central feature of the experimental approach since it is a *condicio sine qua non* for experimenting (and, thus, for knowing). However, not all experiments are the same. On the contrary, different experiments play different roles in the practice of "questioning nature." Accordingly, it is possible to distinguish between *experiments to prove* (Popper 1959, Kadane and Seidenfeld 1990) and *experiments to learn* or *exploratory experiments* (Burian 1997, Steinle 1997). The former are experiments aimed at testing already stated hypotheses, while the latter are those capable of fostering the formulation of new hypotheses. However, this distinction should not be intended in a sharp way (Waters 2007). Different types of experiments differ not just in their supposed independency from theoretical contributions but also in *how* theoretical constraints affect them. Indeed, different experimental strategies enable different *forms of interventions*. Some experiments, the purpose of which is to check something that is guided by the theoretical framework, will rely on on a narrower range of interventions, thus focusing on those ones that seem most promising according to the understanding of the phenomena under investigation. These experiments are "theory-driven" in the sense that their scope and rationale is constructed according to specific constraints shaped by the theory.

Let us clarify this notion of *theory-driven*. The term *theory* could mean (at least) two different things (Franklin 2005). On the one hand, theory may count as general *theoretical background*. This should be seen as a broad conceptual stance grounded on basic well-established empirical findings. For instance, the original hypotheses concerning gene regulation and translation rested on the general idea that protein levels might be controlled by different concentrations of mRNA. On the other hand, theory can mean something more specific than the behavior of objects that are observed and measured. Borrowing a term from Franklin, we call this *local theory*. To illustrate, consider the Western blot technique. Western blot is a method to determine the presence of a protein of interest in a given sample. In the most common protocol, proteins are first separated according to their molecular weight by running them through a specific denaturing gel by electrophoresis. These proteins are then transferred from the gel to a nitrocellulose membrane where the protein of interest is identified by the recognition of a specific antibody. The identification of different bands on the gel, with distinct molecular weights of different proteins, is part of the *local theory* of the Western blot technique. As Franklin argues, a *theory-driven* experiment is not always directed to the test of a specific hypothesis. Rather, it just requires that such an experiment is designed and performed according to specific theoretical constraints. These constraints then shape the possible manipulations, thus narrowing the horizon of expected outcomes. Therefore, the results can meet or fail to meet the expectations of the researchers, eventually leading to confirmations of previous assumptions, methodological problems, or even potential discoveries.

At the same time, there are experiments aimed at "exploring" phenomena. Exploration here means that experimentation, more than questioning nature,

is aimed at "teasing" nature, trying to map how it reacts to material intervention. Of course, such an independence from theory should not be intended in
too strong a sense, as if scientists did their work completely out of the blue.
Methodological freedom here means that experiments are only *theory informed*,
as opposed to *theory driven* (Waters 2007).

To sum up, the differences in the type of intervention should not be understood
in terms of (in)dependence from a theoretical framework but rather in terms of
the degree to which such a dependence is possible and how it is articulated, that
is, how much and according to which modality theory affects practice.

2.2 Hacking's Account Applied to Molecular Biology and Biomedicine

Molecular biology has been often conceived as a science dealing with material
manipulations (Keller 2000). The "experimental" nature of this science and the
standardization of laboratory techniques to tinker experimental systems are, at
least on the received view, distinguished marks of molecular studies in biology
(Rheinberger 1997).

In molecular biology, experiments are generally not done in order to test
theories; rather, they usually serve the purpose of developing and shaping
hypotheses about working models. Most molecular biology articles written since
the mid-1970s proceed by "telling a story," thus emphasizing the importance of
narratives. The underlying idea is the following. First, scientists start from a
general—albeit informed—guess about a biological system; how it is produced,
the effects it has, and so on. Due to the generality of the guess, several—possibly
contrasting—predictions may be derived. In order to test the predictions derived
from the general guess, researchers then devise several experiments *à la* Hacking
to stimulate the experimental system to "reveal" more information. Experiments
not only may be used to corroborate or refute initial predictions; in addition, they
contribute—through additional information they generate—to shape and refine
the hypotheses about the functioning of the experimental systems. This process
of shaping and refining hypotheses through experiments continues, virtually,
ad libitum. This is a sort of progressive and ramified (but not linear) deductive
process, developed by poking and prodding experimental systems.

Molecular studies in biomedicine present a similar structure. For instance, the
development of hypotheses through the manipulation of experimental systems
is the benchmark of molecular studies of cancer (Woodward 2003). Recently,
Robert Weinberg (2014) has outlined a brief history of molecular oncology
wherein he emphasizes the importance of entity manipulation for the advancement of cancer studies. In the 1970s, "the notion that cancer was a disease of
identifiable genes was little more than an attractive speculation" (2014, p 268).
However, through the development of transfection's techniques, biologists have
started to perturb and stimulate cell lines by inserting and knocking out genes
in order to observe phenotypic consequences. Such studies do not just show
the correlation between a gene and an observed phenotype but—again, through

experimental techniques—they try to grasp how such a phenotype is produced, starting from a mutation in a gene. Even today, traditional studies in molecular oncology have this nonlinear, interventionist, and manipulation-based structure. Indeed, the impact of molecular strategies based on material manipulations in biomedicine is also reflected in the shift from "function first" approaches to "target first" approaches (Kell 2013). Roughly, "target first" approaches refer to those strategies in which drugs are developed starting from the detailed knowledge coming from the understanding of biological mechanisms, which is obtained through the typical material manipulations of molecular biology. In particular, the development of new drugs begins with the identification of promising molecular targets; through material manipulation, researchers try to identify components or activities of an experimental system that play a nonnegligible role in the system itself. On the other hand, "function first" refers to the age in which access to molecular details was not possible and drugs were developed starting just from phenotypic screenings.

In short, contemporary biomedical research fits in well with Hacking's conception of experiments and scientific practice. As noted, biological phenomena are studied through experimental systems, which circumscribe and highlight "phenomena" created in the laboratory, where they are materially abstracted from their environment and put into a different context: the experimental context. Knowledge thus obtained is shaped and developed by provoking reactions of the experimental systems through material modifications that conform to excitatory and inhibitory strategies (Bechtel and Richardson 2010). Real-world phenomena are not investigated by means of theorizing over abstract modeling strategies. Real-world phenomena are investigated by playing with them, by exciting or inhibiting parts of experimental systems. This is the kind of access to biological phenomena that molecular biologists favor.

3 Data Mining: The High Throughput Sequencing Revolution and Bioinformatics

The completion of the Human Genome Project (HGP) fostered an allegedly new *way of doing* molecular biology (Pickstone 2001). This is a consequence, at least in part, of technological advancements in data production through sequencing technologies. However, there is more to say on this matter. In 2008, the American magazine *Wired* had on its special issue's cover a provocative title: "The End of Science." Chris Anderson, the former editor in chief of *Wired*, explained that sentence by arguing for the "end of theory" in science. This antitheoretical stance should not be intended along the lines provided by Hacking. According to Anderson, the idea of scientific disciplines being guided by theoretical hypotheses has become obsolete and should be abandoned and replaced by a new picture. The new face of science, according to Anderson, will be shaped by different approaches, new *ways of doing*. Such novelty should be understood in the light of the new challenges provided by so-called Big Data Science.

The label *Big Data* does not simply indicate a big volume of data. If just the amount of data counts, also taxonomy and astronomy could be seen as Big Data Science. Other features then seem to be required to establish what contemporary scientists mean by Big Data Science. Despite the lack of a precise definition, it is possible to select certain signature features of Big Data Science, as characterized by the *quantity* of data (petabytes), the *speed* at which these data are obtained, the ways in which data are *ordered* and *displayed*, the more global/holistic *aim* compared to more traditional statistics, standardized procedures both regarding *resolution* and *identification*, and their relational format, which can be easily expanded or increased in magnitude (Kitchin 2013, 2014). Examples of big data projects are now easy to find in biomedical research. Over the last few years, a strand of molecular research has revolved around the creation of big consortia, such as the Encode Project Consortium, Roadmap Epigenomics Mapping Consortium, and The Cancer Genome Atlas. The aim of these consortia is to generate enormous datasets about specific biological phenomena (the epigenome, the cancer genome, etc.) by means of sequencing technologies and to store these data sets in structured databases (Leonelli 2009, 2011; Ratti 2015).

The purpose of this data-accumulation enterprise is (at least) twofold. The first, comparative, use is quite common, going back to the creation of primary databases such as GenBank.[1] A second use is much more important for the purpose of this chapter. Recently, *biologists have started mining databases of big consortia in order to generate knowledge about biological phenomena*. There are several and straightforward differences with respect to "traditional" manipulation strategies:

1   In these data-accumulation enterprises, molecular biology is focused on the discovery of significant and robust patterns within vast data sets generated through sequencing technologies that, due to their magnitude, only computational approaches can handle.[2]
2   Since all data biologists' need is obtained through a brute-force use of sequencing technologies, there is no need to continuously stimulate experimental systems to squeeze partial information. Though sequencing is a kind of manipulation, data are obtained in "one single shot."
3   Unlike traditional approaches, there is no need to devise new experiments to develop hypotheses, because data obtained by sequencing are all taken to be data that one needs in principle.[3] Data are then subjected to bioinformatics analyses, and these analyses do not need additional experiments to put forth hypotheses.

The aim of these analyses is to uncover patterns or regularities in the data sets of biological databases in order to discover something novel about biological phenomena. This kind of practice is called *data mining*. Setting aside the initial sequencing part, big consortia do not to need, at first glance, robust interventionist strategies. In other words, sequencing technologies *plus* bioinformatics is a kind of molecular biology without the material manipulation aspect.

3.1. Discovering Through the Practice of Mining Databases

If we assume that manipulation, intended as a form of material intervention, is one of the main features of traditional molecular studies, the increasing role of databases in biology has definitely produced a change in the molecular sciences. Although the "comparative approach"[4] has never disappeared from molecular biology (Strasser and de Chadarevian 2011), the creation of biological databases—collections of data computationally administered, and readily accessible—have clearly fostered and enhanced the value of comparative approaches in the generation of scientific knowledge. For instance, UniProt (the greatest repository of protein information) is used to investigate putative functions of unknown proteins by comparing their structure to similar structures of proteins (coming from different organisms, species, etc.) whose function is already/previously known. If structures and molecular contexts are similar, then scientists are in a good position to infer also similar putative functions. Searching for databases is indeed a tool of discovery. The very term *data mining* suggests that, as miners dig into caves for enormous amounts of worthless material in order to find gold and gems, computational scientists penetrate the whole architecture of databases in order to retrieve valuable information. The "logic of discovery" of these procedures differs substantially from "traditional" manipulation strategies. Unlike the search for mechanisms, molecular biology here aims at the individuation of statistically relevant *regularities* within big data sets that only computational approaches can manage.

As an example of the developing strategy of analyzing large cohorts of data stored in structured database to foster biological discovery, consider the data mining of The Cancer Genome Atlas (TCGA) data set (http://cancergenome.nih.gov/). TCGA is a massive project that aims to foster the discovery of somatic mutations and structural variations in cancer and to organize that information toward successful medical interventions. Recently, TCGA fostered the creation of a project called the *pan-cancer analysis* project, namely the reanalysis of data made public by TCGA (The Cancer Genome Atlas Network 2013). Indeed, through the computational analysis of data stored in the TCGA repository, biologists can develop hypotheses about biological phenomena without the traditional "interventionist" (in the sense of experimental) strategy of molecular biology. This is a totally *in-silico* way to access the biological realm.

3.2 The Case Study in Detail

Within the *in-silico* studies promoted by TCGA, we are going to consider the study by Lawrence and colleagues (2013). This study is a paradigmatic example of the flexibility of the practice of clustering and reordering data into meaningful patterns by playing with parameters of data sets. Moreover, in this study, Lawrence and colleagues develop a tentative explanation of a biological phenomenon without any kind of traditional experimental interventionist strategy. By "ordering" a data set, Lawrence and colleagues actually "discover" something of interest for the biological field. This is somehow revolutionary for the experimental culture since scientific discovery (roughly "facts" about natural worlds) is obtained not by poking and prodding nature (through experimental systems) but by looking for patterns of data within the most comprehensive collection of biological samples gathered so far.

In order to understand, in detail, the motivations and the scope of Lawrence and colleagues" study, we should first introduce some facts about cancer genomics.

Contemporary cancer genomics (as well as traditional molecular oncology) has the aim of discovering so-called "cancer genes," that is, genes that, by being suppressed or overexpressed as a result of genomic catastrophic events (e.g., mutations, structural variations), can lead to the development of tumors. Current analytical tools to discover cancer genes are rooted within an evolutionary framework, in the sense that cancer genomics is framed within a selectionist account of cancer development. Indeed, if cancer cells are as such a result of driver somatic mutations, and if those mutations confer a growth advantage to cancer cells, then "driver" mutations are positively selected and they should be more highly conserved than mutations having no effect whatsoever on the fitness of cells. This means that "driver" mutations are more present than other types of mutations. Therefore, the more the sample size is increased, the more one would be able, *in principle*, to discover more of those driver mutations that, by conferring growth advantage to cancer cells, are positively selected. Driver mutations usually are located within "significantly mutated genes," defined as "those genes harboring more mutations than expected given the average background mutation frequency for the cancer type" (Lawrence et al. 2013, p. 214).

However, by analyzing the data set of TCGA, Lawrence and colleagues discover that traditional analytical tools, based on the assumption that bigger sample size will lead eventually to the discovery of new cancer genes, present some problems. They apply such tools to whole-exome[5] sequence data from 178 lung squamous cell carcinoma, and they find out that many recurrently mutated genes could hardly be cancer genes, in the sense that being recurrently mutated is not a sufficient condition for being a cancer gene. For instance, large genes are notably highly mutated. Moreover, olfactory receptor genes (the function of which has nothing to do with lung cancer) are mutated at a suspiciously high rate. *They then decide to see whether taking into account the phenomenon of "heterogeneity" in tumors can make sense of such suspicious cases.*

Heterogeneity in cancer refers to the broad phenomenon that cells within a cancer-cell population can show distinct morphological, genetic, and phenotypic profiles (see also the contributions of Boniolo and Plutinsky in this volume). They decide to analyze heterogeneity in 3,083 tumors samples across 27 tumor types of TCGA. In particular, they analyze three types of heterogeneity: (a) across patients with a given cancer type, (b) across the mutational spectrum of tumors, and (c) the regional heterogeneity across any given genome. Such analyses of heterogeneity at three different levels lead to the observation of "cancer" at the genetic level, from several angles.

Setting technical details aside, they perform a computational analysis for each type of heterogeneity. In technical terms, data have been *clustered*. By *clustering* we mean that they select and group homogeneous bits of data into particular categories. Take for instance Figure 2[6] of the article by Lawrence et al (2013).

The figure is a visual representation of heterogeneity of the mutational spectrum in tumors. Data about somatic mutations are clustered around six important types of mutations (e.g., from C to T or from C to A). While the distance from the origin represents the mutation rate, the angle represents the relative contribution to all spectra.

As with figures derived from clustering data about the two other types of heterogeneity, this analysis does not just advance general knowledge about cancer by showing some interesting and extremely robust correlations. In addition, it explains why there are certain false positives in using traditional analytical tools for discovering cancer genes. For instance, from the whole analysis, Lawrence et al. find out that there is a strong correlation between somatic mutation frequency in cancer and low gene expression levels, in the sense that the more a gene is expressed, the less it is mutated. Moreover, they also observe a marked correlation between somatic mutations and DNA replication timing. Two prominent examples of false positive cancer genes analyzed by Lawrence and colleagues were olfactory receptor genes. Their high mutation rate is explained by the fact that they are late in replication timing, and since they have low expression level in lung tumors, they have a high mutation rate. The same applies to large genes, which in lung cancer are low expressed and are late in replication.

To sum up, the practice of ordering and clustering data in big data sets discussed in this section is an instrument to elaborate explanations and the discovery of new insights on biological phenomena. Mining databases can achieve goals similar to the ones of traditional interventionist and experimental practices. Yet, despite the absence of material manipulations, such a practice is indeed a form of intervention. In the next section, the philosophical details of such a claim will be provided.

4 Is Ordering a Form of Intervention?

In this section, we provide two arguments for why ordering data in biological databases (i.e., data mining) is actually a form of intervention in the sense of manipulation, though not of a material type. The first argument is based on the actual practice of data mining, while the second is more theoretical and stems directly from the notion of *datum*.

4.1 Intervening Without Materially Manipulating

Since data mining does not meet the requirements of material manipulation (because biological databases are not material entities), one might claim that an important strand of contemporary molecular biology lacks the essential features of "intervention" and "manipulation." Thus, we want to ask whether the practice of "ordering" data can be reduced to interventionist strategies.

In a straightforward sense, traditional interventionist strategies are clearly different from the practice of "ordering." Empirical manipulation deals with the "materiality" of living systems. Even though biological systems are built, in the sense that a phenomenon is abstracted from its natural occurrences and "situated" in a different context, still, we argue, the experimental systems are subjected to material manipulation. Following Parke (2014), we might say that when an experimenter wants to study a system (*the object of study*, e.g., lung cancer development in mice) to make inferences about another (*the target*, e.g., lung cancer development in humans), one might intuitively claim that working on the object puts the experimenter in a privileged position because there is a sort of "material" correspondence between the object and the target. Parke calls this *the materiality thesis*. This thesis has been used to argue for the epistemic privilege of experiments over computer simulations, because in experiments there is a material correspondence between object and target, while in computer simulation the relation is a formal one (Guala 2002). Because of the materiality thesis, especially in molecular biology, experiments are supposed to have a remarkable inferential power.

Ordering data in large data sets clearly does not meet the materiality thesis. Data are computational entities with certain features having formal relations with each other. We do not materially modify experimental systems when ordering data. Rather, we play with the parameters of data to cluster data themselves according to a certain aim. But this "playing" is formal, not material. As in computer simulations, no material manipulation occurs in ordering data.

However, the materiality of interventionist strategies is just half of the story. "Intervention" and "manipulation" also imply that we modify the system under investigation by abstracting some features that are of interest to us. From a material point of view, we modify some of its conditions to let certain features emerge. For instance, in the case of disease modeling in animal models, we modify some genetic features of mice (e.g., gene insertion) to see the phenotypic consequences. "Ordering" data implies that we look at a data set just with respect to some of its features in order to see what the data set reveals about them. But this is not mere "observation" of data as we have gathered them. Actually, we *actively* intervene in the data set by changing some parameters to let emerge only what is of interest to us. In the case of Lawrence *et al.* (2013), the data set of TCGA is scrutinized by looking at three different types of heterogeneity. The data set is stimulated by clustering data according to a specific aim. Actually, a data set has been abstracted from its "totality," and they have considered just certain features. The data set observed after the ordering is different from the data set before the ordering. Before ordering, the data set is just the sum of all data. After ordering, the data set is what the data set can tell us about a certain phenomenon because the data set is treated as an experimental system. We therefore "intervene" in the data set because we want to observe only what is of interest to us by stimulating it to reveal its "secrets."

To sum up, "ordering," in the sense of clustering data, is a form of intervention, although it lacks the "material" part of typical interventionist strategies of molecular biology.

4.2 Ordering Data is a Form of Intervention Because of the Nature of the "Datum"

Data mining is also a form of intervention because of the nature of data. Data are meaningful precisely because they are disposed and organized in a particular order. Indeed, data organization is fundamental not just to comprehend data, but also, and more importantly, to consider, conceive, and perceive them for what they are: *data*. Clarifying this point is crucial. Data are such only in relation to other data and to the context of their production and gathering. Let us examine how and why.

Following Luciano Floridi (2008, 2011) it is possible to distinguish several definitions of *datum*. First, data can be intended *epistemically* when conceived as collections of facts. This is probably the closest interpretation to the etymological root of the term.[7] *Data* are then "given" in the sense that they constitute the ground on which further argumentations are constructed. Floridi acknowledges that such an account, although useful, fails to provide an explanation of phenomena as *data compression* and *data cryptography*. Second, data can be equated to *information*. This might be helpful in some practices, but it fails to recognize that the relation is not biconditional, that is, if information is "meaningful and truthful data," then not every data constitutes information. Third, data can be *computationally* conceived as sets of binary elements. This solution is still not completely satisfactory since it conflates data as such with the format in which data are encrypted.

In order to overcome all these issues, Floridi adopts what he calls a *diaphoric interpretation*, claiming that data stand for, basically, *lack of uniformity*. A *datum* is something that can be recognized, perceived, or measured as distinct from the background conditions. However, the relation to the context in which data are produced or gathered should not be simply intended as background. As Floridi writes "[a] white sheet of paper is not just the necessary background condition for the occurrence of a black dot as a datum, it is a constitutive part of the [black-dot-on-white-sheet] datum itself, together with the fundamental relation of inequality that couples it with the dot. *Nothing is a datum in itself.* Rather, being a datum is an *external property*." (Floridi 2008, p. 7, emphasis added). Thus, Floridi suggests the relational nature of data. Indeed, if data are relational entities, thus ordering and reordering a database is formally tinkering with the system. By being clustered in specific manners, data acquire new meanings. Such an operation can be interpreted as manipulation because researchers move bits of data close to others to observe consequences of that clustering.[8] To put it differently, giving order to data is then giving them meaning. Moreover, each order defines a particular epistemic space. Each data organization represents a sort of set of classificatory configurations as much as material interventions delimit experimental conditions.

Assuming that data are relational entities—in the sense that they acquire their meaning only when they stand in specific relations to each other (Leonelli 2015)—then the way scientists associate one bit of data to another makes the difference as to their interpretation. *Contra* Anderson (2008), data *do not* speak for themselves. In order to give meaning to data, we should organize them in a framework. Mining databases is exactly an operation of putting into specific relations different bits of data. Without ordering, a database is just a sum of data with no meaning. Paraphrasing Hacking, in biological databases there is just complexity, and "phenomena" (in the sense of meaningful patterns) emerge only if we intervene in the database by reordering it. If in biological databases there is just complexity, after the operation of ordering, databases themselves look quite different as to the information that we can extract. In this sense, this form of intervention is a kind of (formal) manipulation because the system (the database) is modified. By relating data in a particular way, particular patterns emerge that, strictly speaking, are created by ordering the database. As the Hall effect "does not exist outside of certain kinds of apparatus" (Hacking 1983, p. 226), patterns detected through data mining exist only through the algorithm that we apply to mine the database and through the collection from which we choose to build the database (Leonelli 2009). This is a consequence of the notion of *datum* as Floridi meant it, in the sense that data become meaningful only if put in appropriate relations with each other.

5   Is There a Place for Material Manipulation in Contemporary Biology?

The fact that computational approaches and data mining have changed the practice of molecular biology can mean different things. In a strong sense, one may think that such a "new" scientific venture has diminished, in some cases even eliminated, the role of experimentation within the research in favour of pure bioinformatics efforts. This interpretation is quite inaccurate. It is certainly true that some projects can be pursued mainly due to the reorder of known data. In this sense, mining databases constitutes a legitimate form of creating biological knowledge without any contribution from the experimental side. However, it is pretty obvious that the source of many databases is precisely experimental findings. Experiments still play a central role in contemporary research, but their role has changed. The contribution of experimental work within research has epistemically shifted in scientific practice itself. This means that, against some scientific journalism (e.g., Anderson 2008), data-mining discovery strategies pursued via computational methods will not change biological research toward a progressive reduction of experimental work. On the contrary, experiments still constitute a resource that is not dispensable. Expanding the meaning of possible interventions (i.e., intervention *per se* cannot be equated to material manipulation) results in a different contribution of the experimental side in the process of scientific discovery. If we want to understand how experiments changed their epistemic status within the logic of discovery of new molecular biology, we must first address their role in traditional settings.

In traditional, bench-oriented molecular biology, experiments can be sometimes pursued according to their *exploratory power*. As ancient geographers, molecular biologists were exploring unknown landscapes with no awareness of the configuration of the surrounding areas. It was indeed an exploration. It is not a surprise then that the verb "to explore" has been originally associated with geographical expeditions[9]. However due to technological advancement, such as satellite technology, geographical explorations changed their meaning. As a matter of fact, general mapping no longer requires direct investigation. High-throughput technologies can be seen as the biological counterpart of satellites. However, despite the accuracy of the aerial representations, such an investigation would inevitably leave out some details. These aspects are not fundamental for the global picture, but they can become essential for a more complete description of a location. Thus direct expeditions are now aimed not at exploring, but rather at *fitting the details*. By drawing from this analogy, we might say that many experiments in contemporary molecular biology are designed precisely for a similar purpose. Experiments, in many cases, do not show an intrinsic aim, but rather they present an instrumental value as they serve as *a confirmation tool*. Nowadays, an increasing number of scientific publications perform traditional experiments as confirmations, meaning that empirical results could corroborate the indications provided by high-throughput approaches. For instance, the entire group of results published by the ENCODE project fits such a picture. As recently argued by Germain and colleagues, ENCODE's first step strategy lies precisely in the identification of a "specific subset of biochemical activities (transcription, transcription factor binding, and specific combinations of histone modifications, etc.) which very often contribute and make a difference to the phenomena scientists are interested in" (Germain et al. 2014, p. 816). At a later stage, these activities will be specified, either confirmed or dismissed, via experimentation, which will also clarify their nature and their contribution to the phenomena of interest (by the way, such further efforts do not have necessarily to be conducted by the ENCODE project itself). This means that it is the computational part, not the material manipulation, that has assumed the role of exploration traditionally ascribed to experimentation. *Mutatis mutandis*, experiments in biology now have the same role (i.e., confirmation tools) that, for instance, they have in Poppers so harshly criticized view.[10]

In order to better explain this aspect, let us briefly focus on a recent article (Barozzi et al. 2014), published in *Molecular Cell*, in which traditional experiments are exactly fitting the details, while the general "questioning nature" is entirely based on computational efforts through database consultation and integration. The aim of the article is to show that *transcription factor binding* is somehow coregulated with the *nucleosome occupancy* due to the features of certain DNA regulatory segments (enhancers) shared by mammalians.[11]

Let us briefly examine the rationale of the study from an epistemic point of view. In this article, scientists start from the knowledge that transcription factors (TFs) usually bind sites of regions that previous computational analyses predicted to be with a high nucleosomal occupancy. However, TFs' binding sites are hard to detect since their recognition sequence can be easily repeated just by chance, thus creating a high number of false positives. Next, researchers hypothesized that the same information regulating nucleosome establishment also guides TFs in binding specific regulatory elements and neglecting false positives. From an epistemological point of view, all this starting knowledge has been produced by computational approaches.

These approaches were indeed possible because different specific repositories containing diverse kinds of information (such as factors determining nucleosome occupancy or cell lineage specific enhancers) have been created. Of course this information is as such precisely because it represents the order according to which data coming from experimental findings have been collected and systematized. Coming back to the article and leaving aside technical details, the important epistemic point here is that the choices of the experimental system (i.e., what cell lines to work with, what factors to focus on) are fully determined by the needs of bioinformatics.[12] All the experimental materials in this study are instrumental to the accomplishment of the computational analysis or to corroborate *in vivo* and *in vitro* the discoveries made through bioinformatics tools.

Such a change means that experiments here have changed their epistemic role within scientific enterprise. By that we mean that, contrary to the idea that *data-driven* science has diminished the experimental side of scientific work, such a new way of doing rather changes the epistemic primacy of material manipulation that was the benchmark of traditional molecular biology. In other words, experiments in the sense of "material manipulation" have now more often a confirmatory role rather than a discovery role. To put in a slogan, in biology "experiments do not have any more a life of their own," but rather they are instrumental to corroborate insights emerged from the practice of ordering data.

6 Conclusion

In this chapter, we have argued that bioinformatics analyses of databases (i.e., data mining, data ordering, etc.) are a kind of intervention, although not in the sense of material manipulation. We have provided two arguments in support of this idea. The first is from scientific practice. Through a case study of TCGA, we illustrated how biomedical data in biological databases should be ordered in a specific manner in order to extract useful information and to discover something about biological phenomena. The second argument is more theoretical. Since data could be defined as relational entities, and hence data acquire their meaning only when put in specific relations to other bits of data, then "ordering data" (that is, clustering data according to specific desiderata) is exactly a kind of intervention because we *manipulate* the database by associating bits of data according to specific aims. Finally, through another paradigmatic case study, we showed that the role of material manipulation in molecular biology has changed, passing from being a tool of discovery and hypothesis development to a tool to corroborate hypotheses developed through computational analyses.

Such a change has the potential of being more influential, at the epistemic level, than one may think. Indeed, this new venture seems to challenge the idea, supported by the second phase of philosophy of science (i.e., postpositivism philosophy of science), according to which there is no logic of discovery within sciences, since scientific enterprise cannot be completely reduced to clean and sharp logical steps (Feyerabend 1975). The richness of science, it has been argued, lies precisely in its capacity of going through different paths. Thus epistemic pluralism (Dupré 1993) has been established as the current mainstream view concerning the success of science and also about the intrinsic value of scientific research itself. However, the constraints imposed by consortia running Big Science projects do not pertain just to the economic side of scientific research. By creating a standardization of methods and procedures, Big Science projects are, inevitably and probably unwittingly, creating, from the practical side, a set of criteria for the old *demarcation problem*. If the technical and the epistemic repertoire of molecular studies will be completely subdued to the creation of vast maps of biological knowledge through computational approaches of massive amounts of data, scientific methodological pluralism will pass over. *Pace* Dupré's "disorder of things," it seems that a new order could prevail. If such a situation might occur, at the moment it is just a risk. In particular, this risk is twofold. On the one hand, no one knows whether such a standardization will contribute either to shape a better science (in the sense of more effective) or to impoverish it. For instance, standardization could undermine the creative aspects of scientific discovery, which, quite paradoxically, are sometimes taken as being as important as the rational nature of the scientific work. On the other hand, if such a way of doing should take place more vastly in the research landscape, this would also mean the end of many small labs that will no longer be able to do science. Perhaps the core of the problem rests on the notion of "effectiveness." What does it mean to be effective in science and in particular in biomedicine? Is this a danger for the relation between research findings and their clinical implementation? How is this related to the idea that scientists and philosophers have about what "good science" is? Because of its importance for scientific research, such a concept should be one of the future most important challenges for both science and the philosophy of science.

Notes

1 This aspect would deserve another article, but let us just say that "[r]esearchers routinely compare the sequences they have determined in their laboratories with those in the database, using sophisticated software to infer by analogy the function of genes or the evolutionary relationships between species" (Strasser 2011, p. 63).
2 This access to biological phenomena through the accumulation of data has been perceived in a way as "unbiased," since it is believed they do not rest on traditional "hypothesis-driven" approaches (which can be notoriously affected by several biases). Moreover, these approaches seem not to be biased because phenomena are not created in laboratories by abstracting them from their environment and putting into a different context. Instead, data about phenomena is obtained "directly" through primary samples.
3 Unlike claims to the contrary (Anderson 2008), data do not speak for themselves.
4 By *comparative* we mean that scientific generalizations about observed phenomena are constructed through the comparison among different species, samples, and specimens. This approach was typical of taxonomic work in natural history.
5 The exome is the sum of all exons of the genome.
6 http:\\www.nature.com/nature/journal/v499/n7457/full/nature12213.html#close.
7 "Datum" is the past participle of the Latin verb "dare," which means "to give." Therefore, a datum is a "given."
8 This also shows how classification can be genuinely a form of exploratory science since different classificatory strategies imply different expectations.
9 Etymologically, the verb "to explore" explicitly refers to geographical contexts: "to go to a country or place in quest of discoveries," as reported by the *Online Etymology Dictionary*–www.etymonline.com/index.php?term=explore.
10 Of course, here experiments do not corroborate well-formed theories but rather hypotheses in the form of patterns regularities.
11 Transcription factors (TFs) are proteins that bind to specific DNA sequences thus regulating the rate of transcription of other functional products. Nucleosomes are instead fundamental units of chromatin organization constituted by a core of proteins, called *histones*, around which DNA filaments are somehow wrapped up. Enhancers are DNA regions that favor genetic transcription. Nucleosomes are important also for transcription since they contribute to chromatin conformation, thus either allowing or impeding the possibility of regulatory elements to actually transcribe the genetic information (from DNA to mRNA).
12 Indeed, researchers have chosen to work on primary mouse macrophages and to compare them with several control lines, also because a single specific TF, Pu.1, behaves differently in these diverse cell lineages (it is expressed only in hematopoietic cells).

References

Anderson, C. (2008) "The end of theory: The data deluge makes the scientific method obsolete," *Wired Magazine.*
Bacon, F. (n.d.) *Novum Organum.* London: Bells and Sons.
Barozzi, I., M. Simonatto, S. Bonifacio, L. Yang, R. Rohs, S. Ghisletti, and G. Natoli (2014) "Coregulation of transcription factor binding and nucleosome occupancy through DNA features of mammalian enhancers," *Molecular Cell* 54(5). Elsevier Inc., pp. 844–57. doi:10.1016/j.molcel.2014.04.006.
Bechtel, W., and R. Richardson (2010) *Discovering Complexity: Decomposition and Localization as Strategies in Scientific Research.* Cambridge: The MIT Press.

Burian, R. M. (1997) "Exploratory experimentation and the role of histochemical techniques in the work of Jean Brachet, 1938–1952," *History and Philosophy of the Life Sciences*, 19(1), pp. 27–45.

Dupré, J. (1993) *The Disorder of Things: Metaphysical Foundation of the Disunity of Science*. Cambridge, MA: Harvard University Press.

Feyerabend. *Against Method* (London: 1975). Revised edn. London: Verso, 1988.

Floridi, L. (2008) "Data." In *International Encyclopedia of the Social Sciences*. Macmillan.

Floridi, L. (2011) *The Philosophy of Information*. Oxford: Oxford University Press.

Franklin, L.R. (2005) "Exploratory experiments," *Philosophy of Science*, 72(December), pp. 888–899.

Germain, P., E. Ratti, and F. Boem (2014) "Junk or functional DNA? ENCODE and the function controversy," *Biology and Philosophy*, 29(3), pp. 807–831.

Guala, F. (2002) "Models, simulations, and experiments." In L. Magnani and N. Nersessian (Eds.), *Model Based Reasoning: Science, Technology, Values* (pp. 59–74). New York: Kluver Academic.

Hacking, I. (1983) *Representing and Intervening—Introductory Topics in the Philosophy of Natural Science*. Cambridge University Press.

Kadane, J.B., and T. Seidenfeld (1990) "Randomization in a Bayesian perspective," *Journal of Statistical Planning and Inference*, 35.

Kell, D.B. (2013) "Finding novel pharmaceuticals in the systems biology era using multiple effective drug targets, phenotypic screening and knowledge of transporters: Where drug discovery went wrong and how to fix it," *The FEBS Journal*, 280(23), pp. 5957–80. doi:10.1111/febs.12268

Keller, E.F. (2000) "Models of and models for: Theory and practice in contemporary biology," *Philosophy of Science*, 67(S1), S72. doi:10.1086/392810

Kitchin, R. (2013) "Big data and human geography: Opportunities, challenges and risks," *Dialogues in Human Geography*, 3(3), pp. 262–267. doi:10.1177/2043820613513388

Kitchin, R. (2014). Big Data, new epistemologies and paradigm shifts. *Big Data and Society*, 1(1), 1–12. doi:10.1177/2053951714528481

Lawrence, M.S., P. Stojanov, P. Polak, G.V. Kryukov, K. Cibulskis, A. Sivachenko, G. Getz (2013) "Mutational heterogeneity in cancer and the search for new cancer-associated genes," *Nature*, 499(7457), pp. 214–18. doi:10.1038/nature12213

Leonelli, S. (2009) "On the locality of data and claims about phenomena," *Philosophy of Science*, 76(December), pp. 737–749. doi: 10.1086/684083

Leonelli, S. (2011) "Packaging data for re-use: Databases in model organism biology." In P. Howlett and M.S. Morgan (Eds.), *How Well Do Facts Travel? The Dissemination of Reliable Knowledge*. Cambridge, MA: Cambridge University Press.

Leonelli, S. (2015) "What counts as scientific data? A relational framework," *Philosophy of Science*, 82(5), 810–821.

Parke, E.C. (2014) "Experiments, simulations and epistemic privilege," *Philosophy of Science*, 81(October), pp. 516–536.

Pickstone, J. (2001) *Ways of Knowing. A New History of Science, Technology and Medicine*. Chicago: University of Chicago Press.

Popper, K. (1959/2002) *The Logic of Scientific Discovery*. London and New York: Routledge.

Ratti, E. (2015) "Big Data Biology: Between eliminative inferences and exploratory experiments," *Philosophy of Science*, 82(2), pp. 198–218.

Rheinberger, H.-J. (1997) *Toward a History of Epistemic Things: Synthetizing Proteins in the Test Tube*. Stanford University Press.

Rheinberger, H.-J. (2007) "What happened to molecular biology?" *B.I.F. Futura*, 22, pp. 218–223.

Steinle, F. (1997) "Entering new fields: Exploratory uses of experimentation," *Philosophy of Science*, 64(Supplement. Proceedings of the 1996 Biennial Meetings of PSA), 64–74.

Strasser, B. (2011) "The experimenter's museum—GenBank, natural history, and the moral economies of biomedicine," *Isis*, 102(1), pp. 60–96.

Strasser, B.J., and S. De Chadarevian (2011) "The comparative and the exemplary: Revisiting the early history of molecular biology," *History of Science*, 49(3), pp. 317–336. doi:10.1177/007327531104900305

Waters, C.K. (2007) "The nature and context of exploratory experimentation," *History and Philosophy of the Life Sciences*, 29, 1–9.

Weinberg, R.A. (2014) "Coming full circle—from endless complexity to simplicity and back again," *Cell*, 157(1), pp. 267–271. doi:10.1016/j.cell.2014.03.004

Woodward, J. (2003) *Making Things Happen: A Theory of Causal Explanation*. Oxford: Oxford University Press.

8 Pathways to the Clinic

Cancer Stem Cells and Challenges for Translational Research

Melinda Bonnie Fagan

Abstract

Cancer stem cells (CSCs) have been a focus of research and controversy for over a decade. CSCs are thought to be a small subpopulation of self-renewing stem cells within a tumor or blood-borne cancer, which are responsible for maintaining and growing the malignancy. The concept has profound clinical implications. However, although experimental evidence for CSCs has accumulated, clinical translation of these results has been lacking. This paper examines conceptual and evidential challenges blocking clinical translation of CSC, and proposes a way forward. The solution is to distinguish two CSC concepts with different substantive content, suited to their respective purposes and criteria for success: basic and clinic-oriented. Successful clinical translation requires empirical validation of the latter. I indicate how such evidence could be obtained, building on existing experimental support for the basic CSC model.

1 Introduction

Cancer stem cells, as the name suggests, are a kind of stem cell. Stem cells are defined in terms of two functional properties: self-renewal and differentiation. More precisely, stem cells are unspecialized cells capable of dividing to produce offspring cells that are similarly unspecialized, *and* offspring cells that are or become more differentiated (Ramalho-Santos and Willenbring 2007, p. 35; Melton and Cowan 2009, p. xxiv). Self-renewal maintains or increases the number of stem cells in some tissue, organ, or artificial culture.[1] Differentiation is the process that gives rise to all the specialized cells in the body. Differentiated cells are specialized to play particular roles within a healthy organism, making up its diverse organs and tissues. This general, functional definition encompasses many kinds of stem cell: adult, embryonic, pluripotent, multipotent, induced, mesenchymal, neural, hematopoietic (blood-forming), embryonal carcinoma, and more. In previous work, I have explicated a "minimal" stem cell concept corresponding to the functional definition in terms of four variables: lineage or organismal source, time interval, characters of candidate stem cells, and characters of differentiated cells (Fagan 2013a, Ch. 2). The various kinds of stem cell correspond to different combinations of values of these four variables. Stem cells in general are defined more abstractly as cells occupying the "stem" position in a cell lineage tree organized by reproductive relations between cells exhibiting diverse characters over some time interval of interest (Figure 8.1).

In this chapter, I extend the modeling approach to cancer stem cells (CSCs).[2] As with stem cells more generally, the cancer stem cell concept turns out to be multiple. I argue, further, that different CSC concepts are appropriate for different contexts—notably, basic research and clinical application. The modeling approach to explicating stem cell concepts both helps to clarify the current situation in CSC research and indicates how to move forward. In this way, philosophy of science can contribute, albeit indirectly, to progress in biomedicine. The rest of this introductory section motivates this progressive philosophical project for the case of CSCs.

Amidst the variety of stem cells sketched here is a unifying characteristic: stem cell capacities are *regenerative*. A stem cell, by definition, can produce more stem cells (self-renewal) and also can give rise to specialized cells (differentiation). It follows that stem cells have, in principle, remarkable clinical potential. At least three applications have been proposed. One is to use stem cells directly as renewable sources of healthy tissue. Another is to use knowledge of molecular mechanisms underlying stem cell capacities to direct cells to develop along pathways of our choosing to repair damaged organs and tissues *in situ*. A third potential application is to use stem cells as a renewable source of cells for drug discovery, accelerating the production of new chemical therapies (see Teira, this volume). Stem cell research, accordingly, aims at biological knowledge that can be used to realize these clinical applications. The field's overarching goal is to harness stem cell capacities in order to treat a wide array of pathological conditions, such as heart attack, blindness, cancer, neurodegenerative diseases, spinal cord injury, and diabetes. These clinical aims are central to stem cell science (discussed further in Fagan 2013a, Ch. 10). In this sense, stem cell research is inherently translational.

The idea of a "cancer stem cell" seems to epitomize the translational character of stem cell research. CSCs are thought to be a small subpopulation of self-renewing stem cells within a tumor or blood-borne cancer, which are responsible

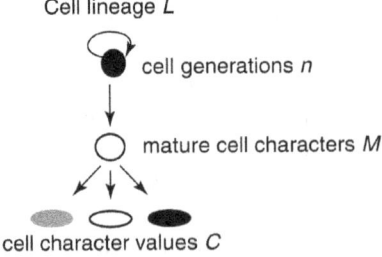

Cell lineage *L*

cell generations *n*

mature cell characters *M*

cell character values *C*

Figure 8.1 Stem cell model, showing a schematic cell lineage tree (after Fagan 2013a, Ch. 2).

for maintaining and growing the malignancy. This idea has significant clinical implications. If the CSC model is correct, the current clinical strategy of seeking to eradicate *all* cancer cells should be revised, to specifically target CSCs. So it is vital to test the model and, if it is confirmed, devise clinical interventions to eradicate CSCs in particular. Conversely, if the model is not confirmed, then clinical efforts should be directed elsewhere. These points are uncontroversial. Yet, after more than a decade of experimental work, the status of the CSC model of cancer remains unsettled. The CSC idea has not been successfully clinically translated nor experimentally disconfirmed. Instead, CSC research appears to be in a holding pattern: experimental evidence accumulates, but progress toward clinical application is lacking. This chapter examines the conceptual underpinnings of this situation and proposes a conceptual innovation that complements scientists' own proposals for moving CSC research forward.

My argument proceeds as follows. The next section briefly sketches the historical and scientific background for current debates over CSCs. Section 3 introduces the CSC model in more detail, applying the same approach used to clarify the general stem cell concept (Fagan 2013a). This explication brings into sharp relief the ambiguities surrounding the notion of CSCs. Section 4 examines "gold standard" experimental assay for CSCs, identifying four challenges for this method of testing the CSC model(s). Section 5 argues that a single CSC concept cannot meet these challenges. This result motivates a distinction between two CSC concepts: basic and clinic-oriented. Section 6 shows how this distinction can be deployed to move CSC research toward clinical translation.

2 Cancer Stem Cells: Background

The CSC concept is central to a developmental model of cancer that views tumors as analogous to organs. On this view, a tumor develops from an enduring population of stem cells, which give rise to more specialized cells in patterns represented by a hierarchical lineage (Figure 8.2). This idea is not new. The notion that cancers are "caricatures" of ordinary tissue, containing a mixture of proliferating stem cells and their more differentiated products, played an important role in 1960s and 1970s cancer research (Pierce and Speers 1988; Kraft 2011; Morange forthcoming). More loosely conceived, a developmental approach to cancer dates back to the first modern characterizations of the disease (Mukherjee 2010).

The developmental approach to cancer was largely displaced by the somatic mutation theory (SMT), which came to prominence in the 1980s. Briefly, SMT states that most cancers arise from a single cell that has undergone mutation, resulting in further genetic instability (Nowell 1976). Cell division then produces a clonal population descended from the initiating cell of origin, with genetic variation arising due to acquired instability (Figure 8.2). Selection within the clone favors "sub-lines" with more rapid rates of division, leading to tumor progression. SMT conceives cancer as a genetic disease maintained by evolutionary processes (see Plutynski, this volume; and Liu, Love, and Travisano, this

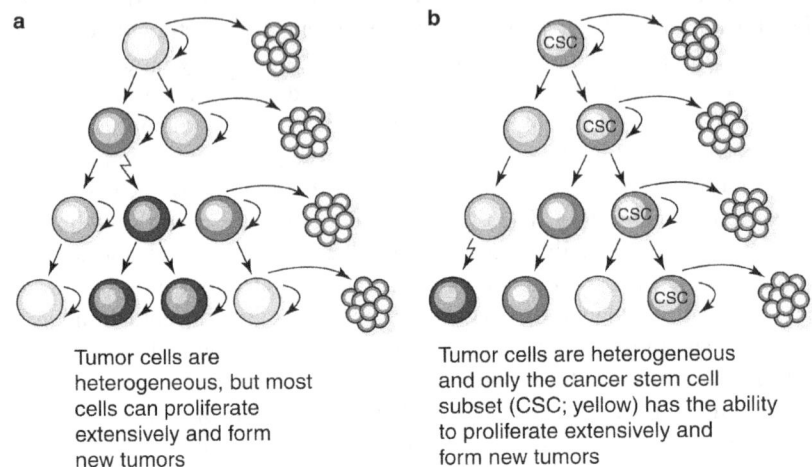

a

Tumor cells are heterogeneous, but most cells can proliferate extensively and form new tumors

b

Tumor cells are heterogeneous and only the cancer stem cell subset (CSC; yellow) has the ability to proliferate extensively and form new tumors

Figure 8.2 Basic concept of cancer stem cells (from Reya et al. 2001, p. 109). (a) The traditional model of cancer. (b) The CSC model of cancer. Reproduced with permission from Nature Publishing Group.

volume). CSCs, and associated developmental ideas, returned to the scene only after the identification of "leukemic stem cells" (Lapidot et al. 1994; Bonnet and Dick 1997) and an analogous cell population in breast cancer, suggesting that CSCs are a general feature of cancers (Al-Hajj et al. 2003).

The CSC model has significant clinical implications. Its key prediction is that most cells comprising a cancer are comparatively benign, with "tumorigenicity" restricted to a small subpopulation, analogous to the rare stem cells that maintain healthy organs and tissues in multicellular organisms. If this is correct, then cancer therapy should target only CSCs (Figure 8.3). But the current practice, informed by SMT, is to try to eradicate *all* tumor cells. So the CSC model is relevant to both basic research on cell development and to clinical cancer treatment. Testing the model requires close coordination between surgeons, laboratory researchers, and clinicians (see Fagan 2013a, Ch. 10). Results of such tests bear on clinical trial design, efficacy criteria, and the dominant conceptual framework of cancer research, as well as strategies for drug discovery. One would therefore expect the study of CSCs to be an exemplar of translational research in stem cell biology.

However, to date the CSC model has produced more controversy and confusion than general empirical results or clinically applicable strategies. The developmental approach remains a minor, and contested, strand of cancer research. While empirical reports of CSCs in various cancers have proliferated (reviewed in Clevers 2011; O'Connor et al. 2014), these results have not translated into clinical success or even early-stage clinical trials.[3] In the absence of breakthroughs, "an attitude of healthy caution seems to be developing in the maturing cancer stem cell community" (Clevers 2011, p. 313). Yet this caution falls short

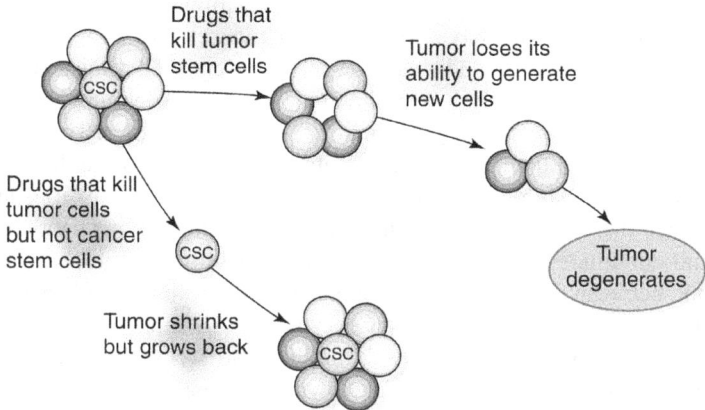

Figure 8.3 Clinical implications of the CSC model (from Reya et al. 2001, p. 110). Reproduced with permission from Nature Publishing Group.

of disconfirmation. There is strong empirical evidence of CSCs, although the model's clinical applicability remains unsettled (Meacham and Morrison 2013). Given the model's obvious clinical relevance, this prolonged situation of controversy, ambivalence, and uncertainty is surprising. Why, after more than a decade, has the CSC model been neither validated by clinical applications nor experimentally disconfirmed? If there is sufficient experimental evidence for the model to maintain it as a focus of inquiry, then what blocks the path to clinical translation? The following sections offer a partial answer to this question and indicate a solution with both conceptual and practical aspects.

To anticipate the conclusion: I argue that the CSC model is ambiguous, encompassing a wide array of properties attributed to the cells in question. Different investigators endorse very different conceptions of "the CSC model." In addition, the "gold standard" experimental method used to identify CSC in practice faces significant evidential challenges, particularly for "thicker" conceptions of the CSC model. Lack of consensus on the essential properties of CSCs compounds the difficulties for interpreting experimental results concerning these cells. The combination of conceptual ambiguity and evidential challenges produces a situation of "epistemic stasis"—certain aspects of the model are confirmed, while others are disconfirmed or not addressed. So repeated experimental tests of the CSC model do not advance toward clinical translation. The situation can be improved by distinguishing two CSC concepts: *basic* and *clinic oriented*. The different purposes and standards of basic research and clinical contexts impose different requirements and constraints on the notion of CSCs. Distinguishing between the CSC concepts appropriate to each context is a prerequisite for characterizing their relation and building a conceptual "bridge" to connect them. Such a conceptual bridge amounts to a blueprint for the clinical translation of CSCs.

One objection to the above must be addressed at the outset. My solution disambiguates the CSC model along traditional disciplinary lines: basic science and clinical medicine. This may seem a step backward, ignoring recent insights about translational research and returning to the discredited "linear model" of basic research followed by clinical application. However, such unwelcome consequences do not follow from my "bifurcated" account of CSCs. To distinguish between concepts and their appropriate contexts is not to separate them utterly. My proposal is compatible with an interactive, nonlinear view of translational research, such that laboratory and clinical contexts influence one another reciprocally and continuously. It is just that this mutual influence involves contact, so to speak, between distinct CSC concepts, instead of a single concept common to both. Understanding which CSC concept is at issue in any particular situation can take us some way toward resolving the controversy and confusion that attend the term; so I shall argue in the following sections.

3   Cancer Stem Cell Model(s)

The simplest way to define CSC is as a variety of stem cell. According to the "minimal model" framework (Section 1), the various kinds of stem cell correspond to different combinations of values of four variables: lineage or organismal source, time interval, characters of candidate stem cells, and characters of differentiated cells. For CSCs, the lineage originates from a cancer cell-of-origin; that is, the original mutated cell that gives rise to a clonal population of cancer cells. The time interval of interest is the "lifespan" of the cancer, which may range from weeks to months or years but is constrained to be less than the natural lifespan of the surrounding organism. Cell characters of interest vary according to cancer type. For CSCs in general, the key stem cell traits are those associated with self-renewal and differentiation capacities, while more differentiated cancer cells lack these traits. This is a "thin" characterization of CSC. The only substantive constraint it adds to the general stem cell model is that the lineage originates from a cancer cell of origin within a multicellular organism, the natural lifespan of which provides an upper limit on the time interval of interest.

   This minimal CSC model can be concisely stated as two assumptions describing essential features of CSCs:

1   Cells comprising a tumor are heterogeneous in phenotype and function.
2   These patterns of variation map onto a hierarchical lineage structure, with more tumorigenic cells giving rise to less tumorigenic, more differentiated, progeny.

The more tumorigenic cells are identified as CSCs. These cells occupy the "stem" position in a cell lineage tree of the form sketched in Figure 8.1. More precisely, the minimal model comprising (1–2) conceptualizes a tumor as a developmentally stratified, or "hierarchical," system of cells, with some cells possessing self-renewal and differentiation capacities sufficient to maintain, expand, or regrow the entire tumor; and others with reduced or negligible capacities for tumorigenesis, such that the latter are the differentiated progeny of the former.[4]

Richer CSC concepts involve further assumptions. A key motivation for these more elaborate CSC models is inference from clinical outcomes. Many cancer patients show a good initial response to therapy, with tumors shrinking past the point of detection, only to relapse later as tumors reappear. Relapses of this sort are particularly common for solid tumors,[5] which account for a large proportion of cancer patients. This phenomenon is in large part why improvements to cancer patient lifespan have been modest overall, despite enormous investment in anticancer drug development and the wide array of chemotherapeutic treatments available (Mukherjee 2010). CSCs are posited to explain these frequent relapses: a rare subpopulation of therapy-resistant tumor cells, which can give rise to all the phenotypically and functionally diverse cells comprising the original tumor (see Figure 8.2).[6] This inferential background supports a model of CSC that includes several further assumptions:

3   CSCs are rare cells within tumors.
4   CSCs are more likely to survive cancer therapy than other tumor cells.
5   CSCs' increased survival rate is due to their either acquiring or inheriting "the molecular armaments of normal stem cells," which protect them from anticancer drugs (Clevers 2011, 313).

The latter assumption, posited as an explanation of (4), calls for further specification. Several alternative, though not mutually exclusive, hypotheses have been proposed in this regard:

6   CSCs are quiescent; that is, they divide at a low rate.
7   CSCs exhibit gene expression patterns associated with pluripotency and long-term self-renewal (the signature abilities of embryonic stem cells) and therefore can be considered cells in a state of "stemness."[7]

Basic cancer research reveals phenomena that are explained by the CSC model, in either "thick" or "thin" versions. It is well-established that cells from a single tumor exhibit heterogeneity in morphology, surface marker expression, proliferative potential, therapy resistance, and capacity for long-term self-renewal and tumorigenicity (see Boniolo, this volume). The CSC model accounts for this intratumor heterogeneity in terms of a cell lineage hierarchy, with CSCs at the apex (Figure 8.2b). The "thicker" explanatory model comprised of (1–7) attributes many additional features to CSCs, including shared molecular mechanisms with normal stem cells.

This clinically motivated model is also positioned as opposed to the prevailing theory of cancer, SMT. Although the two do not conflict outright, the phenomenon occasioning the inference to CSCs exhibiting features (1–7) highlights SMT's shortcomings. So it is natural, though not strictly necessary, to view the thick CSC model as an alternative theory of cancer.

A high-profile example is furnished by the highly successful antileukemia drug imatinib (discussed in Kraft 2011, pp. 208–210). Imatinib specifically targets the molecular mechanism underlying chronic myeloid leukemia (CML). SMT predicts that all CML cells should be destroyed by imatinib; genetic mutations that induce drug resistance should, by the same token, eliminate the CML phenotype.[8] The drug is effective in managing CML: while a patient is taking imatinib, leukemia cells are eliminated below the point of detection. However, if a patient stops taking imatinib, CML cells frequently return; the leukemia regenerates. CSCs are invoked to explain this puzzling result: a few rare CML cells somehow escape the targeted drug and survive to regenerate the cancer when the treatment stops. The mechanism of their drug resistance is, parsimoniously, associated with their regenerative ability; that is, CSCs resist targeted anticancer therapy in virtue of their stem cell properties. Because the thicker CSC model accounts for phenomena that are anomalous for SMT, it appears to be an alternative to that theory. Relatedly, its primary rationale is explanatory power: the thick CSC model "can comprehensibly explain essential, poorly understood clinical events, such as therapy resistance, minimal residual disease, and tumour recurrence" (Vermeulen et al. 2012, p. 83).

The first study to experimentally identify CSC in accordance with this model was of acute myeloid leukemia (AML). However, the conclusion of that seminal study concerned the "cell-of-origin" for AML, that is, the normal tissue cell that serves as the clonal source of the cancer (Bonnet and Dick 1997). The main reported result was that AML originates by mutation of a normal blood stem cell (HSC), rather than from more differentiated blood cells. Because the first experimental identification of CSCs was part of the same study, the "cell-of-origin" question became linked with the CSC model (Figure 8.4). Many researchers thus add a further assumption to the latter:

8   The cancer cell-of-origin is a normal stem cell.

Lack of consensus about which of assumptions (1–8) comprise the CSC model sets the stage for confusion about the model's evidential status.[9] The conceptual situation is further complicated by the fact that some CSC researchers redefine the general stem cell concept, elevating self-renewal to the status of sole essential feature: for example, "the cardinal property of a stem cell is self-renewal, whether normal or malignant" (Kreso and Dick 2014, p. 276; see also Reya et al. 2001, p. 110). The capacity to give rise to differentiated cells of one or more types falls by the wayside in (some) discussions of the CSC model. This conceptual revision is rhetorically useful, as it allows CSC researchers to position "stemness" as central to understanding basic and clinical phenomena of cancer (Figure 8.5). Unregulated cell proliferation is the core of malignancy, and self-renewal is the cellular process involved in proliferation. But in redefining stem cells to showcase self-renewal, CSC researchers open a conceptual gap between their model and core assumptions informing stem cell research.

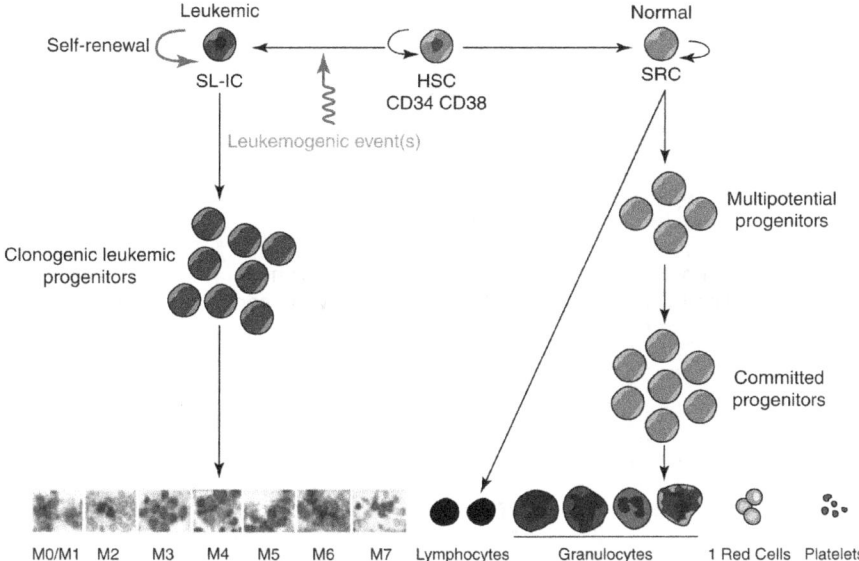

Figure 8.4 Model of cancer etiology integrating somatic mutation theory and cancer stem cell theory. From Bonnet and Dick (1997, p. 735). Reproduced with permission from Nature Publishing Group.

To sum up the analysis so far: "the CSC model" is in practice a family of models, which share a minimal core but include diverse further assumptions. Because "stemness" in CSCs is (sometimes) characterized in terms of self-renewal rather than self-renewal and differentiation, the CSC model is not simply a special case of a more general stem cell model. Different conceptions of the CSC model have different, sometimes overlapping motivations: continuity or analogy with normal stem cells, explanation of clinical phenomena, results of basic research on tumor cells, hypotheses about cancer cells-of-origin and molecular mechanisms underlying cell proliferation, and the redefinition of "stemness" to boost the significance of CSC for cancer research. The diversely motivated variations on the CSC model make for conceptual confusion before the issue of experimental validation is even addressed.[10]

4 Experimental Tests and Uncertainty

In contrast to the multiplicity of CSC models, there is one "gold standard" experimental method for identifying CSC. This is "the xenotransplantation assay," so called because it involves transplanting cancer cells from a human patient to an immunodeficient mouse.[11] The assay was developed by John Dick's research team at the University of Toronto, who used it to identify "SCID leukemia initiating" cells within AML (Lapidot et al. 1994; Bonnet and Dick 1997).[12]

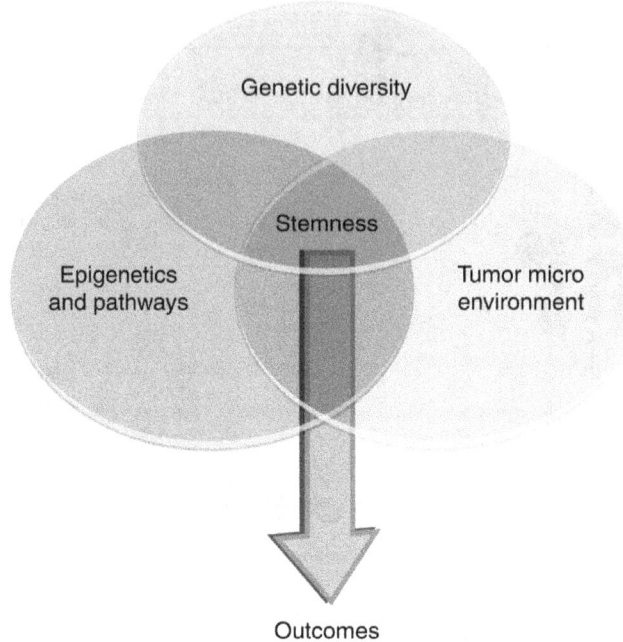

The experimental design is as follows (Figure 8.6). First, cancer cells are
removed from a human patient and sorted into subpopulations based on the
presence or absence of specific molecules on each cell's surface. Next, samples of each cancer cell subpopulation are injected into immunodeficient mice,
sometimes diluted so that each host receives an average of one cell. After some
weeks, transplanted mice are checked for human cancers. Any tumors detected
are removed and their cells characterized by phenotype and function. Cancer
cells showing a candidate CSC phenotype are transplanted into new immunodeficient hosts and the tumor-formation test is repeated.[13] Accordingly, cancer cell
subpopulations that can give rise to new cancers in two successive immunodeficient hosts are identified as CSCs. Cancers that can be sorted into cell subpopulations with different capacities for tumor production, such that one is identified
as CSC and others as less tumorigenic, are said to "fit the CSC model," that is,
to exhibit developmental hierarchy.

This experimental design is a special case of the general design of experiments
aiming to identify and characterize stem cells (described in Fagan 2013a, 2015).
The resemblance is deliberate: Dick and colleagues modeled their original CSC
assay on earlier stem cell methods (discussed in Fagan 2013a, Chs. 2 and 8).

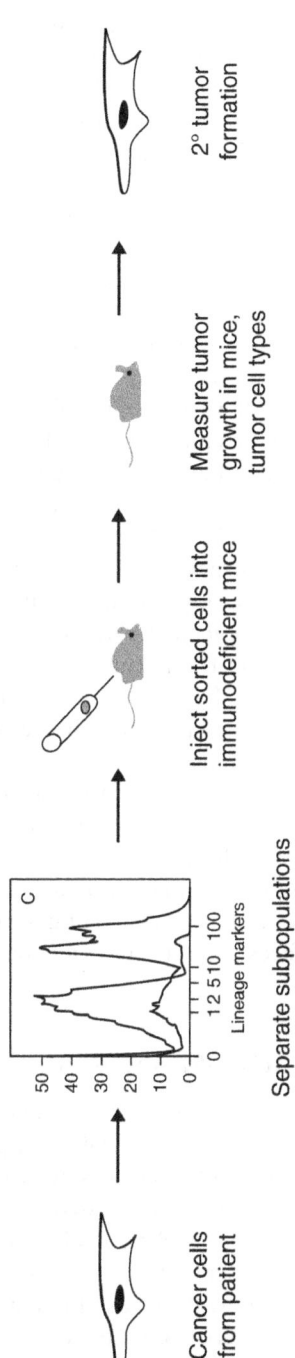

Cancer cells
from patient

Separate subpopulations
by characters C

Inject sorted cells into
immunodeficient mice

Measure tumor
growth in mice,
tumor cell types

2° tumor
formation

Figure 8.6 Basic design of the xenotransplantation assay to identify CSCs.

These canonical stem cell experiments exhibit a basic structure of three stages. First, cells are extracted from an organismal source and placed in a new environment in which candidate stem cell characters are measured. The measured cells are then moved to another environment, in which capacities for differentiation can be realized. Finally, characters of differentiated cells are measured. For the CSC assay, the organismal source is a human cancer, the first environment is the cell sorting apparatus, and the second environment is an immunodeficient mouse.[14] The characters on the basis of which cancer cells are sorted define different subpopulations, which are transplanted to different hosts. At the end of the assay, correlations between subpopulation identity and tumor-forming ability are recorded. If the assay is successful, tumor-forming ability is restricted to or enhanced in one subpopulation over others. Self-renewal is inferred from the ability of transplanted cells to propagate and clonally expand in two successive new hosts. The capacity for differentiation is inferred from regrowth of the original cancer in a mouse, with all its phenotypic and functional heterogeneity.

The xenotransplantation assay is the primary method used to evaluate the CSC model.[15] As any particular assay is restricted to one cancer from one patient, the general applicability of the CSC model emerges from performing the experiment on a wide variety of cancer types across many patients. Although a comprehensive survey of all cancer types in the assay is still lacking, results to date indicate considerable developmental variation across cancers and even across patients with the same cancer (reviewed in Clevers 2011; Magee et al. 2012; Meacham and Morrison 2013). I discuss concerns raised by these variable results below. For now, however, I will put aside this feature of emerging CSC results to focus on the xenotransplantation assay itself. The purpose of the assay is twofold: to identify and characterize CSC, and to evaluate the CSC model. The relation between these two goals seems straightforward: accomplishing the first offers positive support in regard to the second. However, the relation between CSCs, as identified by the xenotransplantation assay, and the CSC model of cancer, is complex and tenuous. As a means of testing the model's assumptions, the xenotransplantation assay faces at least four evidential challenges. I discuss each in turn.

(i)   *Some assumptions of the CSC model are not tested*

Most obviously, several assumptions of the CSC model are not tested by the experimental design described above. Because only transplanted cells that give rise to tumors are "read-out" in the results, the possibility of CSC quiescence (6) is unexamined. Therapy resistance (4) is not assessed, as immunodeficient mice are not a good proxy for human patients. A clinical follow-up to test (4) is possible in principle: track CSC populations identified by the xenotransplantation assay in patients, and correlate CSC numbers with stages of treatment. But to do this we need unambiguous identification of robust CSC populations—which turns out to be a difficulty for many cancers (see below). For the same reason, experimental tests of shared molecular "defenses" (5) and gene expression pathways (7) with normal stem cells are problematic. Testing these assumptions requires a robust CSC population *and* a corresponding stem cell population to compare the molecular characters of each. But stem cell populations are also various and in many tissues remain poorly characterized (Fagan 2013a). Except in cases where assumption (8) has been experimentally validated, rigorous comparisons between CSCs and normal stem cell populations are not possible, even if robust CSC populations were to be had. But the xenotransplantation assay does not test the idea that the cancer cell-of-origin is a normal stem cell. So the xenotransplantation assay cannot, on its own, test assumptions (4–8) of the CSC model.

Moreover, there are concerns about assumption (3)—that CSC are rare cells within a tumor. Although the xenotransplantation assay is quantitative, there is reason to think it systematically underestimates the number of CSCs in patient samples. The environment of an immunodeficient mouse differs significantly from that of a tumor "in situ" (reviewed in Magee et al. 2012, p. 288). Different signaling molecules, cellular microenvironments, tissue architectures, and cell–cell contact mechanisms could all affect transplanted cells' tumor-forming ability—most likely negatively. Furthermore, immunodeficient mice do not entirely lack mechanisms of immune defense. These mechanisms destroy some fraction of the transplanted cells, leaving their tumorigenic potential untested. So results indicating that CSCs are rare in patient cancers may be misleading. The xenotransplantation assay therefore reliably tests only the assumptions that cancers are developmentally heterogeneous (1) and contain a subpopulation of tumor-forming cells, CSCs (2). That is, the primary method of CSC research tests only the "minimal" model, not thicker conceptions of CSCs.

(ii) Stem cell capacities are context dependent

The phenomenon of clinical interest for CSC research is the different regeneration and tumor-resistance abilities of cancer cells within human patients. Obviously, the xenotransplantation assay measures these cells' tumor-forming abilities in a very different environment—within the bodies of immunodeficient (and highly inbred) mice. I have already discussed how these differences are reasons for concern about tests of CSC rarity. In addition, extrapolation from animal models to human diseases faces well-known challenges (e.g., Steel 2008). These challenges are particularly acute for the CSC case because stem cell capacities in general are notoriously context dependent. Different cellular microenvironments and chemical signals can dramatically alter stem cells' abilities to self-renew and differentiate (see Fagan 2013a, 2015). The xenotransplantation assay involves a drastic change in cancer cells' environment. So the cells that exhibit stem cell capacities (tumor formation) in the host tissue environment *may not be the same cells* that would exhibit such capacities in the original context of the tumor. And if they are not, then cell populations identified as CSCs by the assay will be clinically irrelevant.

The problem of context dependency is related to an evidential challenge faced by all stem cell experiments, which I have raised elsewhere (Fagan 2013a, 2013b, 2015). CSC researchers have recognized it as well: "transplantation of any stem cell can reveal the potential of the stem cell under the particular assay conditions, but it cannot reveal the actual fate of the transplanted cell in its original tissue or tumor" (Clevers 2011, p. 315).[16] Transplantation experiments show whether or not the transplanted cells can self-renew and differentiate within the body of the host. But these results do not tell us what those cells would do in another environmental context.[17] A stem cell's differentiation potential is revealed by placing it in an environment conducive to differentiation, then measuring its descendants to see whether these exhibit specialized features of mature cells; and similarly for self-renewal. For stem cells, different balances of self-renewal and differentiation are elicited by different environments. So we cannot tell what transplanted cells might give rise to in a different range of environments than those actually included in the experiment. Xenotransplantation experiments, therefore, do not include a rigorous basis for comparison across cell capacities across environments, including those unmanipulated by experiment.

(iii) Cell surface markers may not identify robust, specific, stable CSCs

The xenotransplantation assay aims to prospectively identify CSC populations that could be tracked and measured in clinical contexts. A prerequisite for achieving this goal is that CSC markers validated by the assay are robust, specific, and stable (see Nathan, this volume). But emerging evidence suggests that surface markers correlated with CSC capacities in the xenotransplantation assay are highly variable, lack specificity, and may be unstable over time. I discuss each point in turn.

Accumulating results from multiple CSC experiments suggest that CSC markers may vary among patients diagnosed with the same cancer, and perhaps even within a single patient's cancer (Magee 2012; Meacham and Morrison 2013). Such variation would undercut the clinical applicability of CSC phenotypes identified by the assay. At the very least, it is too soon to say whether CSC markers are robust across a wide range of patients. More systematic study of patients diagnosed with the same cancer would answer the question—but the answer may be unwelcome.

Specificity here refers to the association between cell surface markers and CSC behaviors. Ideally, all and only CSCs should display the surface marker phenotype, such that assay results show a perfect correlation between surface markers used to sort cancer cells and tumor-forming ability. But few experiments have produced such results. For a given surface marker phenotype positively correlated with CSC activity, assay results indicate that there are CSCs that do not exhibit that phenotype and/or cells with the phenotype that are not CSCs. Such "mismatch" is unsurprising, because current CSC markers are primarily chosen as robust, heterogeneously expressed FACS markers that allow the faithful sorting of marker-positive and marker-negative populations; however, they are not selected on the basis of a deep understanding of the underlying stem cell biology of the pertinent tissue from which the cancer originates.

(Clevers 2011, pp. 315–316)[18]

That is, CSC markers are chosen because they give clean results in a particular experimental set-up, not for any biological reason connected with CSCs. Ideally, surface markers would be chosen based on good understanding of the biology of the stem cells in question. But only in leukemias are the blood cell lineage relationships and stem cell phenotype well enough understood to make these judgments.

This epistemic situation is another aspect of the "stem cell uncertainty" noted above (Fagan 2013b, 2015). We do not know the characters of CSCs in advance, cell surface markers or otherwise. So we cannot know, when selecting cell surface markers to sort cancer cells into subpopulations, that those characters are the right ones for identifying CSCs for the cancer in question. The same experimental design using a different set of surface markers might very well produce different results. Yet experiments with this basic design are our only way of identifying stem cells' properties and capacities. There is no predetermined list of phenotypic characters that pick out all and only stem cells of any variety; experimental researchers must work this out for themselves. This is done via repeated use of the experimental method using different combinations of cell surface markers, so as to gradually reduce mismatch with CSC developmental capacities. However, only for the leukemias has this "whittling down" process taken place. Other CSC phenotypes confirmed by the xenotransplantation assay should be treated as provisional at best.

Even if specific CSC phenotypes are identified, they may not be stable within a single patient's cancer. The xenotransplantation assay is effectively a "snapshot" of a patient's tumor cells at the time of sampling. Its results cannot tell us whether those cells, in situ, can change their surface phenotype and/or regenerative capacities (discussed in Clevers 2011; Meacham and Morrison 2013). That is, the xenotransplantation assay cannot determine whether the CSC phenotype identified in any particular experiment is stable, in the environment of clinical interest. Evidence from other methods (see Section 5) suggests that cancer cells do shift between CSC and non-CSC phenotypes, depending on their cellular microenvironment (reviewed in Vermeulen et al 2012; Meacham and Morrison 2013).[19] That is, the same individual cancer cell, depending on what other cells it contacts, can behave as CSC or a more differentiated, less tumorigenic cancer cell. If this is generally the case, then CSCs are inherently unstable. In terms of the developmental model of cancer, it would then follow that the cell lineage hierarchy is not fixed and unidirectional but reversible. The assumption that cell development within a cancer follows a fixed hierarchical order is at the core of the developmental approach. If that assumption does not hold, then the CSC model must be modified accordingly.

To sum up: CSC populations identified by the xenotransplantation assay may not be robust across patients, specific for tumorigenic activity, or stable within a single patient. While it would be hasty to conclude at this point that robust, specific, and stable CSC populations are not to be had, for most cancers their existence has not yet been demonstrated.

(iv) Alternative explanations for cancer cell heterogeneity are not considered

Results indicating fit with the CSC model do provide experimental support for the minimal CSC model. This model posits only that cancer cells have different developmental capacities and that these variations map onto a "developmentally defined hierarchy of heterogeneous phenotypes derived from a small subset of "cancer stem cells" (Clarke et al. 2006, p. 9339). Results of this sort have been obtained for a variety of cancers, including leukemias, breast cancer, several forms of brain cancer, pancreatic cancer, colorectal cancer, and ovarian cancer. Other cancers tested with the xenotransplantation assay (notably melanoma) do not fit the CSC model, while for many others there is insufficient evidence to say one way or the other (in large part due to the evidential challenges discussed above). As Meacham and Morrison conclude in their 2013 review, "[w]e do not yet know what fraction of cancers follows the stem-cell model" (p. 335).

However, cancers that do fit the CSC model face a further evidential challenge. "Fit with the CSC model" demonstrates only that the results of a xenotransplantation experiment can be interpreted in terms of a developmental hierarchy emanating from a CSC population. But there are other ways of interpreting these results. Clevers (2011) summarizes the situation concisely:

> a marker or marker combination is found to be expressed in a heterogeneous fashion in a certain tumor type. On the basis of this marker heterogeneity, subpopulations of cells are sorted from primary tumors and transplanted into immunodeficient mice by limiting dilution, after which tumor growth is scored some weeks or months later. Different capacities for tumor initiation between tumor cell subsets can be interpreted as evidence for the presence of CSCs in the primary tumor, and it is then often said that the tumor adheres to the CSC model.
>
> (p. 314)

This approach conspicuously ignores alternative explanations for heterogeneity among cells of a tumor.

Two alternatives in particular are relevant for CSCs: somatic mutation and diverse microenvironments. The somatic mutation theory (SMT, see Section 1) predicts that genetic instability will produce "sub-lines" within a clone of cancer cells. Different "triggering" mutations can produce cancer cells with different tumorigenic potential. If SMT is correct, then the different developmental capacities revealed in the xenotransplantation assay could be caused by different somatic mutations arising within the same cancer, and not cells occupying different positions in a developmental hierarchy. To rule out that alternative, genomes of more and less tumorigenic cells should be sequenced, to determine if somatic mutations correlate with tumor-forming ability. Research along these lines is in early stages.[20] Another possible source of variation in cancer cells is microenvironmental variation within a tumor. Like stem cells in general, cancer cells are sensitive to their particular "niche"; subtle differences in microenvironment can make for large differences in tumor-forming ability. CSC capacities might be responses to environmental cues in the immediate vicinity of particular cancer cells, not the behavior of a fixed cell developmental hierarchy.[21]

Because most uses of the xenotransplantation assay are not combined with studies of genetic variation within a growing tumor, or examination of tumor cell microenvironment and its effects, these experiments do not test the CSC model in relation to relevant alternatives. Therefore, the quality of the evidence provided is less than if those alternatives were considered. Importantly, the different sources of variation for cancer cells are not mutually exclusive causes; all could be operating.[22] But unless each source is considered and variation across cancer cells partitioned so as to determine what difference each makes to the overall outcome, even the minimal CSC model cannot be said to have been rigorously tested.

Summary of Section

To sum up: as a test of the CSC model, the xenotransplantation assay has serious limitations. Most assumptions in thicker versions of the model are not tested. Extrapolation of results to human patients is questionable, given the known context dependence of stem cell capacities and attendant obstacles to extrapolation across environments. Assay results often lack specificity, as cell surface phenotypes correlated with CSC capacities are uncertain and provisional. Furthermore, accumulated results to date indicate that phenotypes associated with CSC capacities vary widely across cancers, across patients diagnosed with the same cancer, and even within a single patient's cancer. Finally, claims that a particular cancer "fits the CSC model" based on results of the xenotransplantation assay seldom rule out (or even consider) alternative sources of cancer cell heterogeneity, such as somatic mutation and microenvironment effects.

5 Two CSC Concepts

Let us take stock. There is considerable ambiguity concerning the CSC concept and there are diverse conceptions of "the CSC model." The experimental method used to identify CSCs and evaluate the model faces significant evidential challenges. Together, these problems synergize to create a situation of uncertainty about CSCs, which blocks efforts at clinical translation. A number of CSC researchers, noting the limitations of the xenotransplantation assay, have argued that it should be supplemented with other experimental methods: genomic sequences, cell lineage tracing within cancers in situ, and characterization of tumor microenvironments (e.g., Clevers 2011; Magee et al. 2012; Valent et al. 2012; Meacham and Morrison 2013). As noted above, genome sequencing of more and less tumorigenic cells could determine if somatic mutations correlate with tumor-forming ability. Similarly, lineage tracing of individual cancer cells in specific microenvironments would distinguish effects of the latter from CSC developmental hierarchy. These experimental methods could complement the limitations of the xenotransplantation assay and so address the concerns above.

However, alongside this methodological pluralism, conceptual pluralism is also warranted. Tacit commitment to a "unified" CSC concept, which spans basic science and clinical contexts, contributes to the current translational impasse. We have seen that different CSC researchers conceptualize CSCs differently (Section 2). Yet they plausibly concur that, however CSCs are characterized, this characterization will be the same in basic science and efforts toward the clinic. This section argues that such a monist view of CSCs impedes clinical translation, given our current evidential situation.

Suppose the monist CSC concept is "thick," corresponding to a model including all or most of assumptions (1–8) in Section 2. This CSC model does not have much experimental support. Many of its assumptions are not even addressed by the xenotransplantation assay, and for those that are, support is equivocal. The environment in which cancer cells express their tumorigenic capacities in the assay differs in many ways from the environment of clinical interest—cancer in situ, within a human patient. The assay's "snapshot" access to a patient's cancer leaves unanswered questions about CSC phenotypic stability, while cell surface markers used to identify CSCs are uncertain proxies for cancer cells' developmental abilities—lacking the desired specificity. Results that have accumulated within these limitations suggest that CSC phenotypes may not be robust across patients with the same cancer. Moreover, neglect of alternative hypotheses to explain cancer cell heterogeneity leaves CSC research open to dismissal by those committed to the SMT exclusively, while studies of cell microenvironment undermine key assumptions of the developmental approach to cancer. Taking all these considerations into account, the thick CSC model appears a poor prospect for clinical translation. So efforts in this direction do not flourish.

Suppose instead that the CSC concept corresponds to the minimal model, consisting of assumptions (1–2) and perhaps (3). This minimal CSC model conceptualizes a tumor as an ordered cell lineage hierarchy, with cells at different developmental stages possessing different capacities for self-renewal and differentiation. The lineage stems from CSCs, which have capacities sufficient to maintain, expand, or regrow the entire tumor while descendant cells comprising the bulk of the tumor have reduced or negligible capacities for tumorigenesis.

For some cancers (though not all), this CSC model has good empirical support. But these experimental results are disconnected from clinical contexts. Tests of this model do not address the motivating phenomenon of thick CSC models: patterns of recovery and relapse in human patients. Clinically relevant hypotheses must be experimentally assessed by correlating frequency of CSCs with patient outcomes. A precondition for such tests is the identification of robust, specific, and stable CSC populations for particular cancers, which could be tracked within patients. Yet features of the xenotransplantation assay, and emerging evidence from it and other methods of studying cancer, raise doubts as to whether such CSC populations can be found. Therefore, experimental results showing that a particular cancer fits the minimal CSC model do not suffice for extrapolation to clinical contexts.

Although intermediate versions of the CSC model are possible, none alone are adequate to the task of moving CSC research toward clinical translation. The current evidence is too equivocal on too many issues relevant to embarking on clinical trials for research on any one CSC model to enter a translational phase. Overall, there is a profound disconnect between results of the xenotransplantation assay (and accumulated experimental evidence more generally) and clinically relevant features of CSCs. If the CSC model includes the latter, then its empirical basis is weak. But if it does not, then its empirical support does not pave the way for clinical translation. If we assume one CSC concept common to both contexts, clinical translation will not move forward. Multiplying experimental methods, without accompanying conceptual change, will not resolve the problem.

The solution is to distinguish between two CSC concepts, one appropriate for basic research (including but not limited to the xenotransplantation assay), the other suited to clinical translation. The "basic CSC concept" consists of assumptions (1–2) only—it is, in brief, the minimal CSC model discussed above. All that is required for a given cancer to fit this model is that its cells vary such that these patterns of variation map onto a cell lineage hierarchy, with CSCs occupying the apex position (schematically shown in Figure 8.1). Cancers that show this pattern include leukemias (e.g., Figure 8.4), breast cancer, several forms of brain cancer, pancreatic cancer, colorectal cancer, and ovarian cancer. If alternative sources of heterogeneity (mutation and microenvironment) are ruled out and cell lineage tracing experiments corroborate the developmental hierarchy, then the xenotransplantation assay provides good evidence that that the "clonogenic core of the tumour" is a cell population "driving tumour growth and progression" (Vermeulen et al. 2012, p. 83). These multiple methods—xenotransplantation, genomic sequencing, and cell lineage tracing, used in coordination—should also help to reveal robust, stable, and specific CSC populations (should any exist). However, the basic CSC concept does not presuppose any such populations or any specific CSC phenotype. All its experimental validation shows is that the cancer in question undergoes developmental processes resulting in variation among cancer cells. "Basic CSCs" are just those lineage-initiating cells, whatever their other properties.

A substantively different CSC concept is appropriate to the context of clinical translation. This "clinic-oriented" CSC concept is more restrictive than its basic science counterpart. One further requirement is that CSCs are a rare cell population within the cancer (assumption 3). Otherwise, there is little difference between targeting all cancer cells (the current practice) and targeting CSCs.[23] A second constraint is stable CSC phenotype; that is, the cell developmental hierarchy is not reversible, at least regarding CSCs and other tumor cells. If non-CSCs can become so, then the strategy of targeting CSCs for clinical intervention will be undercut in practice—as long as any cancer cells remain, more CSCs can appear. Although one can imagine successful clinical interventions targeting rare or unstable CSCs, these would not be *more effective* than interventions targeting all cells of the cancer. Clinical trials, under the current regulatory regime, require that new treatments be shown to be more effective than existing treatments. So the clinic-oriented CSC concept includes at least two features not required of its basic science counterpart: *rarity* and *stability*. Importantly, these features can also be tested for by the multiple methods noted above. Further assumptions often made about CSCs (Section 3) are not required by the clinical context.

6 Pathways to the Clinic

In the section, I show how the two concepts distinguished above can help move CSC research beyond its current "holding pattern" vis-à-vis clinical translation. The core idea is simple: distinguishing between the two CSC concepts allows us to recognize and consolidate experimental support for the basic model and to pinpoint what further research is needed to obtain empirical support for the more restrictive clinic-oriented concept. The preceding sections show that the basic CSC concept has been (provisionally) experimentally validated. It can therefore serve as a starting point to articulate further requirements for the clinic-oriented CSC concept. Having thus clarified the latter concept, we can see what experimental results would experimentally validate it. If these results in fact emerge for one or more cancers, then clinical translation of CSCs should proceed. While treating the CSC concept as unitary helps sustain the current impasse to clinical translation, distinguishing between basic and clinic-oriented CSC concepts draws attention to the question of how they are related. Explicating this relation provides a "blueprint" for clinical translation of CSCs.

The model-based approach taken in this chapter has particular advantages in this regard. For one, it is *minimalist*: both CSC concepts include as few prior assumptions as possible. This has two benefits. First, it simplifies the conceptual terrain, reducing the potential for ambiguity for each CSC concept. Second, minimal models are easier to confirm experimentally, all else being equal. The fewer the assumptions requiring validation, the more cases the models are likely to fit, and the easier such fit will be to experimentally demonstrate. Both CSC concepts proposed here are "stripped-down" for efficiency.

Another advantage of the model-based approach to CSCs is that models are related to one another in ways that can be stated explicitly. In keeping with the minimalist character noted above, the clinic-oriented CSC concept is limned by contrast with the basic CSC concept. Requirements for a CSC concept suited to clinical translation are inferred by asking: what more should be added to the basic CSC concept, which simply posits a developmental hierarchy of cells comprising a cancer, to accommodate the standards and purposes of clinical research? In answering this question, one charts a path from the laboratory to the clinic, so to speak, making explicit what is required for a CSC concept to be suitable for clinical translation. It is important to note also that these conceptual moves in no way attempt to prejudge experimental results. Rather, the aim is to clarify what kind of experimental results are needed to move CSC research toward clinical translation.

The obstacles to such progress should not be underestimated. But the model-based approach can also help anticipate these, so that efforts to ameliorate them can be undertaken in advance. For example, the requirement that clinically relevant CSCs be rare in turn necessitates a modification of clinical standards for cancer research. Currently, regulatory agencies take significant decrease in tumor size to be a key benchmark in assessing efficacy of new interventions.[24] But if CSCs comprise a small proportion of tumor cells, then a new benchmark is needed, which measures disruption of defining CSC activities: long-term self-renewal and tumor-forming ability. Making the rarity requirement for clinic-oriented CSC explicit motivates and justifies efforts to modify regulatory standards for clinical translation of these cells.

The cell lineage visualization of CSC models suggests a perspicuous way of characterizing the relation between basic and clinic-oriented CSC concepts. Any given process of development operating within a cancer produces a cell lineage tree with a specific topology, representing developmental and reproductive relations between cancer cells. CSC models constrain the space of possible topologies for cancer cell lineage trees. If cancer cells undergo development, such that developmental change is correlated with tumor-forming ability, then the basic CSC model applies. This CSC concept can be realized by "cell lineage trees" exhibiting many different topologies: deep or shallow, directed or reversible, narrow or "bushy." The only constraint is that the lineage tree emanate from a single stem. The clinic-oriented CSC concept further constrains the space of possible cell lineage trees, requiring that the single stem be rare within the overall cancer, and that lineage relations be irreversible. More abstract characterizations of the basic–clinical CSC relation, making use of graph theory, are possible. While it is not obvious that formalizing CSC concepts in this way would contribute to clinical translation, the option is available.

To summarize: the model-based account of basic and clinical CSC concepts offers a simple, articulated sketch of a path toward clinical translation, which clarifies the experimental results that are needed to move forward. It also indicates where accompanying regulatory changes will be needed if the hoped-for experimental results (validating the clinic-oriented CSC concept) are in fact observed. In this way, the conceptual argument of this chapter complements recent recommendations of some CSC researchers to require multiple methods for identifying and characterizing CSC rather than a single "gold standard" assay. The methodological pluralism those scientists (rightly) advocate will more effectively segue to clinical translation if accompanied by conceptual pluralism, of the sort defended here.

These points are reinforced, and extended, by a recent example of successful clinical translation of stem cell research (McNeish et al. 2015). Working with clinical scientists and industry groups, Kevin Eggan's laboratory at the Harvard Stem Cell Institute brought new results in basic stem cell research to Phase IIa clinical trials in under 2 years. Eggan's group took cells from patients with amyotrophic lateral sclerosis (ALS), a neurodegenerative disease; produced pluripotent stem cell lines from these cells; and directed these stem cells to differentiate into motor neurons in culture. This *in vitro* system reproducibly exhibited physiological defects that resembled ALS presentation in patients. The genetic and biochemical basis of the observed defect was known, so Eggan and colleagues screened drugs targeted to these features of the system. One of these, ezogabine, alleviated the disease phenotype and was already approved for patient use for other neurological conditions. This stem cell–derived treatment rapidly progressed to the stage of clinical trials.

Eggan and colleagues attribute their success in this case to several factors—some fortuitous (e.g., ezogabine's previous history), but others the result of careful planning (MacNeish et al. 2015, pp. 8–9). The latter include (i) a well-established stem cell technology, yielding robust results according to clear criteria; (ii) an *in vitro* defect characterized at molecular genetic and physiological levels; (iii) stem cells relegated to a "discovery" role, identifying a chemical treatment that alleviates physiological symptoms; (iv) use of known clinical biomarkers in the in vitro assay; (v) establishing robustness, with cell lines from patients of various genetic backgrounds and disease forms; and (vi) a research plan including collection of samples from patients to produce more (diverse) human stem cell lines to "grow the test," establishing relevance to an increasingly wide potential patient population.[25] The last is especially important, as it produced a resource for future basic and clinical research on ALS, whatever the ezogabine trial's outcome.

This example, although concerned with another kind of stem cell, offers some useful guidelines for facilitating CSC clinical translation in the event that empirical results warrant it. Feature (i) corresponds to multiple integrated experimental methods for rigorously confirming the basic CSC model and the additional assumptions of rarity and stability required for the clinic-oriented model. If the latter is experimentally confirmed, then counterparts of Eggan's (v) and (vi) become relevant. General or robust patterns of CSC behavior in patients with cancers conforming to the clinic-oriented CSC model should be the next focus of translational research. The way to detect such patterns is to pursue counterparts of Eggan's features (ii) and (iv): detailed study of the cancer cell growth and metastasis in physiologically-relevant contexts, with CSCs characterized by markers that are already in clinical use. Here the CSC models' neutrality with respect to surface phenotype is important. Efforts should be made to characterize clinic-oriented CSCs using markers that have been previously used in patients. Robust results within and across cancer patients, correlating CSC capacities with previously characterized markers, would smooth the path to clinical translation, as for Eggan's group. If CSC capacities can be modeled *in vitro*, then feature (iii) could be incorporated into CSC research as well.

Eggan's example indicates ways in which the two-concept account could more efficiently point toward clinical translation of CSC. Although CSCs are not yet ready for clinical translation, the pathways by which they could reach that stage can be discerned more clearly with the two-concept framework.

7 Conclusion

Stem cell research is motivated by the hope of treating a wide range of pathologies for which current therapies are inadequate, including spinal injury, neurodegenerative diseases, heart attack, and diabetes. CSCs seem to epitomize the translational character of stem cell science. The basic idea, which has profound clinical implications, is that cancers exhibit a developmental hierarchy analogous to normal organs, with tissues grown and continually regenerated from a stable population of stem cells, which give rise to more specialized (less tumorigenic) descendants. However, after more than a decade of experimental research and heated debate, the CSC model has not entered a phase of clinical translation. In this chapter, I have diagnosed the problem in terms of ambiguity in the meaning of *cancer stem cell* and evidential challenges for the method used to identify cells to which the term applies. To move beyond the current "epistemic holding pattern," I argue that we should distinguish between two CSC concepts: one associated with models of biological development, the other with clinical intervention. In this two-concept framework, progress toward clinical translation can be made. The basic CSC concept provides an empirically supported platform to which further conditions are added in response to requirements of the clinical context. In addition, the model-based approach to CSC concepts offers a simple, explicit "blueprint" for moving CSC research toward clinical translation, which is both amenable to rigorous formalization and reinforces lessons from a successful example of clinical translation of a stem cell–derived therapy. This treatment of the CSC case demonstrates one way that philosophical approaches can be useful for biomedicine.

Acknowledgments

Many thanks to editors Marco Nathan and Giovanni Boniolo, both for the opportunity to contribute to this volume and for helpful comments on an earlier draft. This chapter has also benefited from discussions with Allan Beke, Mike Clarke, Sara Green, Lucie Laplane, Kirstin Matthews, Sean Morrison, Leila

Perie, and Irv Weissman. Early research contributing to this chapter was funded by the Humanities Research Center at Rice University's Collaborative Research Fellowship (2009–2010), and Faculty Innovation Fund (2010–2012). Funding for later stages of the project was provided by the Department of Philosophy at the University of Utah.

Notes

1   Cells reproduce by binary division; in a division event, a parent cell divides to yield two offspring cells. Eukaryotes exhibit two modes of cell division: mitosis and meiosis. Stem cell phenomena involve mitosis; the term *cell division* here refers to that mode only. Cell division may be *symmetric* (two similar offspring) or *asymmetric* (offspring different from one another). Self-renewal with symmetric division increases the stem cell population; self-renewal with asymmetric division maintains stem cell numbers.

2   For an alternative approach to CSCs, emphasizing "cell ontologies" and theories rather than models, see Laplane (2014a, b). Interestingly, Laplane's conclusions about CSCs are consonant with those of this paper. This rebuts Laplane's (2014a) criticism of my model-based account, that my account "hides a problem" that her approach can clarify. In fact, both our approaches yield similar conclusions (albeit with different purposes in view; see Fagan 2015).

3   A search of ClinicalTrials.gov, the public database of medical trials in the U.S., yields only 20 studies of cancer stem cells (as indicated by title and intervention; accessed 9/6/2015). Of these, three are classified as status "unknown," having not been verified in over 2 years; two have been terminated; seven are in progress; and eight are completed. Of the studies in progress, two are tests of anti-CSC activity by specific drugs, four are preclinical studies of CSC biomarkers in particular cancers, and one (sponsored by the University of Michigan Cancer Center) is testing efficacy of a CSC-targeted drug in preventing relapse of gynecology-related cancers. Only the last falls into the category of clinical cancer research, and it is still in early stages. Of the completed studies, seven are from a single sponsor (Fuda Cancer Hospital, Guangzhou, in collaboration with the University of Michigan). Although listed as Phase 1 and 2 trials, no results have been reported for any of the seven. The other completed study is a preclinical feasibility study for characterizing biopsies of head and neck cancers (sponsored by Stanford University). This is an extremely modest clinical output, to say the least, for a model that has been a focus of study for over a decade.

4   Examples of CSC researchers endorsing this minimal conception include Magee et al. (2012), Meacham and Morrison (2013), and Vermeulen et al. (2012).

5   Cancers are classified by the organismal location in which they (are thought to) originate. The term *solid tumor* generically refers to abnormal tissue masses within an organismal body (see *NCI Dictionary of Cancer*, www.cancer.gov/publications/dictionaries/cancer-terms). Contrasting forms of cancer are comprised of cysts or liquids (e.g., leukemias).

6   E.g., the "explanations it provides for several poorly understood clinical phenomena" are "a major attraction of the CSC concept" (Clevers 2011, p. 313).

7   Other hypotheses are also proposed to explain assumption (4); however, (6) and (7) have received the most attention to date in the CSC literature.

8   There is evidence that some resistance-conveying mutations change the conformation of the binding pocket so the drug cannot bind and destroy the tumor cell without eliminating the CML phenotype (Vermeulen et al. 2012). However, these mutations do not account for the robust relapse phenomenon when imatinib treatment ends, nor the return to remission when the treatment is resumed.

9 For example, the American Association of Cancer Researchers workshop on CSCs discussed all the assumptions noted above (Clarke et al 2006, p. 9339). In contrast, Morrison's group endorses the minimal CSC model, consisting of assumptions (1) and (2) only (Meacham and Morrison 2013). Clevers, the leader of another influential research group, defines CSC in terms of (1–3) only (2011, p. 313). Dick and colleagues take assumptions (1–2), (4–5), and (7) all to describe necessary attributes of CSCs (Kreso and Dick 2015, p. 276). Laplane (2014b) characterizes CSCs as a theory comprised of three diagrammatic models, which correspond to assumptions (1–2), (4), and (7), together with their biological and therapeutic implications.

10 The problem has not gone unnoticed. The "Year 2011 Working Conference on CSCs," comprising 22 specialists from hospitals and research institutes in 11 countries, was convened to explicitly to address "lack of consistency in the terms used for these cells and how they are defined" (Valent et al. 2012, p. 767). However, their proposal, based on current classification practices for cancer, has not, to my knowledge, been widely taken up.

11 *Xenotransplanation* refers to the crossing of species boundaries in the transplantation from human to mouse. Because the assay induces tumors in the host, ethical constraints forbid within-species transplantation in this case. Immunodeficiency is required because a normal host immune system would destroy the transplant, all the quicker for the species difference.

12 *SCID* refers to the "severe combined immunodeficiency" status of mice used in the experiment.

13 Self-renewal is operationalized as the ability to give rise to tumors through two cycles of transplantation, which is why this further step is necessary.

14 The self-renewal requirement for CSCs mandates two such immunodeficient hosts, but this elaboration does not alter the basic experimental design.

15 *In vitro* assays using cell culture surrogates for tumor formation are also used, but these are given less evidential significance and are considered more a preliminary study than a rigorous test of the CSC model (e.g., Valent et al. 2012).

16 The quotation is based on remarks in Shackleton et al. (2009); see Clevers (2011) for the original reference.

17 The evidential problem I have discussed in previous work concerns the experimental identification of single stem cells (see references above). Here the problem is inferring stem cell capacities across environments. So the challenge is not exactly the same. Nonetheless, some of the same arguments apply to both cases.

18 *FACS* refers to fluorescence-activated cell sorting, the method used to sort living cells into subpopulations based on the presence or absence of specific molecules on their surfaces (see Fagan 2013a, Ch. 8 for discussion of FACS technology's significance for stem cell research).

19 E.g., "The CSC phenotype is much more fluid than anticipated and is strongly regulated by the tumour-cell environment" (Vermeulen et al. 2012, p. 83).

20 For example, Anderson et al. (2011) and Piccirillo et al. (2015) show that CSCs for leukemia and glioblastoma include different mutational "subclones."

21 Tumor microenvironment has also been implicated in treatment resistance (Valent et al. 2012, p. 767) and instability of CSC phenotype ("plasticity"; Meacham and Morrison 2013).

22 Magee et al. (2012) "the cancer stem cell and clonal evolution models can be interacting, or independent, sources of heterogeneity depending on the cancer" (p. 291).

23 Here I concur with Clevers (2011) that a low frequency of CSCs is not essential for a tumor to "adhere to the CSC model"—in the basic science case. The clinical significance of such CSC results is undermined, however.

24 The current Phase II standard for efficacy is "Response Evaluation Criteria in Solid Tumors" (RECIST), where *response* is defined as "cumulative longest diameters of target lesions decrease by more than 30%" in patients receiving the treatment

(Vermeulen et al. 2012). This radiological standard tacitly assumes that bulk reduction of the tumor is the intent of the proposed therapy. Therapies targeting rare CSCs require different benchmarks.

25 "[T]he strong scientific foundation we had produced using iPSCs, a clear clinical question that could be tested using an established biomarker, and a compound with strong pharmacological properties were each essential pieces in the puzzle of organizing partnerships between academics, clinicians, patient advocacy groups, and industry that were needed to mount a clinical trial" (McNeish et al. 2015, p. 9). Other key factors were sociological: working within the infrastructure of a pre-existing clinical network and building a diverse funding consortium.

References

Al-Hajj, M., S. Wicha, A. Benito-Hernandez, S.J. Morrison and M.F. Clarke (2003) "Prospective identification of tumorigenic breast cancer cells," *Proceedings of the National Academy of Sciences, USA 100*, pp. 3983–3988.

Anderson, K., C. Lutz, F.W. van Delft, C.M. Bateman, Y. Guo, et al. (2011) "Genetic variegation of clonal architecture and propagating cells in leukaemia," *Nature 469*, pp. 356–361.

Bonnet, D., and J.E. Dick (1997) "Human acute myeloid leukemia is organized as a hierarchy that originates from a primitive hematopoietic cell," *Nature Medicine 3*, pp. 730–737.

Clarke, M., J.E. Dick, P. Dirks, C. Eaves, C. Jamieson, et al. (2006) "Cancer stem cells—perspectives on current status and future directions: AACR workshop on cancer stem cells," *Cancer Research 66*, pp. 9339–9344.

Clevers, H. (2011) "The cancer stem cell: Premises, promises and challenges," *Nature Medicine 17*, pp. 313–319.

Fagan, M.B. (2013a) *Philosophy of Stem Cell Biology*. London: Palgrave Macmillan.

Fagan, M.B. (2013b) "The stem cell uncertainty principle," *Philosophy of Science 80*, pp. 945–957.

Fagan, M.B. (2015) "Crucial stem cell experiments? Stem cells, uncertainty, and single-cell experiments," *Theoria* 30, pp.183–205. (Special Section: Philosophy of Experiment)

Kraft, A. (2011) "Converging histories, reconsidered potentialities: The stem cell and cancer," *BioSocieties 6*, pp. 195–216.

Kreso, A., and J.E. Dick (2014) "Evolution of the cancer stem cell model," *Cell Stem Cell 14*, pp. 275–291.

Lapidot, T, C. Sirard, J. Vormoor, B. Murdoch, T. Hoang, J. Caceres-Cortes, M. Minden, B. Paterson, M.A. Caligiuri, and J.E. Dick (1994) "A cell initiating human acute myeloid leukaemia after transplantation into SCID mice," *Nature 367*, pp. 645–648.

Laplane, L. (2014a) "Stem cell epistemological issues." In P. Charbord and C. Durand (eds) *Stem Cell Biology and Regenerative Medicine*. River Publishers.

Laplane, L. (2014b) "Identifying some theories in developmental biology: The case of the cancer stem cell theory," In A. Minelli and T. Pradeu (eds) *Toward a Theory of Development*. Oxford University Press, pp. 246–259.

Magee, J.A., E. Piskounova, and S.J. Morrison (2012) "Cancer stem cells: Impact, heterogeneity, and uncertainty," *Cancer Cell 21* (Special Issue), pp. 283–296.

McNeish, J., J.P. Gardner, B.J. Wainger, C.J. Woolf, and K. Eggan, (2015) "From dish to bedside: Lessons learned while translating findings from a stem cell model of disease to a clinical trial," *Cell Stem Cell 17*, pp. 8–10.

Meacham, C.E., and S.J. Morrison, (2013) "Tumour heterogeneity and cancer cell plasticity," *Nature* 501, pp. 328–337.

Melton, D., and C. Cowan (2009) "Stemness: Definitions, criteria, and standards." In R. Lanza et al. (eds.) *Essentials of Stem Biology, 2nd edition*. San Diego, CA: Academic Press, pp. xxii–xxix.

Morange, M. (forthcoming) "Is there an explanation for ... the diversity of explanations in biological studies?" In P.-A. Braillard and C. Malaterre (eds.) *Explanation in Biology. An Enquiry into the Diversity of Explanatory Patterns in the Life Sciences*. Dordrecht, Springer.

Mukherjee, S. (2010) *The Emperor of All Maladies*. New York: Scribner.

Nowell, P.C. (1976) "The clonal evolution of tumor cell populations," *Science* 194, pp. 23–28.

O'Connor, M.L., D. Xiang, S. Shigdar, J. Macdonald, Y.Li, T. Wang, C. Pu, Z. Wang., L. Qiao and W. Duan (2014) "Cancer stem cells: A contentious hypothesis now moving forward," *Cancer Letters* 344, pp. 180–187.

Piccirillo, S.G.M., S. Colman, N.E. Potter, F.W. van Delft, S. Lillis, M.J. Carnicer, L. Kearney, C. Watts, and M. Greaves (2015) "Genetic and functional diversity of propagating cells in glioblastoma," *Stem Cell Reports* 4, pp. 7–15.

Pierce, G.B., and W.C. Speers (1988) "Tumors as caricatures of the process of tissue renewal: Prospects for therapy by directing differentiation," *Cancer Research* 48, pp. 1996–2004.

Ramalho-Santos, M., and H. Willenbring (2007) "On the origin of the term 'stem cell'," *Cell Stem Cell* 1, pp. 35–38.

Reya, T., S. Morrison, M. Clarke, and I.L. Weissman (2001) "Stem cells, cancer, and cancer stem cells," *Nature* 414, pp. 105–111.

Steel, D. (2008) *Across the Boundaries: Extrapolation in Biology and Social Sciences*. Oxford University Press.

Valent, P., D. Bonnet, R. De Maria, T. Lapidot, M. Copland, J.V. Melo, C. Chomienne, F. Ishikawa, J.J. Schuringa, G. Stassi, B. Huntly, H. Hermann, J. Soulier, A. Roesch, G.J. Schuurhuis, S. Wöhrer, M. Arock, J. Zuber, S. Cerny-Reiterer, H.E. Johnsen, M. Andreeff and C. Eaves (2012) "Cancer stem cell definitions and terminology: The devil is in the details," *Nature Reviews Cancer* 12, pp. 767–775.

Vermeulen, L., F. de Sousa e Melo, D.J. Richel, and J.P. Medema (2012) "The developing cancer stem-cell model: Clinical challenges and opportunities," *The Lancet Oncology* 13, pp. e83–e89.

9 Counterfactual Reasoning in Molecular Medicine

Marco J. Nathan

Abstract

Counterfactual reasoning is just as commonplace in scientific theory and practice as it is in our ordinary lives. Understanding the significance of such inferences requires paying close attention to the nature and foundations of conditionals, an important—albeit thorny—endeavor that philosophers of science typically (and happily) delegate to metaphysicians, philosophers of language, and linguists. This, I argue, is a mistake. The aim of this essay is to discuss the role of counter-factual reasoning in science, focusing specifically on molecular medicine. I suggest that subjunctive conditionals play a key part in the diagnosis and prognosis of molecular diseases and other pathological conditions. Yet, understanding the prominence of these sui generis statements in medical decision making requires some radical departures from traditional semantic analyses.

1 Introduction: Tough Choices

Momentous advancements in the biomedical sciences have provided contemporary physicians with an unprecedented array of theoretical and tech-nological tools, which allow the treatment and, perhaps even more importantly, the prevention of pathological conditions that just a few decades ago would have been incurable. Yet, the paradigm shift brought about by molecular medicine—illustrated by the essays included in this collection—carried along a plethora of concepts and practices that still await appropriate philosophical foundations.

Consider, for example, the familiar process of decision making in clini-cal practice. When it comes to health, making the right decisions is vital (pun intended). Yet, many choices are far from obvious or straightforward. For start-ers, a patient needs to realize that there is or might be a problem, or that it is time for a preventive check-up. Next, the appropriate physician must be selected for a preliminary scrutiny, including a medical history (*anamnesis*) of the patient. Once the doctor has completed her examination and provided a *diagnosis*, together with the patient, she must decide on a course of action or *therapy*, which requires—among other things—the evaluation of different *prognoses*. In addition, serious conditions often require ethical counseling, patients need to plan their life after treatment, and lifestyle adjustments are often required to make sure that the problem does not recur (*secondary prevention*). Given this multifaceted decision-making process, it is hardly any wonder that figuring out the optimal or suboptimal path often turns into a daunting task.

In determining the appropriate course of action, one cannot merely focus on what *actually* is, or will be, the case. In addition, it is also important to consider what *could have happened* or what *could happen* under different circumstances that did not occur or might not occur. But why should this be so? Once we can safely dismiss the occurrence of a past or future event, why is knowledge that such event could be, or could have been, relevant to assessing, explaining, and understanding what actually is, was, or will be? Answering this question requires delving into the nature of a specific linguistic class of conditionals in the subjunctive mood with a false antecedent. Statements of this form—such as "Had the tumor not been removed, Jones would not have lived through the month," where, in fact, the tumor has been removed and Jones is alive and well—are known in the philosophical literature as *counterfactual conditionals* or *counterfactuals*, for short.[1]

Counterfactual conditionals are by no means confined to medical and scientific reasoning or linguistic analysis but pervade virtually every aspect of our lives, including attributions of responsibility ("Had Sue known his intentions, she wouldn't have helped him"), financial decisions ("Had Jill followed my investment advices, she would have made more profit"), and more frivolous scenarios of dubious epistemic status, dear to many sports fans ("Had our coach been fired earlier in the season, we would have won the championship"). In addition, these statements play a prominent role in philosophical analyses of science, where virtually every foundational concept—including causation, probability, explanation, propensity, lawfulness, and law-likeliness—has, at some point or another, been analyzed counterfactually. Indeed, as noted early on by Goodman (1955, p. 3), "if we lack the means for interpreting counterfactual conditionals, we can hardly claim to have any adequate philosophy of science."

The commonplace nature of counterfactual claims and their widespread presence are potentially deceptive. Humdrum as they may appear, these statements give rise to notorious puzzles, and an uncontroversial analysis is yet to be found. Despite their extensive usage, scientists and philosophers of science typically (and happily) delegate the thorny task of providing a precise account of subjunctives to metaphysicians, philosophers of language, and linguists. This, I maintain, is a mistake. As Goodman promptly noted, understanding the semantic, pragmatics, and epistemology of these statements is a crucial step toward grasping their central role in science. To the extent that we are interested in the structure, validity, and justification of scientific reasoning—and, by extension, biomedical reasoning—counterfactuals ought to take the center of the stage.

This chapter focuses on the role of counterfactual thinking in medicine, with particular emphasis on the changes fostered by the ongoing "molecular revolution." I begin by showing that counter-to-the-facts reasoning plays a crucial role in two fundamental medical inferences, namely *diagnosis* (§2) and *prognosis* (§3). Next, I argue that much traditional philosophy of science is guilty of conflating two independent scientific goals—*truth* and *explanation*—and, consequently, traditional analyses of subjunctives have trouble explaining the role of counterfactual reasoning in medicine (§4). In the final part of the essay, I present (§5) and defend (§6) an alternative perspective that, I maintain, fares better on this score.

Before moving on, a brief note about the intended scope of this chapter. I should immediately warn more empirically-oriented prospective readers that my essay develops a principled methodological point concerning the role of an important class of conditional statements in clinical medicine. A detailed discussion of medical applications and normative implications must be set aside for a different occasion. Nevertheless, I hope that even practicing physicians and bioethicists will find some interest in a general philosophical analysis of concepts, such as diagnosis and prognosis, that are central to clinical practice and yet are frustratingly hard to define accurately.

2 Diagnosis

Suppose that a physician suspects that a patient is suffering from a specific condition, say, pathology *P*. How does the doctor go about inquiring into whether the patient, in fact, suffers from *P*? In general, there seem to be two routes: a direct one and an indirect one. The *direct* strategy is to seek the presence of *P* via observation—possibly aided by technological enhancements—of *P* itself or some constitutive characteristic. For example, a dark spot on an x-ray scan could reveal the presence of lung cancer, and the unnatural shape of a limb might be the direct consequence of a broken bone. Similarly, one could confirm the occurrence of an infection by detecting the presence of bacteria or other invading agents, since "infection" is defined as the invasion of bodily tissues by disease-causing agents. In short, direct verification is the most straightforward and evidentially secure way of diagnosing a condition. However, in many cases, such a strategy is unavailable because the nature of pathology *P* is yet unknown, because of lack of appropriate probing technology, or because a direct test would be overly dangerous or invasive for the patient. When this is the case, we may follow a different, *indirect* strategy, and look for an "indicator" of *P*, that is, signs, symptoms, or associated conditions that would likely be observed if it were the case that the patient was, indeed, suffering from *P*.[2]

These symptomatic conditions are generally known, across the biomedical sciences, by the name *biomarkers*. More precisely, a biomarker can be defined as a measurable indicator that can be experimentally employed to evaluate certain characteristics about the respective source: a normal biological process, a pathogenic process, or a pharmacological response to therapeutic intervention. For instance, a body temperature of 97.7–99.5°F indicates a normal (healthy) physiological state, whereas a brain computed tomography (CT) scan that shows specific signs of bleeding or other kinds of damage can be a diagnostic marker of a stroke. A biomarker at the molecular level is often referred to as a *molecular biomarker*. To illustrate, mutations in human genes that produce tumor suppression proteins, such as BRC1 and BRC2, are molecular biomarkers that indicate an increased risk of pathologies such as breast or ovarian cancer. The simplicity of these examples, however, is not representative of all pathologies. A condition as complex as Parkinson's disease may have a whole array of biomarkers of different kinds, as diverse as sleep abnormalities, radiolabeled tracer imaging, and cerebrospinal fluid and serum tests, including α-synuclein, and DJ-1 (Chahine and Stern 2011). Finally, laboratory procedures aimed at detecting specific molecular markers that allow the assessment of the probability that an individual is affected by a particular disease or condition are called *predictive* or *presymptomatic molecular tests*.

Three features of molecular markers ought to be emphasized. First, most biomarkers are *probabilistic* indicators. In medicine, signs or symptoms that indicate the presence of a particular disease or pathological condition beyond any reasonable doubt are called *pathognomonic*. For instance, a vesicular rash is a pathognomonic sign of chicken pox.[3] Yet, this situation is the exception rather than the rule. Most biomarkers fall short of being sure-fire indicators; rather, they are probabilistic risk factors that increase the likelihood that the individual is or will be affected by the condition without approaching certainty or near certainty. Thus, detection of a mutated MYH7 gene raises the probability that a subject is affected by hypertrophic cardiomyopathy (HCM), but this inference falls short of the "beyond any reasonable doubt" status.[4]

Second, molecular biomarkers—or, more generally, biomarkers—are *indicators* of diseases, disease states, or disease rates. As such, they should not be confused with *causes* of pathological conditions. For instance, a high level of prostate-specific antigen (PSA) is a reliable indicator of a situation that could be causally linked with cancer; however, a high level of PSA, *per se*, is not a cause of prostate cancer. A marker that a subject is more or less prone to have or develop a certain pathological condition does not suggest how to intervene on, remove, or avoid the condition itself, precisely because it fails to provide any kind of causal knowledge. To be sure, causes, especially when it takes them a long time to trigger their effects, can sometimes be used as markers of diseases. For instance, un unstable expansion of a CAG-repeat in the *huntingtin* gene is necessary and, unfortunately, sufficient for predicting that the subject will develop Huntington's Disease at a specific stage in life, typically midlife. However, the genetic condition is not merely an indicator of the (future development of) the disease; the mutation plays a critical role in triggering the cascade of effects leading to serious neural degeneration in affected patients.[5] Whether causes can be considered as *bona fide* indicators is an significant question that, however, I shall not address here. The important point, for present purposes, is that "indirect" means of detecting a disease—such as signs, symptoms, and biomarkers—should not be confused with more "direct" means, such as the detection of constitutive traits or causes.

Third, biomarkers should not be flatly identified with diagnoses *tout court*. A biomarker is an indicator that a subject might be affected by a disease and, as such, it can be used to diagnose the disease in the subject. Yet, a complete and reliable diagnosis typically requires other, independent evaluations.

These considerations help dispelling a potential confusion. It is tempting to assume that a *diagnosis* of a condition corresponds to an *explanation* of that same condition, whereby the presence (or absence) of a pathology is established based on the evidence of relevant symptoms or signs. This, however, is a mistake. To illustrate, consider the following example. It is well-known that certain genotypes may increase the chances of developing certain pathological conditions. For instance, a germline mutation called TP53 is associated with Li-Fraumeni syndrome, which indicates an increased risk (up to 85 percent) in early adulthood of developing tumors such as bone and soft-tissue sarcoma, premenopausal breast carcinoma, leukemia, brain cancer, and adrenocortical carcinoma. Suppose that a patient who has tested positive for TP53 mutation exhibits some symptoms of brain cancer. Intuitively, the genetic background supports the diagnosis that the patient is, in fact, affected by brain cancer. However, it would be preposterous to claim that the genotype, by itself, *explains* the disease. To gloss the main philosophical accounts of scientific explanation, even a full account of the genetic condition would still constitute, at best, a very incomplete description of the mechanism triggering brain cancer or of the underlying causal network and would lack significant unificatory power.

In short, diagnoses provide inadequate representations and poor explanations for the reason that diagnoses are meant to be neither explanatory nor causally or descriptively accurate. What is diagnosis, then? Inspired by Popper (1935, 1963), one could understand diagnoses as hypothetical statements or conjectures regarding possible pathogenic states of a patient. Following a related yet alternative route, I conceive of diagnoses as a particular kind of inductive argument that is commonly known in philosophy of science as *abduction* or *inference to the best explanation* (IBE) where one infers from the available evidence to the hypothesis that would, if correct, best explain such evidence (Lipton 2004). To illustrate the main idea underlying IBE, consider a detective who concludes that the butler murdered the victim because this is the hypothesis that best explains the facts: the victim was shot, no one else was inside the mansion, the butler's fingerprints are on the gun, and the victim was blackmailing the butler with compromising data. Instances of IBE are abundant across the sciences as well. To focus on an example from evolutionary biology, Darwin rightly contended that his hypothesis of natural selection best explained observed patterns of plant and animal distribution, and that is what makes it superior to alternative accounts. Note that this is precisely the kind of reasoning underlying medical diagnoses: the conclusion that the patient is affected by condition P best explains the evidence that the patient does display sign or symptom S, and S is a diagnostic marker of P. On the other hand, if R were also a biomarker of P, as strong as or stronger than S, then the fact that the *patient* is affected by S but not R might suggest that a better diagnosis for P might be available.[6]

It is important to note that an IBE is an essentially comparative intellectual endeavor that requires considering not only what *is* the case, but also what *could have been* the case—that is, the array of possible explanations. To wit, even if condition *P* is the actual cause of symptom *S*, such a conclusion requires judging what would have happened had *P* not occurred and had alternative conditions *Q*, *R*, *T*… occurred instead. As mentioned at the outset, counterfactual conditionals are the most natural and efficacious way of expressing these hypothetical situations. Hence, we now begin to see why nonactual conditions are so important in understanding actuality: counterfactual reasoning plays a crucial role in IBE (Edgington 2008) and in all sorts of related hypothetical inferences underlying diagnosis. Consequently, lacking an appropriate foundation for subjunctives, we *ipso facto* lack a foundational basis for medical diagnosis.

3 Prognosis

Let us set diagnosis aside, for the moment, and focus on a different—albeit equally important—medical inference. Imagine that a physician has diagnosed a patient with condition *P*, and suppose that such diagnosis is, in fact, correct. Now, the doctor and the patient, together, need to decide how to intervene in order to resolve the problem. This decision requires evaluating different modes of intervention on the basis of *P*'s *prognosis*, that is, the likely outcome of the patient's condition, including the expected duration and (dys)function of the disease, together with a description of its course, such as expected progressive decline, intermittent crisis, sudden unpredictable crisis, and so on. Note that deciding on a course of action requires weighing various prognoses, for alternative therapeutic paths are likely to alter the expected course of a pathological condition.

In many cases, there is only one reasonable course of action. If my doctor tells me that removing my appendix is the only safe and effective way to resolve my problems with appendicitis, then that's the obvious way to go, and all that's left for me to do is make the necessary arrangements. However, there are more problematic cases where the choice is not as simple, as illustrated by the following example, based on a real-life scenario (from Boniolo and Sanchini 2016). Consider a young woman who is diagnosed with a dangerous form of breast cancer during the initial stages of her pregnancy. She is offered three therapeutic paths. The first option is standard chemotherapy, which is most effective against the disease but, unfortunately, is incompatible with fetal development and, consequently, would entail interrupting the pregnancy. The second option is to undergo surgery, followed by adjuvant chemotherapy. While this strategy is compatible with fetal development, it has a lower response rate to treatment than standard chemotherapy; hence, it lowers the chances of recovery. Finally, the third option is to undergo immediate surgery and to postpone all other forms of treatment until the delivery of the baby. Clearly, this last option minimizes the effects of therapy on the fetus but, delaying other forms of treatment, it also jeopardizes the mother's recovery more than the alternative solutions. Here, the (sub)optimal course of action is far from obvious, involving various unknown variables—what, exactly, is the impact of delaying therapy until delivery of the baby on the probability of the mother's full recovery?—as well as painful trade-offs: for example, should one compromise the chances of the mother's survival to protect the fetus? In such cases, the value of a reliable prognosis is not merely predictive; in addition, it provides information to doctors, patients, and family members allowing them to make a conscious and informed decision on how to proceed. In sum, a prognostic judgment may have a substantial impact on a patient by influencing her choices.

Modern prognoses, powered by theoretical, statistical, and technological innovations, are not confined to the proximate future but can extend quite far on the temporal scale. To wit, suppose that a patient is diagnosed with a form of cancer, is successfully treated, and appears to be fully recovered. What are the chances that she will still be alive in one, five, or ten years? In order to answer this crucial question, one can perform a *survival analysis* to obtain the relevant *survival rate*, that is, the percentage of individuals with a certain type and stage of cancer who have survived for a certain period of time after the condition has been detected. For instance, the *five-year absolute survival rate* specifies the number of individuals who are alive at least five years after the diagnosis.[7] To be sure, survival rates are by no means easy to obtain. Correctly plotting the development in time of a disease in a group of individuals requires careful and subtle distinctions between types of diseases (sites, histology, etc.), periods of diagnosis, gender, stages of the disease, ethnicity of patients, and various other relevant variables.[8] In addition, survival analyses face significant limitations. Survival rates are statistical analyses and, as such, cannot predict what happens to particular individuals. Each patient is unique, and both treatments and responses vary greatly across subjects; hence, statistical analyses can only provide an indication of what occurred to relevantly similar patients in that same condition. Moreover, survival rates cannot be used to determine the effectiveness of different forms of treatments, an important endeavor that requires properly conducted clinical trials (see Teira, this volume). To further complicate the situation, medical decisions under uncertainty or risk are not independent of values and preferences of patients, such as their aversion to risk. Appropriate counseling is thus required not only to understand the objective risks that a patient is facing and her chances of recovery, but also to articulate the patient's preferences regarding the consequences, for instance, of the trade-offs between the benefits of early detection and the losses derived from years of unnecessary treatment, if a condition is overdiagnosed. Despite all these challenges and limitations, survival analyses, when available, provide precious prognostic information.

What kind of inference, exactly, is a prognosis? Just like diagnoses, prognoses convey virtually no causal or mechanistic knowledge. In gathering information regarding the expected course of a disease, one is providing neither a description of the underlying causal nexus nor an explanation of the disease itself. A prognosis is akin to a prediction of the outcome of a condition under certain circumstances. Given that, for every pathological condition, there likely are various therapeutic paths, and each path can be subject to distinct predictions, the process of prognosing a disease consists in selecting, for each alternative, the most plausible or likely prediction. In a sense, if diagnosis is a kind of inference to the best explanation, prognosis can be conceived as an *inference to the best prediction*, where one infers from the available evidence to the outcome that is most probable in the course of a disease or in its aftermath.

Much like diagnosis, prognosis requires counterfactual reasoning. In forecasting the future condition of a patient, selecting the optimal or suboptimal outcome, and determining a course of action, one cannot merely focus on what actually will happen; one must also consider what could happen were circumstances to change. Going back to the above example, suppose that undergoing immediate surgery and postponing all other forms of treatment until delivery maximizes the five-year survival rate of both the mother and the child. Still, given how high the stakes are, it is important to consider what *could* go wrong, no matter how far-fetched it is, and what could be done under those unfortunate circumstances. Once again, conditional thinking plays a role in such calculations. The subjunctive "Had chemotherapy been administered earlier, it would have prevented the spreading of metastasis," and the corresponding indicative "If chemotherapy is not administered early, the spreading of metastasis cannot be prevented," are a simple and effective way of conveying the relevant information.

Taking stock, in the first part of this essay, I argued that counterfactual thinking and hypothetical scenarios play a prominent role in medicine, focusing on two common and important kinds of inferences: *diagnosis* and *prognosis*. IBE and prediction are concepts that have long been at the center of both medical practice and philosophical reflection. It would be a mistake, I maintain, to argue that these notions play a different role in molecular medicine compared to their function in classical or premolecular physiology. Now, surely, molecular explanations are novel in the sense that they often appeal to previously unknown molecular properties, as opposed to, say, morphological or physiological structure. Yet, molecular medicine does not deliver new *kinds* of explanations. The true paradigm shift, illustrated well by the essays collected in this volume, consists in the application of molecular techniques that allow us to identify, predict, and explain pathological and otherwise abnormal conditions in much greater detail and at a much more extended temporal scale.[9] A clear example of how the "molecular revolution" has extended the reach of prediction and explanation in medicine comes from the much debated field of *genetic testing*—molecular assays that employ genetic data to diagnose and prognose conditions that the individual could experience at various stages of her lifetime. In short, the reason why counterfactual reasoning gains increasing importance in the context of molecular medicine is not because it underlies a specific class of molecular explanations; there is no such thing as a *sui generis* molecular type of explanation. Molecular markers, survival analyses, genetics tests, bioethical counseling, and various other statistical and technological advancements make the process of decision making extremely powerful and, at the same time, complicated. This makes a careful analysis of counterfactuals even more pressing. I now move on to discuss the foundations of this important class of statements.

4 The Truth (Still) Doesn't Explain Much

In an influential article aptly entitled "The truth doesn't explain much," Nancy Cartwright (1980) argued that many philosophers are guilty of conflating two functions of science that should be kept distinct: *truth* and *explanation*. More specifically, explanation is commonly viewed as a byproduct of truth: scientific theories are taken to explain by dint of the description they give of reality. It follows that descriptive accuracy is all that science is really after; explanatory power is a (welcome) free lunch. This mistake was fostered by the "covering law model," which presupposes that all explanation requires is knowledge of laws of nature and initial conditions in addition to some basic logic and probability theory (Hempel 1965). Cartwright goes on to argue that there are perfectly good explanations that are not covered by any law. At best, they are (more or less explicitly) prefixed by *ceteris paribus* generalizations which, read literally, are not only false, but known to be false; read with the *ceteris paribus* modifier, they only cover those few cases where the conditions are right. Either way, this is bad news for the covering law model of explanation.

Thirty-five years after the publication of Cartwright's article, the covering law model is now virtually dead and gone. Few philosophers (if any) still believe that explanation can be fully reduced to the logical derivation of an explanandum from a complex explanans that contains laws of nature together with a specification of initial conditions. Yet, the conflation of truth and explanation into a single, indivisible goal of inquiry is still commonplace in philosophy of science and is presupposed, often implicitly, in many scientific discussions. For instance, advocates of the so-called "new mechanistic philosophy" typically suggest, more or less explicitly, that scientific explanation is a matter of providing descriptions of mechanisms (Bechtel 2011; Craver and Darden 2013). Whether such descriptions should be as accurate as possible, or whether abstractions and idealizations constitute important and irreducible features of such representations, is still an open question (Love and Nathan 2015). Yet, there is a widely shared presupposition that mechanistic depictions will take good care of both endeavors at once: a single model, sufficiently rich in detail, can provide both accurate descriptions and illuminating explanations.

The conflation of truth and explanation is not only confined to science and related philosophical discussions but also extends to other areas of philosophy, such as philosophy of language and metaphysics. More importantly, for our present concerns, it affects traditional accounts of subjunctive conditionals. Counterfactuals first surfaced in contemporary philosophy early in the twentieth century as a solution to the problem of dispositions, which could not be analyzed in terms of the truth-functional material conditional of classical logic, and eventually turned into a self-standing research topic. Setting a few complications aside, the "standard" possible-world analysis of counterfactual conditionals, boosted by Kripke's (1963) groundbreaking possible-world semantics for modal logic and developed independently by Stalnaker (1968) and Lewis (1973), goes as follows.[10] Let us formalize the statement "Had the tumor not been removed, Jones would not have lived through the month" as

$$\neg T \; Ł \rightarrow \neg L$$

where "¬T" stands for the decision not to remove the tumor on a specific date, and "¬L" is the event of Jones's death within the subsequent four weeks. Such statement is true, according to the analysis, if and only if, in the possible world(s) closest to the actual world in which the antecedent is true—where the tumor is not removed—the consequent is also true; in those worlds, Jones does not live through the month.

The Stalnaker-Lewis analysis is arguably the most popular and influential semantics for counterfactuals, providing a clear and compelling account of an important class of linguistic statements that, for a long time, appeared to be intractable. At the same time, this analysis is confronted by an array of objections. Just to mention some of the better-known problems, standard possible-worlds accounts have trouble delivering the intuitively correct analysis of counterfactuals involving disjunctive antecedents, impossible antecedents, probable antecedents, and conditionals describing big changes consequential upon small changes (Fine 1975; Bennett 2003). In this essay, I shall not be concerned with these (or other) objections, important issues that, however, transcend our present purposes. The point that I want to emphasize is that, once again, all the focus is on *truth*: while the analysis is tailored toward what makes a counterfactual right or wrong, correct or incorrect, their predictive and explanatory value has been largely ignored. To be sure, the extensive literature on possible-world semantics does touch upon epistemological and pragmatic issues. Yet, epistemology and pragmatics are typically subordinated to semantic analyses. To wit, from an epistemological perspective, the main problem is how one comes to know whether a given counterfactual is true or false. Similarly, a central question regarding the pragmatics of counterfactuals is how contextual and conversational cues affect the truth value of a conditional statement. In short, truth and falsity are constantly at the center of philosophical analyses of counterfactuals, even with regards to their epistemic and pragmatic value.[11] My main claim is not that any specific view is wrong, rather, I argue that verocentric views fail to provide adequate epistemologies that explain the role that counterfactual reasoning plays in prognosis, diagnosis, and other important scientific inferences. In the rest of this essay, I develop an parallel analysis that, I maintain, fares better on this score and can thus integrate extant theories of counterfactuals.

Before moving on, three important disclaimers are in order. First, in what follows, I shall remain as agnostic as possible regarding the metaphysics and epistemology of models, possible worlds, and explanations—significant questions that, however, lie beyond our present scope. In this essay, I use *model* and *possible world* interchangeably to refer to interpreted structures that represent features of the world (Giere 1988, 2004; Weisberg 2013).[12] Second, I take it for granted that models and representations play a central role in explanatory practices (Cartwright 1983; van Fraassen 2008). Third, and finally, I adopt a pluralistic stance regarding explanations, which may come in various forms: D-N derivations, unifications, causal attributions, mechanistic sketches, and so on. However, I will not attempt a general specification of necessary and sufficient conditions for explanation. Following a common practice in contemporary philosophy of science, explanations are individuated by glossing practice: successful explanations specify models that provide a satisfactory answer to questions posed by the appropriate portion of the relevant community; appeal to their language, concepts, and methodology; and meet their standards.

5 Counterfactuals as Placeholders

At the outset of this chapter, I introduced subjunctive conditionals by asking why anyone would ever use counter-to-the-facts statements to discuss factual, empirical matters. To address this question, let us begin by considering a simple, mundane situation. Imagine being in a place where it would be dangerous or inconvenient to light a flame, and trying to explain to a young child how matches work. Suppose she asked you about the match and, in response, you told her:

> (1) "If the match were scratched, it would light."

I want to set aside the (important) question of what makes a statement such as this true or false and focus instead on what this conditional conveys. Obviously, in uttering it, you are not providing information about a possible world, near or far, where the match actually lights. What you are conveying is the *flammability* of the match, that is, its capacity or potential to light when struck. Two points should promptly be emphasized. First, in uttering (1) you are not explaining the disposition, any more than Molière's notorious *virtus dormitiva* explains the sedative power of opium. "Flammability" is a restatement, not an explanation of the capacity. Second, and equally important, ascribing flammability to a match, in and of itself, does not explain the actual or potential behavior expressed by the conditional; again, it is a paraphrase. A flammable match is a match that, *ceteris paribus*, would light if struck.[13]

Note that the same point straightforwardly applies to scientific and biomedical conditionals. Imagine that a young child, Sue, has learned that her friend Jim could not go skiing in the Rockies because he has a disease called sickle-cell anemia (SCA). Sue asks what is wrong with Jim and, in response, you tell her:

(2) "Had Jim been at altitude, he would have suffered respiratory crises."

In uttering (2), you are not providing information about a possible world, near or far, where Jim is sick. What you are conveying is Jim's pathological condition, that is, his predisposition to experience respiratory crises when exposed to low-oxygen conditions. Statement (2) is neither an explanation of SCA nor an explanation of the actual or potential behavior expressed by the conditional; (2) is a partial description of the effects of the relevant pathology: drepanocytosis.

In sum, utterances like (1) or (2) explain neither dispositions or pathological conditions, nor the associated behaviors. Does this mean that, in uttering a subjunctive conditional, you are somehow misleading your young interlocutor or failing to provide her with relevant information? Of course not. While restating dispositional properties, such as flammability or drepanocytosis, in counterfactual terms hardly explains anything, it conveys a commitment to the possibility, in practice or in principle, of providing an explanation of the underlying causal mechanisms.[14] To illustrate, (1) can be used to convey the existence of some kind of mechanism that explains the capacity of a match to light when scratched. (2), in turn, stands in for a fascinatingly complex story involving a network of molecular, developmental, and environmental events. Oversimplifying a bit, the details will describe how a point mutation in the *hemoglobin* gene leading to a single amino-acid substitution in the encoded molecule, under conditions of low oxygen, may cause erythrocytes to assume a rigid crescent shape, reducing their flexibility and elasticity and affecting their ability to dilate capillaries to facilitate their passage.

To clarify the main point, it might be useful to compare (1) and (2) with utterances of the opposing conditionals with the contradictory consequent:

(3) "If the match were scratched, it would *not* light."

(4) "Had Jim been at altitude, he would *not* have suffered respiratory crises."

Again, I shall set aside, for the time being, whether (3) and (4) are true or false and focus instead on why anyone would utter statements of this kind. (3) effectively conveys that a particular match, at this point in time, lacks the capacity to light, if scratched. As in the case of (1), (3) does not explain what will or might interfere with the match, preventing it from lighting; yet it does commit the speaker to the presence of *some* mechanism or condition that would prevent the lighting of the scratched match. Thus, if the speaker had no reason to believe that the match is faulty or humid or that oxygen is lacking, then an utterance of (3) would be inappropriate or misleading because it would suggest the possibility of providing an explanation that, however, cannot be produced, even in principle, because of the lack of appropriate conditions. In contrast, if the speaker suspected that some interfering conditions are actually in place, then the relevant explanation would be available, making an utterance of (3) felicitous and, correspondingly, an utterance of (1) inappropriate.

This point about explanation can be extended, *mutatis mutandis*, to prediction. (1) and (3)—or, better, the corresponding indicatives "if the match is scratched, it will (not) light"—describe what will happen if the match gets scratched. Note that the conditionals do not ground the prediction in anything; they do not provide any reason to believe that the scratched match will light as opposed to not lighting. However, the appropriateness of uttering (1) versus (3) presupposes that there is an underlying fact, reason, or mechanism that makes one prediction more plausible than the other. With all of this in mind, consider the case of drepanocytosis. Statement (4) commits the speaker to the possibility of explaining or predicting why Jim would not have suffered respiratory crises despite being at altitude. If one knows that Jim, or someone on his behalf, is taking measures to avoid the drastic effects of his condition, then (4) is felicitous. However, if the speaker is unaware of any relevant fact, reason, or mechanism, then the appropriate utterance would be (2).

We can summarize the entire discussion, so far, as follows: *counterfactual conditionals are placeholders* that stand in for predictions and explanations that a speaker commits to producing in principle, regardless of whether she (or anyone else) is capable of producing them, in practice. To emphasize, *qua* placeholders, counterfactuals themselves do not provide the explanation or justify the prediction; they simply commit the speaker to the existence of some explanation or justification that gets the job done.

Interestingly, the idea that counterfactuals are placeholders harks back to the pioneering intuitions of Frank Ramsey and Nelson Goodman. In an influential paper, Ramsey (1931) claimed that, in order to evaluate a subjunctive conditional *C*, a subject *s* adds the antecedent to her stock of beliefs, making whatever adjustments, if any, are required to maintain consistency. Next, *s* considers whether or not the consequent would hold, under those circumstances. Essentially, Ramsey provides a recipe for determining whether or not a subject should *believe* in *C*, without a corresponding specification of truth conditions for *C*. Similarly, Goodman (1955) framed the problem of counterfactuals as the task of defining the circumstances under which a given counterfactual holds while the contradictory conditional (the subjunctive with the same antecedent and the negated consequent) fails to hold. "Holding" is commonly identified with "being true." This, I maintain, is a mistake. As discussed above, the circumstances under which a statement is true and those in which it is explanatory, assertible, or useful need not—and often do not—coincide. Indeed, the emphasis on truth came later, with possible-world semantics. However, all this virtually exclusive hype on truth values might be unjustified. For one thing, asking whether a model (or a possible world) is "true" or "false" is a category mistake. Models are (logically) *consistent* or *inconsistent*, *similar* or *dissimilar* (in certain respects, to their targets), *useful* or *useless* (with respect to a given goal), *explanatory* or *nonexplanatory* (relative to an explanandum), and so on. But they are neither "true" nor "false." Thus, if counterfactuals stand in for models (or collections thereof), focusing solely on their truth value is equally misleading. Another important question regarding counterfactuals concerns not their logic or semantics, but their *pragmatic assertibility*, that is, their informativeness, their contribution to arguments or conversations, their diagnostic or prognostic value. In the following section, I develop this notion of pragmatic assertibility, focusing specifically on its role in diagnostic and prognostic judgments.

6 Semantic Truth or Pragmatic Assertibility?

The observation that the literal meaning of a statement needs not (and frequently does not) correspond to the information it conveys is old news in linguistics and philosophy of language (Grice 1989). My suggestion, developed in the previous section, is essentially that this old adagio be applied to conditionals of both the indicative and the subjunctive kind. A counterfactual *per se* does not explain anything; it is a placeholder that stands in for explanatory hypotheses regarding a real or alleged similarity between a theory, model, or hypothesis and bits and pieces of the real world. The contribution of a counterfactual to a conversation has little or nothing to do with its truth value. In uttering a subjunctive conditional, a speaker pragmatically implicates that the model (possible world) in question can be used to predict or explain some significant feature of reality. This is why counterfactuals have diagnostic and prognostic value in medical reasoning. I now elaborate on this claim.

In the first part of this essay, I argued that diagnoses and prognoses require one to consider not only what actually is (or has been) but also what could occur (or could have occurred). This information is effectively conveyed through conditionals of both the indicative and subjunctive kind. As noted in §2, diagnoses provide very limited causal accounts and poor explanations, and this should be hardly surprising, as diagnoses are meant to be neither explanatory nor descriptively accurate. Diagnoses are inferences to the best explanation, where the conclusion that would best explain the symptoms is inferred on the basis of available evidence and counterfactual considerations. Note how nicely this fits in with the account of counterfactuals as placeholders. By uttering "Had the tumor not been removed, Jones would not have made it through the month," I am not explaining why the tumor was removed, why Jones is currently alive and well, or anything else. Rather, I am indicating what the key to the explanation is: if you want to explain what is going on with Jones, you had better look at what happened to his tumor. Needless to say, different contexts will require distinct explanatory models of varying depth. While Jones's family members might rest content knowing that the operation went well and Jones is now safe and sound, surgeons should "dig deeper," making sure that the situation is truly under control and that Jones is on his way toward full recovery.

An analogous point can be made regarding prognoses. In §3, I argued that a prognosis is a kind of inference to the best prediction, where a physician selects the optimal or suboptimal outcome of a condition and suggests the best paths towards "actualizing" it. Once again, the placeholder view fits in nicely with the role of counter-to-the-facts statements in this kind of inference. As noted, assessing the likelihood of various scenarios can be extremely hard, and a precise quantification of probabilities often exceeds our current grasp. Conditionals, of both the indicative and subjunctive kind, provide an effective means of conveying the relevant information in a quick and efficient way. To wit, nobody is likely able to quantify exactly Jones's chances of full recovery (90 percent? 91.3 percent?). Yet, knowing that, if the tumor is not removed, he will not survive the month is likely to suggest the optimal or suboptimal course of action regardless of the actual underlying probabilities, of either the subjective or objective ilk.

At this point, some readers might be inclined to reply that my attempt to move away from truth conditions is a red herring. Regardless of which semantic analysis we adopt, it is the *truth value* of a counterfactual, the objection runs, that makes it valuable in science. To wit, "Had the tumor not been removed, Jones would not have lived through the month" gestures toward the *correct* explanation of Jones's current well-being in virtue of its truth. Similarly, "if the tumor is not removed, he will not survive the month" constitutes the correct prognosis of what will happen to Jones precisely because such a statement is true.

This substantial objection ultimately misses the point, and it is important to see why. Take the statement "Had the tumor not been removed, Jones would not have lived through the month" and, despite its plausibility, assume that it is false. Perhaps Jones would have been particularly responsive to belated treatment, perhaps a really skillful surgeon would have been able to save his life anyway, or perhaps a miracle would have occurred: pick out your favorite scenario and suppose that the statement is false. If you adhere to the standard possible-world semantics, the reason will likely be that worlds in which the tumor is not removed and Jones dies shortly thereafter are farther away from actuality than worlds in which the tumor is not removed and Jones survives; if you buy into a different account, you will likely provide a different story, but it doesn't matter. The important point is that, provided that Jones's condition is serious enough, the conditional statement is perfectly assertible even on the assumption that it is false. There are many ways for a statement to be false, and many of them do not affect assertibility. Suppose, for instance, that Jones would not have died within a month had he not received treatment, but that he would have been in serious danger, or he would have died shortly thereafter anyway (say, eight days later), or that he would have been seriously disabled throughout the rest of his life. Under all these circumstances, the conditional is false, but it still serves the function for which it was uttered in the first place: it is a convenient and efficacious way to convey the seriousness of Jones's condition, to bring attention to the importance of surgically intervening on Jones, and so on. In short, the truth value of a conditional is not what underlies its scientific value. Indeed, the truth of many counterfactuals is way too fragile, subjective, and context dependent to base upon them something as vital as medical decisions. We are much better off grounding decisions about our life in empirical statements about the actual world, where truth and falsity do play a decisive role.

In sum, the placeholder view, which sets aside the semantic content of a conditional in favor of its pragmatic implicatures, provides a simple and convenient framework for understanding why (and how) counterfactual reasoning is so central to medical theory and practice. Questions concerning the *truth value* of subjunctive conditionals should therefore share center stage with the assertibility of a conditional in a given context.

Before moving on, two final philosophical remarks on the placeholder view of counterfactuals (more empirically oriented readers are encouraged to skip directly to the following section). First, suppose that, at some point, we were to achieve a complete causal explanation of a physiological condition, pathological or otherwise. Under such circumstances, would we be able to completely dispense with conditionals? The short answer is: in principle, yes; in practice, no. In principle, if we had complete causal knowledge of a disease, then counterfactuals would become disposable since, given the capacity to explain and predict the condition, there would be no need to employ placeholders. However, in practice, we might still employ conditionals for the sake of brevity and simplicity. To illustrate, chemistry has already provided us with a reasonably accurate and comprehensive explanation of flammability and solubility. Still, in most contexts, conditionals (or the associated dispositions) provide a much more efficient and convenient way to "explain" and "predict" the relevant behavior by pointing to the underlying mechanism while leaving it unspecified. A complete causal explanation of a pathological condition (assuming, for the sake of the argument, that such the notion of "complete explanation" is indeed coherent and meaningful) is likely to be exponentially more complex than the mechanisms underlying matches or salt. Hence, although in principle, we could completely dispense with conditional reasoning, in practice, it is extremely unlikely that we will ever do so; counterfactuals will remain useful and practical linguistic shortcuts.

Second, my emphasis on representation indicates that, like virtually all other authors writing on counterfactuals, I am assuming that possible worlds (models) can be assessed in terms of their similarity to actuality. What characterizes my proposal, distinguishing it from extant accounts, is the claim that this similarity relation needs not have significant effects on the truth value of counterfactuals. The similarity metric presupposed here affects the pragmatic assertibility of conditionals—the appropriateness or felicitousness of its utterance, in a given context—which depends on whether the counterfactual picks out a class containing (at least some) models of explanatory value. In other words, the notion of similarity presupposed here is not a metaphysical relation of closeness between worlds but an explanatory relation between a real system and various representations, including pictorial, verbal, analogical, graphic, computer-simulated, and hybrid ones. Making sense of this world–model relation is one of the hardest and more important problems in contemporary philosophy of science—the problem of scientific representation.[15] Yet, the practice of selecting appropriate models on the basis of relevant similarities with the world is commonplace in both scientific and more "mundane" investigations. While philosophers of science have typically assumed that fundamental representational resources are *semantic* (e.g., truth and reference), both classic and recent work has begun to shift attention to *pragmatic* accounts of representation as a kind of activity (Hacking 1983; Giere 2004).

7 Conclusion: Tough Choices

Making the "right" decisions, when it comes to health, is not only vital, it is also challenging. Medical decisions presuppose that patients are able to successfully comprehend and process a swarm of probabilistic data provided by physicians and other medical specialists. While this might sound obvious and unproblematic, recent studies suggest that we should not be too hasty in jumping to conclusions. Contemporary psychology reveals an unflattering picture of human beings as remarkably prone to mistakes, misunderstandings, and misinterpretations when it comes to statistical reasoning (Gigerenzer 2008; Kahneman 2011). To mention a famous example from a classic study, almost 90 percent of subjects judged it more likely that a fictional character, introduced in an experimental context, was a *feminist* bank teller than a bank teller *simpliciter*, an obvious and overt violation of basic probability theory, since the set of feminist bank tellers is a proper subset of the set of bank tellers (Tversky and Kahneman 1983).

Fortunately for most patients, physicians are there, ready to support those of us who have had a hard time crunching numbers since grade school. But is help really on the way? Further studies point out potential problems. An influential experiment, for instance, suggests the disconcerting conclusion that the majority of American physicians might not fully grasp the meaning of statistical concepts they employ on a daily basis (Gigerenzer et al. 2007).[16] The consequences of these kinds of mistakes are significant, as misinterpreting clinical data might lead to patients who are not in need of treatment ending up receiving it or, even more dangerously, patients not being administered appropriate treatment. Indeed, some authors have recently pointed out that the percentage of misdiagnoses in current medical practice could be as high as 38 percent (Sadegh-Zadeh 2011).

There are different approaches for avoiding such errors or, at least, minimizing their impact. An obvious recommendation would be to improve the statistical literacy of both physicians and patients. Unfortunately, this strategy is hard to implement: improvements are often limited, depending on the level of education of the subjects, and the relative importance of context and form in which the information is conveyed is subject to controversy.[17] In addition there are subtler and more complex kinds of probabilistic biases underlying medical decision, such as *lead time bias* in survival analyses, which further complicate the picture. In short, communicating risks to patients and guiding their decisions is a challenging task for which there are no clear standards yet.

As emphasized in the first part of the chapter, accurate, reliable, and comprehensive diagnoses of pathological conditions and prognoses of their courses are important medical goals, albeit hardly simple ones. Physicians and patients sometimes need to act promptly and make important decisions even if the available predictions and explanations are inaccurate or incomplete. When this is the case, conditionals can be a fruitful tool to process and convey complex statistical information in a simple and yet effective fashion. To illustrate, consider how much mechanistic and statistical knowledge is required to assess precisely the chances that a patient with a specific condition (such as a mutated MYH7 or BRC2 gene) will be affected by a serious condition in the course of her life. In addition, much relevant information is still lacking; for instance, the precise impact of various environmental and lifestyle choices on these predispositions. Rather than flooding concerned patients with complicated probabilistic data, which are prone to be misinterpreted by both physicians and patients, under many circumstances, it might be more useful and productive to provide simple "rules of thumb" expressed in the form of simple indicative or counterfactual conditionals, such as "if symptom X appears, make sure to consult a specialist" or "had the scan reported a dark spot, that would have been a bad sign." Obviously, simple counterfactuals by themselves cannot replace the accurate diagnosis and prognosis of complex pathologies, which require painstaking lab work and much ingenuity. Similarly, conditional reasoning is not the panacea to the probabilistic biases underlying human rationality, which constitute an extensive topic of research in psychology, economics, philosophy, and many related fields. In order to understand this important role, however, it is necessary to move away from semantic discussions that focus almost exclusively on the truth value of conditionals toward an approach—such as the "placeholder view" presented here—that emphasizes their pragmatic role in predictions and explanations.

At a general level, the present discussion illustrates the mutual relevance of scientific and philosophical research. On the one hand, clinical practice provides a wealth of examples that should inform, constrain, and inspire the work of metaphysicians, philosophers of language, and philosophers of science. On the other hand, philosophical analysis plays an important role in systematizing such knowledge and providing solid conceptual and methodological foundations. This essay attempted to emphasize the importance of pursuing both. Counterfactuals constitute an important class of linguistic statements that underlie important scientific inferences and, hence, deserve more attention and respect than they have been accorded so far by either scientists or philosophers of science.

Acknowledgements

I would like to express my gratitude to Giovanni Boniolo, Mike Brent, Jeff Brown, Serafino Garella, Maël Lemoine, Lorenzo Magnani, and David Teira for constructive comments on various versions of this article. Earlier drafts have been presented at the Konrad Lorenz Institute in Klosterneuburg, at the University of Urbino, and at the University of Sassari. All audiences have provided helpful feedback.

Notes

1   Philosophers sometimes distinguish between *subjunctive conditionals* and *counterfactual conditionals* in that the latter must have false antecedents. In this article, I set these subtleties aside and use the terms *counterfactual* and *subjunctive conditional* interchangeably.

2   In general, it is customary to distinguish between *symptoms*—medically relevant complaints reported by the patient or others familiar with the patient—and *clinical signs*, that is, objective conditions ascertained by direct medical examination. For the sake of simplicity, in the following discussion, I shall leave it up to the reader to determine whether a condition is better understood as a "sign," "symptom," or an altogether different kind of marker.

3   When the chance that an individual bearing the sign or symptom is (or will be) affected by the pathological condition is (virtual) certainty, it is tempting to dub the indicator "deterministic." However, such label is slightly misleading since, while a pathognomonic sign or symptom, by definition, has a very high *specificity*, it needs not have high *sensitivity*. (Any test is associated with two values that we can call *sensitivity* and *specificity*. The sensitivity of a test t, testing for condition T, is the probability that a subject s tests positive given that s is affected by T. The specificity of t is the probability that s tests negative, given the absence of T.) Hence, the presence of a pathognomonic sign or symptom is virtually sure-fire indicator of the condition, but the occurrence of the pathology does not necessarily imply that its pathognomonic signs or symptoms will also be present. For all these reasons, the relation between signs, symptoms, and their underlying conditions is better understood in terms of probabilities and difference-making rather than determination (Waters 2007; Nathan 2012).

4   The probabilistic nature of biomarkers is further complicated by the truism that virtually no real-life test is infallible and, consequently, results are seldom conclusive. When a test is not perfectly reliable, a fraction of the results may be *false positives* (the subject tests positive without having the disease) or *false negatives* (the subject tests negative while having the disease). While *sensitivity* and *specificity* are features of the test, positive and negative *predictive values*—the probability of having the disease given the positive outcome of the test and the probability of not having the disease given the negative outcome of the test—depend on the prevalence of the disease in the population of interest. Hence, a clinician prescribing and evaluating a test should be aware of both its sensitivity and specificity as well as the correlated false positive and false negative fractions (Boniolo and Teira 2016).

5   Indeed, Huntington's Disease has a whole host of symptoms—such as adult-onset personality changes, generalized motor dysfunctions, and cognitive decline—and biomarkers—including imaging studies, metabolomic, proteomic, and transcriptomic approaches; and biomarkers built on hypothesis-driven experiments—that help us diagnose, prognose, and monitor the occurrence of the disease without playing any significant causal role in its development (Zuccato et al. 2010).

6   Before moving on, a few clarifications are in order. First, while some authors draw subtle distinctions between various kinds of diagnoses (Sadegh-Zadeh 2011) and abductive inferences (Magnani 2001), for the sake of simplicity, I shall treat *diagnosis* and *abduction* as monolithic concepts. Second, one could draw a distinction between "induction" and "abduction"; yet, in classifying abduction as a particular kind of inductive argument, I am simply drawing attention to the fact that such inference falls short of deductive certainty. Third, the analogy between diagnosis and inference to the best explanation, I surmise, fuels the mistaken assumption, discussed above, that diagnosis is a sort of explanation. After all, isn't an IBE a particular kind of explanation? The answer, unfortunately, is negative. In a diagnostic inference, the presence of S, coupled with the conditional premise "If S, then P" provides evidence for P; yet it does not explain P.

7 It is important not to confuse the five-year *absolute* survival rate with the five-year *relative* survival rate, which describes the percentage of patients who are alive five years after the diagnosis, divided by the percentage of the relevant population with similar characteristics of gender, age, ethnicity, and so on that is alive after five years.

8 Yet another complicating factor is due to the fact that members of the study group are seldom observed for the same amount of time. For instance, patients diagnosed toward the end of the study period may provide incomplete data. In addition, the status of patients observed at the beginning of the study period may be unknown by the end of the experiment, as some may have moved, changed physician, died for "irrelevant reasons," and so on (Boniolo and Teira 2016). Patients observed until the endpoint of interest are sometimes called *uncensored cases*; patients who survive beyond the end of the follow-up period or who are lost to follow-up are dubbed *censored cases*.

9 These considerations are part of a more general problem, namely, how to conceptualize the momentous change brought about by the molecular study of medicine. While it is tempting to straightforwardly define *molecular medicine* as the study of the molecular basis of disease, as Boniolo (this volume) persuasively argues, such a simplistic characterization would involve both historical and methodological misconceptions (see also Gadebusch Bondio and Spöring, this volume; and Russo and Vineis, this volume).

10 Here I ignore some technical details and salient differences between Stalnaker and Lewis's approaches. A comprehensive account can be found in Sider (2010, Ch. 8).

11 I should mention that, over the last few years, some authors have attempted to move away from this "vero-centrism." Dorothy Edgington (1995, 2008), for instance, applies to counterfactuals the so-called "suppositional account." On this view, a conditional statement, of either the indicative or subjunctive kind, is not a categorical assertion—true or false, as the case may be. Rather, a conditional is a statement of the consequent under the supposition of the antecedent. Originally developed by Adams (1975), the suppositional view is broadly accepted in the case of indicatives; however, its extension to subjunctives is much more controversial. Many philosophers reject a unified account of indicatives and subjunctives on the grounds that expressing a conditional of the former kind in terms of the latter, or vice versa, can affect the truth value. To cite a somewhat trite example, the indicative "If Oswald didn't kill Kennedy, someone else did" is intuitively true, whereas the corresponding subjunctive "Had Oswald not killed Kennedy, someone else would have" is intuitively false. But, Edgington says, the suppositional view provides a straightforward explanation for the discrepancy: one can consistently be confident that someone else did it, on the supposition that Oswald didn't, while judging it unlikely that someone else would have done it back in 1963 supposing that Oswald hadn't. The details of Edgington's proposal need not interest us here. The important point, for present purposes, is that, if conditionals are understood in terms of probabilities, then they cannot be understood in terms of truth conditions. This is because conditional probabilities are not measures of the truth of propositions; a famous result proves that there is no proposition X such that, necessarily, the probability that X is true is the conditional probability $p(C|A)$. Edgington suggests that, in order to assess a counterfactual, we take a probability distribution over the relevant worlds and figure out how likely it is that we have a C-world, given that we have an A-world. This conditional probability is the chance, at a time where A still had the chance of coming about, of $C|A$ together with any relevant, causally independent, subsequent facts that causally bear on C.

12 While not taking a stance on the nature and ontological status of these models, I do want to make explicit two important provisos. First, *representation* should be understood in the broadest and most ecumenical sense, including not only mathematical axiomatization, which is common in formalized fields such as theoretical physics

and (parts of) economics, but also linguistic descriptions, diagrams, and various kinds of pictorial depictions, more typical of sciences such as biology, psychology, and neuroscience. Second, it is common among philosophers to treat possible worlds as "total" in the sense that, for any proposition P, either P itself or its negation $\neg P$ must be true at any given world (Adams 1974). In keeping with the analogy with scientific practice, I shall relax this "totality requirement," which poses unnecessary strictures on the notion of representation. Thus I allow a model (possible world) to be "partial" or "selective" in the sense that a given model (world) might realize or make true some proposition P without committing to the truth value of other independent propositions $Q, R, S\ldots$ Thus, I can have a model where Jones has Asperger syndrome and is diabetic, but such model says nothing about, say, Jones's immunization record.

13 To be sure, providing a systematic account of the *ceteris paribus* modifier is no easy task; indeed, it is precisely what makes dispositional properties such a longstanding and frustrating philosophical conundrum. Since the realization that dispositions cannot be straightforwardly reduced to subjunctive conditionals, because of problems involving *finking, masking,* and *mimicking* (Martin 1994), various solutions have been provided, from replacing the simple subjunctive conditional with a more sophisticated one (Prior et al. 1982; Lewis 1997; Mellor 2000) to exploring nonconditional analyses (Fara 2005) or abandoning altogether the quest for analysis in favor of nonreductive explanations of dispositions (Bird 1998; Molnar 1999). In a recent article, I suggested a different route for saving the simple conditional analysis from Martin's problems and its variants (Nathan 2015), but I shall not rehearse it here.

14 The "or in principle" qualifier is important because I will often not be able to rehearse the relevant explanation off the top of my head. Alternatively, the complete story might be obvious and presupposed or unnecessarily complicated, depending on whether I am talking to experts or novices. Finally, the relevant explanation might not have been discovered yet. Still, in all these cases, there is an explanation to be provided, at least in principle.

15 Importantly, finding the appropriate model is not simply a matter of providing the most accurate representation for, as noted, the most accurate representations are not necessarily the most explanatory ones. Indeed, the distinction between explanation, description, and prediction plays an important role in recent discussions of idealization and abstraction in scientific representation (Weisberg 2013; Love and Nathan 2015).

16 To establish this point, the experimenters presented the following scenario to 160 gynecologists. Given a breast cancer screening of a mammogram with a sensitivity of 90 percent, a false positive rate of 9 percent, and a disease prevalence in the relevant population of 1 percent, if a woman tests positive, what are her chances of really having breast cancer? Of the responding physicians, 60 percent were quite far from the correct result (approximately 10 percent), wrongly believing that 80–90 percent of women who tested positive would actually have the condition. Admittedly, this exercise is not as straightforward as the "bank teller example," especially since the probability of obtaining a positive result in the test cannot be trivially inferred from the provided data. Still, the answer can be derived through a direct application of Bayes's theorem, a well-known formal tool that guides the determination of the conditional probability of hypothesis h given evidence e (intuitively, how to update your degree of belief in a hypothesis given new evidence):

$$P(h|e) = \frac{P(e|h)P(h)}{P(e)} = \frac{P(e|h)P(h)}{P(e|h)P(h) + P(e|\neg h)P(\neg h)}$$

In our case, *P(h|e)* corresponds to the probability of having the disease, given that the test positive; *P(e|h)* is the conditional probability of testing positive given that you have the disease; $P(h)$ is the prior probability of having the disease; and $P(e)$ is the probability of testing positive. In order to obtain the correct result, the sensitivity of the test, that is, the probability of having breast cancer given a positive mammogram (90 percent), is multiplied by the prevalence of the disease (1 percent), and the result is divided by the probability of obtaining a positive in the test (10 percent).

17 Whereas psychologists such as Tversky and Kahneman tend to be more pessimistic regarding the prospects of enhancing our capacity to process probabilistic data, evolutionary psychologists, such as Gigerenzer, argue that substantial improvements can be obtained by presenting the data in the form of natural frequencies. Consequently, they advocate the use of simple decision rules (*heuristics*) that allow physicians to handle uncertainty with a success rate similar to a complex statistical algorithm (*fast-and-frugal trees*).

References

Adams, E.W. (1975) *The Logic of Conditionals*. Dordrecht: Reidel.

Adams, R. (1974) "Theories of actuality," *Nôus* 8, pp. 211–31.

Bechtel, W. (2011) "Mechanism and biological explanation," *Philosophy of Science* 78, pp. 533–57.

Bennett, J. (2003) *A Philosophical Guide to Conditionals*. Oxford: Oxford University Press.

Bird, A. (1998) "Dispositions and antidotes," *The Philosophical Quarterly* 48, pp. 227–34.

Boniolo, G., and V. Sanchini (eds.) (2016) *Ethical Counseling and Medical Decision-Making in the Era of Personalized Medicine*. Heidelberg: Springer.

Boniolo, G., and D. Teira (2016) "The centrality of probability." In *Ethical Counselling and Medical Decision-Making in the Era of Personalized Medicine*, ed. G. Boniolo and V. Sanchini, chapter 6. Heidelberg: Springer.

Cartwright, N. (1980) "The truth doesn't explain much," *American Philosophical Quarterly* 17(2), pp. 159–63.

Cartwright, N. (1983) *How the Laws of Physics Lie*. Oxford: Clarendon.

Chahine, L.M., and M.B. Stern (2011) "Diagnostic markers for Parkinson's disease," *Current Opinion in Neurology* 24(4), pp. 309–17.

Craver, C.F., and L. Darden (2013) "In search of mechanisms." In *Discoveries Across the Life Sciences*. Chicago: University of Chicago Press.

Edgington, D. (1995) "On conditionals," *Mind* 104, pp. 235–329.

Edgington, D. (2008) "Counterfactuals," *Proceedings of the Aristotelian Society* 108(1), pp. 1–21.

Fara, M. (2005) "Dispositions and habituals," *Nôus* 39(1), pp. 43–82.

Fine, K. (1975) "Review of *Counterfactuals* by David Lewis," *Mind* 84, pp. 451–58.

Giere, R.N. (1988) *Explaining Science. A Cognitive Approach*. Chicago: University of Chicago Press.

Giere, R.N. (2004) "How models are used to represent reality," *Philosophy of Science* 71, pp. 742–52.

Gigerenzer, G. (2008) *Rationality for Mortals: How People Cope with Uncertainty*. New York: Oxford University Press.

Gigerenzer, G., W. Gaissmaier, E. Kurz-Milcke, L.M. Schwartz, and S. Woloshin (2007) "Helping doctors and patients make sense of health statistics," *Psychological Science in the Public Interest* 8(2), pp. 53–96.

Goodman, N. (1955) *Fact, Fiction, and Forecast*. Cambridge: Harvard University Press.

Grice, H.P. (1989) *Studies in the Way of Words*. Cambridge: Harvard University Press.

Hacking, I. (1983) *Representing and Intervening. Introductory Topics in the Philosophy of Natural Science*. Cambridge: Cambridge University Press.

Hempel, C.G. (1965) *Aspects of Scientific Explanation and Other Essays in the Philosophy of Science*. New York: Free Press.

Kahneman, D. (2011) *Thinking, Fast and Slow*. New York: Farrar, Straus and Giroux.

Kripke, S.A. (1963) "Semantical considerations on modal logic," *Acta Philosophica Fennica* 16, pp. 83–94.

Lewis, D.K. (1973) *Counterfactuals*. Oxford: Blackwell.

Lewis, D.K. (1997) "Finkish dispositions," *The Philosophical Quarterly* 47(187), pp. 143–58.

Lipton, P. (2004) *Inference to the Best Explanation* (2nd ed.). London: Routledge.

Love, A.C., and M.J. Nathan (2015) "The idealization of causation in mechanistic explanation." *Philosophy of Science* 82, pp. 761–74.

Magnani, L. (2001) *Abduction, Reason, and Science: Processes of Discovery and Explanation*. New York: Springer.

Martin, C.B. (1994) "Dispositions and conditionals," *The Philosophical Quarterly* 44(174), pp. 1–8.

Mellor, D.H. (2000) "The semantics and ontology of dispositions," *Mind* 109, pp. 757–80.

Molnar, G. (1999) "Are dispositions reducible?" *The Philosophical Quarterly* 49(194), pp. 1–17.

Nathan, M.J. (2012) "The varieties of molecular explanation." *Philosophy of Science* 79(2), pp. 233–54.

Nathan, M.J. (2015) "A simulacrum account of dispositional properties," *Nôus* 49(2), pp. 253–74.

Popper, K.R. (1963) *Conjectures and Refutations. The Growth of Scientific Knowledge*. London: Routledge and Kegan Paul.

Popper, K.R. (2002 [1935]) *The Logic of Scientific Discovery*. London and New York: Routledge.

Prior, E.W., R. Pargetter, and F. Jackson (1982) "Three theses about dispositions," *American Philosophical Quarterly* 19(3), pp. 351–57.

Ramsey, F.P. (1931) "General propositions and causality." In *Foundations of Mathematics and Other Logical Essays*. New York: Routledge and Kegan Paul, pp. 237–55.

Sadegh-Zadeh, K. (2011) "The logic of diagnosis." In F. Gifford (ed.), *Philosophy of Medicine*, pp. 357–424. Amsterdam: Elsevier.

Sider, T. (2010) *Logic for Philosophy*. New York: Oxford University Press.

Stalnaker, R.C. (1968) "A theory of conditionals." In Nicholas Rescher (ed.), *Studies in Logical Theory*, Volume 2, pp. 98–112. Oxford: Blackwell.

Tversky, A., and D. Kahneman (1983.) "Extensional versus intuitive reasoning: The conjunction fallacy in probability judgment," *Psychological Review* 90(4), pp. 293–315.

van Fraassen, B.C. (2008) *Scientific Representation*. New York: Oxford University Press.

Waters, C.K. (2007) "Causes that make a difference," *The Journal of Philosophy* 104(11), pp. 551–79.

Weisberg, M. (2013) *Simulation and Similarity: Using Models to Understand the World*. New York: Oxford University Press.

Zuccato, C., M. Valenza, and E. Cattaneo (2010) "Molecular mechanisms and potential therapeutical targets in Huntington's disease," *Physiological Review* 90, pp. 905–81.

Part IV
Inference

10 Forms of Extrapolation in Molecular Medicine

Pierre-Luc Germain and Tudor Baetu

Abstract

Biomedical research is built upon inferences transposing knowledge across systems–be it across species, between an experimental system and another, or from controlled clinical studies to routine healthcare contexts. While the notion of extrapolation received increasing attention in the recent philosophical literature, extrapolative inferences having distinct epistemic aims are often conflated together, thus making it difficult to assess the validity of such inferences. In this paper, we begin by characterizing the general form of extrapolations, whose structure and components allow a careful and systematic dissection of distinct types of extrapolations, highlighting the specific aims, methods and challenges associated with each type of extrapolation. Finally, we show how some contemporary research practices can challenge the boundaries of this classification, pointing to the need to consider extrapolations in their general form.

1 Introduction

In biological and biomedical research, the notion of extrapolation is intimately linked to the problem of generalizing or applying findings and results obtained by studying one biological system to other biological systems. Perhaps the most widely recognized and discussed type of extrapolation in the philosophical literature refers to the translation of the results of basic science to clinical practice (Bolker 2009; Howick et al. 2013; LaFollette and Shanks 1996; Shanks et al. 2009; Steel 2007). For instance, basic science can demonstrate that an antiviral treatment reduces the viral load in simian immunodeficiency virus (SIV)–infected rhesus monkeys, that an anticancer drug kills human hepatocytes *in vitro*, and that some commonly used food additive is carcinogenic in mice. Given this information, medical practitioners, pharmaceutical companies, and healthcare policy makers must make a decision whether to use the antiviral drug as a treatment for AIDS, issue a toxicity warning for the anticancer drug, and prohibit the use of the additive in the food industry. The most informative data for making these decisions would be to repeat the experiments of basic science in humans. However, for ethical and practical reasons, such experiments are often not possible. The alternative is to assume that what is true of one biological system is also true or approximately true of another; for instance, one can conclude that a substance is likely to be toxic in humans on the basis of preclinical data from animal models. Such practices qualify as extrapolations insofar as they extend a claim from the system in which it was actually documented to a different system. We can therefore characterize extrapolation as the attempt make a claim about a *target system* based on experimental results obtained in a different *source system*. The source system is often said to serve as a *surrogate model* of the target, that is, as a more manageable experimental setup for studying a phenomenon used as a substitute for an experimentally less manageable, but biologically, physiologically, or clinically more relevant target setup. The hope is that the findings generated by the investigation of the surrogate can be safely extrapolated to the target despite known and unknown differences between the two (Baetu 2014, 2015; Bolker 2009; Steel 2007).[1]

Extrapolated results are evaluated along two distinct axes (Fletcher et al. 1996). First, it is important to establish the *internal validity* of a study,[2] that is, the extent to which the study supports claims about the source system. Assuming that there are no reasons to doubt the internal validity of the study, *external validity* refers to the extent to which the results of the study and the claims they support can be legitimately extrapolated to targets others than those directly tested, such as different organisms, different experimental conditions, or different interventions and outcomes; or, most commonly, to targets involving a combination of such differences.

Despite the central importance of the concept of external validity, clinical researchers and practitioners have long complained that trial design guidelines or evaluation committees made little or no explicit mention about external validity, and that there were no "accepted guidelines on how external validity of RCTs [Randomized Clinical Trials] should be assessed" (Rothwell 2005, p.82). Although efforts have been made in that direction, this remains largely true today. As we will argue, two factors have contributed to this issue. The first is a reliance on relatively ill-defined concepts to discuss and assess external validity. The second is the fact that validity of extrapolation cannot be ensured by purely statistical means; it requires an assessment of relevant similarities which necessarily emphasizes, selectively, some features over others—and thereby escapes strictly statistical solutions. A third issue addressed in the paper links to the fact that external validity, generalizability, "transferability," and applicability tend to be used "with overlapping meanings" (Dekkers et al. 2010, p.90), and although to some extent conceptual analysis and definitional rigor might disentangle this situation, we contend that such a laxity in language is the consequence of the fact that these concerns, like the forms of extrapolations they address, are seldom independent.

We begin by characterizing the general form of an extrapolation before using it to disentangle the different threats to its validity. Considering the nature of and relationship between the different pairs of variables characterizing an extrapolation, we show, can guide a clearer and more precise discussion of its validity.

2 The General Form of Extrapolation

Critics of the use of animal models in biomedical research have construed extrapolation of causal claims as a type of inference where the premise "[i]n an animal test, X led to Y" grants the prediction that "X will lead to Y in humans also" (Shanks et al. 2009), where X and Y are assumed identical in both statements. Likewise, talking of "transferring causal generalizations" (Steel 2007, p. 3) or asking whether "the same effect" is present in both the model and the target system seems to suppose an identity of the claim extrapolated from one system to another. In practice, however, the extrapolated claim seldom remains the same when transferred from source to target. For instance, a mouse is never given the same doses of a drug that a human would be given; a surgeon operating in a clinical trial is not the average surgeon; the endpoints of a clinical trial are often not the same as the clinical outcomes sought after in routine health-care (Howick 2011; Howick et al. 2013). The success of the extrapolation will depend on the nature of the source system in which the claim is established, the nature of the target system to which it is to be extrapolated, the relationship between the two systems, and the relationship between the established and the extrapolated claims. Therefore, the general form of an extrapolation should be:

(i) X1 causes Y1 in system P1 (under setting S1), therefore X2 is likely to cause Y2 in system P2 (under setting S2).

The validity of the extrapolation will depend not only on the overall relation between biological systems P1 and P2, as is most often emphasized in the philosophical literature (Ankeny and Leonelli 2011; Schaffner 2001; Weber 2005), but also on the relations between X1 and X2, Y1 and Y2, and S1 and S2. In the example of Figure 10.1, the potential for extrapolation depends not only on the overall anatomical, physiological, or phylogenetic similarity between the animal model and the human targets (the relationship between P1 and P2), but also on how interventions are adjusted in order to produce a measurable effect in humans, such as a higher dosage for humans (the relationship between X1 and X2); on whether cortical spreading depression is a causal determinant of migraine in humans or, in more general terms, how accurately an outcome in the source system matches key attributes of the outcome in the target (the relationship between Y1 and Y2); and on whether differences between lab and routine health-care conditions are likely to affect the outcome by introducing confounding variables (the relationship between S1 and S2).

It is perhaps useful to make the following clarification. In some cases, research relies on a more or less linear, step-wise compounding of distinct

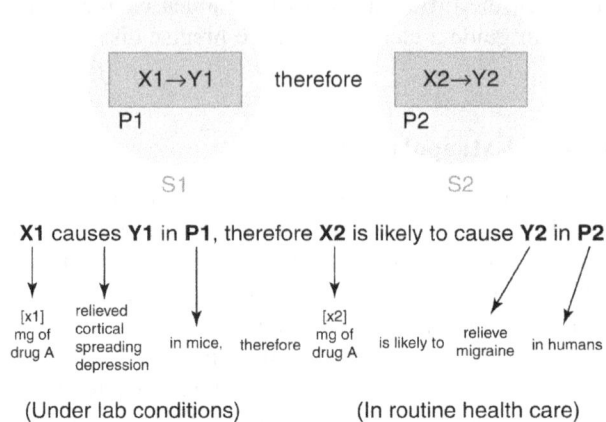

Figure 10.1 Illustrates this structure with an example.

extrapolative inferences. For instance, results obtained in laboratory models are first extrapolated to humans and are then tested in clinical trials before being generalized to the general population and finally applied to specific patients. Thus, one may legitimately take the extrapolation exemplified in Figure 10.1 to be a compounded extrapolation that, for analytical clarity, could be represented as a series of distinct extrapolations. For instance, one may argue that "1 mg of drug A relieves cortical spreading depression (CSD) in mice, therefore 10 mg of drug A is likely to relieve CSD in humans, and therefore 10 mg of drug A is likely to relieve migraine in humans." In other words, cortical spreading depression (Y1) is taken to be an experimentally tractable causal component of migraine (Y2), since the presence or absence of a "migraine" is difficult to assess in the mouse model. There is an obvious practical value in decomposing extrapolative claims in this way, such as troubleshooting faulty extrapolations by raising the degree of similarity between the two outcomes.

At the same time, it is equally important to keep in mind that not all extrapolations can or should be decomposed in this way. To begin with, failing to recognize the extent of similarity between these inferences downplays the importance of assessing, for all forms of extrapolations, relevant similarities between the systems, and it prevents us from transposing epistemological insights and scientific strategies from one form of extrapolation to the other, which can be particularly fruitful given that some features of extrapolation or means of supporting it are more noticeable in some contexts than in others. Second, decomposing extrapolations may obfuscate the fact that similarity between one pair of variables (such as Y1 and Y2) is dependent on the other variables (such as P1 and P2). Consider the following example:

> For example, rearing on the hind legs is a prominent component of stimulant-induced stereotyped behavior in rats, but not in primates, whereas the

reverse is true of scratching (Randrup and Munkvad, 1970). The physical topography of these behaviors is quite different; nevertheless, we are able to say that they are homologous across species.

Willner (1991, p. 14)

We cannot just say that "scratching" (Y1) is equivalent to "rearing on the hind legs" (Y2)—it is not, and for many species there is no relation between the two. However, it so happens that scratching is what primates do when stimulated, and it so happens that rearing on hind legs is what mice do on similar cues. Therefore, you cannot say whether Y1 or Y2 are similar without knowing in what species they obtain. In such cases, it is necessary to consider the problem of extrapolation in its general form.

Nevertheless, the decomposition is useful insofar as it reveals threats to validity, because each pair of variables in (i), namely interventions (X1 and X2), outcomes (Y1 and Y2), populations (P1 and P2), and settings (S1 and S2), brings with it specific issues. In a first, analytical step, we will therefore show how extrapolations and corresponding concerns of validity, commonly conceived as of different kinds, map onto these variables. This framework will provide more robust distinctions than the categories traditionally used to discuss extrapolation and its validity.

3 Forms and Dimensions of Extrapolation

3.1 Generalization

Generalization refers to the extrapolation of a result obtained in a sample of a population to the entire population. Such an inference is generally accomplished by means of statistical analysis and is accepted provided that the size of the sample is sufficient in relation to the variability of the phenomenon. However, such a statistical inference rests on the assumption that the sample is *representative* of the sampled population, in other words, that the distribution of confounding variables is similar between the the control and the test groups, which can be approximated using large, randomized[3] samples. However, this condition is almost never met in biomedical research.

There are a number of reasons why randomization is difficult to achieve. To begin, participation in a trial is dependent on patients' consent (more or less forced by poverty and lack of access to health care), which can represent an immense sample bias in some circumstances. For instance, Fisher and Kalbaugh (2011) noted that some groups are underrepresented in therapeutic trials and overrepresented in "higher risk or lower benefit research" such as phase I trials (where two-thirds of participants are male and nearly two-thirds are non-White). In addition, representativeness often conflicts with another desideratum of experiments, namely the *taming of variability*. Experimental design generally involves major efforts at reducing within-group variability, which in turn increase precision and power but can lead to decisions reducing representativeness. For example, clinical trials generally exclude participants presenting comorbidities (other medical conditions on top of that under study), because these would greatly increase the variability in symptoms or outcomes and would most often prevent true therapeutic effects from being detectable. Likewise, pregnant women, children, or the elderly are all systematically excluded from most trials.[4] While these practices enhance the internal validity of the study by reducing variability and hence improving precision and power, they jeopardize generalization insofar as aging, pregnancy, and diseases do not wait in line to strike us one at a time. Finally, an additional difficulty in sampling (or assessing the quality thereof) is that the nature and boundaries of the population are often unclear (even the "whole human population" changes with time). For instance, given the participants in a trial, one might wish to generalize the results to the population for which it is representative, for instance "non-pregnant young adults" (see Teira, this volume). In practice, however, clinicians have to work with the population they encounter. As some epidemiologists have put it, "The problem is that in clinical practice different doctors may want to apply the same research evidence to different target populations" (Dekkers et al. 2010, p. 90).

The less a sample is statistically representative of the population, the more the extrapolation departs from a generalization, and hence a purely statistical approach will be insufficient. Whenever this happens—and there are reasons to think that, in biomedicine, it nearly always does to some degree—a direct assessment of the relevant similarities and dissimilarities between the sample and the population will be necessary. In this context, so-called generalizations from clinical trials can be considered as between-population extrapolations.

3.2 Extrapolation Across Populations

Extrapolation from a population to a different population is seen very differently depending on whether the two populations belong to the same species. Extrapolation from Caucasian to Asian human populations (whatever these are supposed to be) is deemed much more trustworthy than from mice to humans (Asian or not). The distinction is most commonly made in terms of humans and mice being two distinct kinds or types. However, given that humans and mice belong to a same taxonomic group (mammals), such a distinction has to rest on some ground for preferring one taxonomic unit over another.[5] Although species are not natural kinds in the way electrons are thought to be, they are useful insofar as they instantiate homeostatic property clusters (Boyd 1999), and as such are likely to represent, as groups, some kind of local optimum between size and within-group similarity.

However, not all similarities are equally important, and high "overall" similarity might well hide few but critical dissimilarities. Indeed, for specific purposes, animal models may be a better surrogate for a given patient than most other humans. For instance, for most relevant purposes, a genetically engineered mouse model of cystic fibrosis will be more similar to a human patient suffering from the disease than a healthy human would be. Likewise, a mouse model of cancer harbouring a tumor with the same oncogenic mutations as those of a patient is in many respects a better model than other patients (Chen, Cheng et al. 2012). This is especially critical in the context of molecular medicine, and we will therefore look at these examples in more detail in Section 3.

What matters to the extrapolation is not, therefore, whether or not the two populations belong to the same type or species, but once more whether they are similar in relevant respects—and similarity naturally comes in degrees and flavors.

3.3 Application

Applicability generally refers to a variety of concerns (Howick et al. 2013). The first kind is when the intervention (X1) or the setting (S1) of the experimental/ trial/model system differs significantly from those of the target system (X2 and S2). There are concerns, for instance, that a given trial intervention might not be applicable in routine health care, or that routine practice might not allow the same degree of supervision, care, and nursing support that was possible under the idealized trial setting. We can call this kind of concern *applicability across contexts*, and it is sometimes referred to as the *ecological validity* of a finding.

The second kind of concerns about applicability arise because, as discussed earlier, the trial sample (P1) often differs in relevant respect from the general population (P2). As mentioned, "real" patients often have features that are not represented in trials due to exclusion rules: a typical patient is more likely to be older, with comorbidities, and on multiple medications. In this context, applicability to an individual can be framed as an extrapolation between populations: between the idealized trial sample and either the whole population or a subgroup thereof which, although it was excluded from the sampling, would include an actual patient (e.g., old men, children, pregnant women, diabetic patients). Therefore, we can call this a concern of *applicability across populations*.

To address these issues, *pragmatic trials* have been designed to better represent "real world" practice and patients (Schwartz and Lellouch 1967; Godwin et al. 2003).[6] Pragmatic trials are said to be "less-perfect experiments" than traditional (or explanatory) trials, as they "sacrifice internal validity to achieve generalizability" (Ware and Hamel 2011, p.1686). Indeed, as mentioned earlier, a major drawback of such an approach is that it increases the intragroup variability to a degree that can easily hamper the detection of significant effects and requires larger sample sizes. In addition, pragmatic trials often omit some epistemic features of explanatory trials such as blinding, because they "alter the 'ecology' of care" (Ware and Hamel 2011, p. 1686).

However, even when patients do not present obviously relevant differences to the trial sample, extrapolating from a group to a specific individual is problematic because of unexplained interpatient variability: participants in a trial might respond very differently to a treatment and in unpredictable ways. Biological systems, and especially human patients who have long medical histories and live in a variety of environments, are highly heterogeneous. Because of this variability, even if a claim can be successfully extrapolated from the studied sample to the general population, this does not guarantee that the claim is applicable to a particular individual (P2) within the population (La Caze 2008; 2009; Howick et al. 2013; Khorsan and Crawford 2014). As Mant (1999) puts it, "The paradox of the clinical trial is that it is the best way to assess whether an intervention works, but is arguably the worst way to assess who will benefit from it." (p. 744). To distinguish this concern from the previous ones, we can call its applicability to an individual, and in analogy to mathematics and to highlight the fact that the individual is part of the sampled population, we could tentatively call this form of inference *interpolation*.

In principle, the same issue could arise when inferring, from a sample of interventions (X1), that a specific intervention (X2) should work (all other things being equal). For pragmatic reasons—in general, there is no relevant difference between different batches of the same pills—such issues are not often discussed. Nevertheless, it is not inconceivable that a clinician could, for instance, refrain from prescribing a given treatment on the grounds that the surgeon of the local area is not very apt at administering it.

3.4 Dealing with Heterogeneity: From Single-Subject Trials to Subgroup Analysis

The rationale behind pragmatic trials is to improve generalizability by drawing heterogeneous samples that are more representative of the diversity of the population. While this might yield generalizations that are to some extent satisfactory for public health purposes, it does not address the issue of applicability to a specific individual, which the personalized medicine agenda has made a prime concern. The general strategy for addressing this issue is to find more-or-less homogeneous subpopulations within a population and study them independently. This can be performed with different degrees of granularity but with increasing homogeneity with smaller group size, ultimately leading to fully individualized treatment evaluation.

This individualized approach is implemented in N-of-1 trials, or single-subject cross-over experiments (Langreth and Waldholz 1999; Lillie et al. 2011; Barr et al. 2015). Such a design is especially common for chronic or relatively stable conditions for which different treatments are available and there is substantial heterogeneity in patients' response to treatments, such as chronic pain and psychiatric disorders (similar designs are also widespread in behavioral research). A common example of such procedures is a "sandwich" experimental design, where a condition is measured in the same patient before a treatment is introduced, during the treatment, and after withdrawal of the treatment. An important limitation of this approach is that it assumes that the patient and his/her condition remain stable, while people and diseases progress and are affected by treatment history (e.g., carry-over). This is especially problematic in cancer research, where tumors evolve in response to treatment; or when testing multiple treatment options, as is commonly done for major depression. Therefore, N-of-1 trials have limited internal validity and limited potential for generalizability, and they are restricted in scope to stable conditions (Howick 2011).

Most often, however, the patient population is divided into larger subgroups for both epistemic and economic reasons. Importantly, the identification of these subgroups and the classification of patients according to them necessarily involve placing a selective emphasis on some similarities and dissimilarities over others. The subgroups can be defined beforehand on the basis of groups that often respond differently to treatments (e.g., men/women) or of pathophysiological rationales, such as the expression, in tumors, of the protein targeted by a treatment. Such analyses are generally considered valid provided that hypothesis tests are appropriately corrected[7] and that the study was sufficiently powered (Lagakos 2006). Alternatively, subgroups can be defined *post hoc*, that is, after a trial was performed and on the basis of its results. Like most experimental designs not testing a predefined hypothesis, this is considered a particularly dangerous method easily amenable to data fishing. Moreover, because the relevant subgroups are unknown, a large number of hypotheses are tested more or less formally, yielding to difficulties in assessing the rate of false discoveries (see Footnote 9). For these reasons, results from *post-hoc* subgroup analyses are generally considered exploratory, with promising results being replicated in an appropriately-designed trial. In fact, *post-hoc* subgroup analyses are considered so problematic from an evidential point of view that when the scientists of the ISIS-2 trial were pressed by the editor publishing their results to include subgroup analyses, they insisted on putting on the top of the list the significant correlation between astrological sign and the treatment's effect so as to warn their readers of the dangers of such analyses (ISIS 1988; Horton 2000).

Whether predefined or *post hoc*, subgroup analysis presents important challenges. The more fine grained the groups are (and the more hypotheses are tested), the larger a sample is needed to achieve sufficient statistical power, thereby limiting the widespread applicability of the method. For an evaluation of the statistical limitations of subgroup analysis, see Assman et al. (2000).

Subgroup analysis, or patient stratification more generally, can address the issue of applicability to specific individuals by reducing the variability in treatment responses within each subgroup and finding the subgroup to which the specific patient corresponds, thereby reducing the width of the distribution of probability of response. In principle, the same strategy could be implemented *ad hoc* by assessing whether trial participants more similar to the specific patient to be treated responded better or worse than other trial participants or at the level of trials when data from multiple trials is available). Obviously, such an approach faces the same problems presented by *post-hoc* subgroup analysis. In this context, similarity between patients is evaluated on the basis of pathophysiological rationales (La Caze 2011; Steel 2013). Finally, even in the absence of trial participants sufficiently similar to the patient (as if often the case with comorbidities) or when the data is unavailable, the same pathophysiological rationales, or mechanistic reasoning more generally, can be used to assess whether the patient presents features likely to affect his/her response. Although such a strategy was explicitly supported by the Evidence-Based Medicine (EBM) working group (1992), there is no clear methodology as to how clinicians should use such reasoning to guide their judgment.

The previous sections have highlighted the fact that although extrapolation can be performed and interpreted in very different ways, these different forms of extrapolation and the concerns they raise are seldom entirely distinct or insulated from each other. For instance, so-called "generalizations" seldom are, at least in biomedical research, purely statistical inferences from truly representative samples, and when they are framed as such (for instance when restricting the scope of the extrapolation by redefining the target population), they inevitably give rise to applicability concerns analogous to between-populations extrapolations. As a consequence, statistics itself cannot tell us how to perform "real life" extrapolations, which one way or another require the selection of *relevant* similarities between individuals and/or populations (Ankeny 2001; Burian 1993; Cartwright and Munro 2010; Leonelli 2007). The key question therefore remains how these relevant similarities and dissimilarities are to be identified— a question that molecular medicine purports to address.

4 Extrapolation in Molecular Medicine

4.1 Mechanisms and the Molecular Metric of Similarity

Molecular medicine and the set of *-omics* technologies backing it often purport to provide a complete and unbiased picture of the phenomena (Chen, Mias et al. 2012), raising the possibility of a fundamental, molecular metric for similarity assessment. However, it faces the same traditional problems. First, even if these technologies were providing an exhaustive list of characteristics, there would still be different ways of measuring the similarity between two profiles (e.g., Euclidean distance, correlation, distribution-relative scores, nonparametric measurements). Second and more critically, overall similarity might be misleading when the phenomenon of interest is particularly determined by few variables. Not all similarities are equal, and there is no straightforward way of identifying relevant similarities.

It has often been argued that the most relevant similarities—especially when the extrapolation targets are specific qualitative and quantitative aspects of a phenomenon rather than general approximations—are those concerning *mechanisms* causally productive of the phenomenon of interest (Cartwright 1989; Craver 2007; Schaffner 2001; Steel 2007; Weber 2005; Wimsatt 1976). The general argument is that surrogate models that share mechanistic features with their targets are more likely to generate the phenomena of interest via the same causal pathways and to respond in similar ways when these pathways are disturbed. By contrast, surrogates that do not share causal features might generate similar phenomena by means of different mechanisms and therefore may behave very differently when subjected to similar experimental interventions. This approach is indeed documented in the biomedical practice of animal model validation. For example, in order to successfully extrapolate from an animal model of human disease, empirical evidence must be provided that the model accurately matches specific attributes of human disease (Bolker 2009; Cardiff et al. 2004; Piotrowska 2012). Crucially, these attributes include not only descriptive features of the phenomenon (e.g., symptoms of AIDS in humans as compared with simian AIDS or immunodeficiency in mice), but also known or potentially relevant mechanisms (e.g., AIDS pathogenesis as compared with simian AIDS caused by SIV or immunodeficiency caused by HIV in humanized mice). Molecular medicine can be seen as advancing that view insofar as it offers mechanisms that are (at least in an experimental setting) easy to track, validate, and manipulate, thereby allowing an evaluation of mechanistic similarities.

The notion that models can be validated as surrogate objects of investigation based on mechanistic similarities has been criticized on the grounds that those extrapolations are justified only if there are no relevant disanalogies between surrogate and target. Since the absence of relevant disanalogies can only be ascertained if one already has access to a relatively advanced knowledge of the mechanisms at work in the two systems, extrapolators are trapped in a vicious circle (analogous to the experimenter's regress—see Collins 1981), whereby establishing the suitability of a system as a surrogate model would require already possessing knowledge of the mechanisms at work in the target system, in which case the extrapolation would be unnecessary (Howick et al. 2013; LaFollette and Shanks 1996).

This objection can be circumvented if one takes into consideration that, in practice, researchers consider mechanistic similarities in conjunction with other criteria, such as phylogenetic, genetic, structural, and symptom/phenomenological similarities when assessing the overall suitability of a surrogate model. Typically, the validation of surrogate models relies on a criterion of double similarity—symptom similarity at the level of the phenomenon under investigation and structural/causal similarity at the level of the physical system in which the phenomenon is documented (Cardiff et al. 2004). In addition, model validation strategies often exploit the fact that many biological mechanisms are decomposable in causally modular stages that can be investigated on an independent basis (Steel 2007; Woodward 2002). In turn, modularity makes it possible to adopt a "divide and conquer" strategy, whereby any given surrogate model is judged to be more or less suitable for drawing certain extrapolations depending on which, how many, and to what degree attributes associated with a specific stage or mechanistic module are matched. For example, different stages in pathogenesis are often studied in different surrogate models, each fine-tuned to accurately match a particular stage of pathogenesis.[8]

Although these approaches offer genuinely productive means of assessing the credibility of extrapolations, they are limited in that the support they provide comes from a mosaic mechanism itself drawn through a number of extrapolations across experimental systems. As a consequence, it can never ensure extrapolation, yet it can significantly strengthen it and offer "fallback" methods of troubleshooting it (Baetu 2014, 2015).

4.2 Conflating Extrapolations

Molecular medicine does not change the fundamental issue behind extrapolation. Nevertheless, it—and in particular molecular oncology—disrupts the modalities and categories with which extrapolation and its validity were traditionally discussed. The key reason is that these categories were formed in a context where they held a more or less defined relationship with each other due to the way we used to understand and act upon biomedical phenomena. However, the more a phenomenon, such as cancer (including the therapeutic approaches to it), is dependent on molecular characteristics that vary across patients, the more it becomes possible for models to become, for a specific patient, more relevant than they would be for other patients. Two examples will help illustrate the implications of this with particular salience.

A first interesting example is the "murine co-clinical trial" reported in Chen, Cheng et al. (2012). A clinical trial on KRAS-mutated lung cancer, designed primarily to test the overall efficacy of a treatment, also attempted to identify biomarkers predicting differential treatment response (see Nathan, this volume). As mentioned previously, such subgroup analyses run the risk of identifying spurious correlations. Therefore, as the first candidates came out, bench scientists designed a parallel "co-trial" in genetically engineered mice harboring some of the markers for differential response, thus testing by controlled intervention whether the marker really made a difference to the response (it did). Beside the noteworthy point that this involved a translation from the clinical context to the laboratory, what this example shows is that, insofar as groups are based on molecular characteristics, they are amenable to schemes of extrapolation that bypass the traditional limitations of subgroup analysis and that conflate extrapolations normally thought of as distinct. Indeed, the final finding of the study rests upon interdependent extrapolations ranging from animal, cellular, and clinical studies. In particular, the conclusion relied on two main lines of evidence: a causal claimed established in the mouse model,[9] and a correlation observed in the human trial population.[10]

It is obvious that both inferences, when evaluated on their own, have a high risk of error, and because of this interdependence, they cannot be evaluated individually (Germain 2014a). Together, they yield very confident knowledge regarding the target system, but the fact that they ought to be evaluated together also implies that concerns of generalizability, transferability across species, and applicability cannot be treated independently either.

Xenografting, or xenotransplantation—the transplantation of cells, tissues, or organs across species—offers another interesting example (Maugeri and Blasimme 2011; Germain 2014b). In cancer research, this generally means the transplantation either of pieces of human tumors or of cellular populations derived from them into an immunodeficient mouse host (to avoid the host from mounting an immune reaction to the foreign tissue). Notwithstanding the discussion as to the extent to which transplanted tumors resemble the original human tumor (e.g., Creighton et al. 2003; DeRose et al. 2011), these xenograft models have long been and continue to be widely used in research precisely because they conserve (better than alternative models) genetic and epigenetic characteristics of cancer cells, which are highly relevant for the cancer and its differential response to treatment. For this reason, and given the great heterogeneity between tumors, a xenograft harboring your very tumor can be a better model, for the purpose of predicting the therapeutic effect of some treatment, than most often human patients would be (Hidalgo et al. 2011). Unsurprisingly, there are now companies in the U.S. that propose (for about 15,000 USD) to study drugs on mouse models xenografted with your own tumor.[11] Once more, this example bypasses traditional limitations of personalized treatment evaluation such as N-of-1 trials; because xenografts can be performed in simultaneous replicates, they can have a high internal validity. These models address—in the lab—concerns of applicability to an individual by introducing other validity concerns raised by extrapolation across (host) species. This shows the inaccuracy of a linear picture in which results are first extrapolated from laboratory models to humans and then tested in clinical trials before being generalized to the general population and finally applied to specific patients.

These two examples illustrate the interdependence, in contemporary research practices and in particular in molecular medicine, of some compounded extrapolations and how some threats to the validity of an extrapolation can be addressed by exposing oneself to a different threat. As such, these examples illustrate the analytical value of the general form of extrapolation in exposing and discussing real-life extrapolations. Moreover, they also highlight the necessity, when dealing with some concrete extrapolations, of considering them in their general form to properly evaluate its credibility.

5 Conclusion

Extrapolations are most often complex inferences translating claims across not one but several relevant disanalogies, which often cannot be completely disentangled. We proposed to characterize and discuss these using what we termed the general form of extrapolation, namely:

(i) X1 causes Y1 in system P1 (under setting S1), therefore X2 is likely to cause Y2 in system P2 (under setting S2).

We argued that this general framework allows a more accurate characterization of any given extrapolation than traditional typologies. We also suggested that the nature of and relationship between the different pairs of variables allowed a more precise discussion of the threats to the validity of the extrapolation than traditional categories (generalizability, applicability, etc.). This is especially important in molecular medicine and in contemporary biomedical research more broadly, where traditional distinctions between types of extrapolations and their mapping to issues of validity are regularly conflated.

However, it is important to note that the general form of extrapolation we propose is not an all-encompassing or *a priori* formulae, nor is it entirely determined. For instance, features such as lifestyle, nutrition, and exposure to pollutants could well be understood as properties of a population (P), of the setting in which it dwells (S), or even of the intervention (X) or outcome (Y). The answer to this question will obviously depend on the research context, but it is also changing with scientific developments (see Meloni and Testa 2014) and has clear political meanings (Levins 1998). Similarly, there might come (or have been) a time or there might be actual situations in which characterizing an extrapolation with these variables will be inefficient. The structure of the general form of extrapolation is largely the consequence of how we have come to construe and construct biomedical phenomena. Our proposal, we argue, captures and articulates the most relevant axes of extrapolations in contemporary biomedical research and thereby allows a more precise and productive discussion of the issues it encounters and of the strategies available to address them.

Finally, our analysis of extrapolation highlighted the fact that, barring idealized cases, all forms of extrapolation are based on the assessment of relevant similarities—a problem that keeps resisting philosophical analysis. For although mechanism-based approaches offer productive means of identifying relevant points of similarity, they are themselves the product of an elaborate complex of extrapolations (Baetu 2015).

Notes

1   Extrapolation differs from enumerative induction in that it does not presuppose the homogeneity of tokens across the type concerned and explicitly acknowledges the potential for relevant differences between source and target (Fuller and Flores, 2015).

2   *Validity* is defined as the degree to which a result from a study is likely to be true and free from bias or error (Cochrane Collaboration, www.cochrane.org/glossary/5#letterv).

3   Randomization is not a goal in itself but has the purpose of avoiding biases (Worrall 2007). Random allocation of trial participants to treatment groups will most often result in an even distribution of potential confounding factors across groups. Likewise, random selection of a sample (i.e., trial participants) has the purpose of reducing the risk of the sample diverging from the population in causally relevant respects. In other words, random sampling is an imperfect but highly efficient means of ensuring similarity between the sample and the population.

4   Sweeney et al. (1995) note that some important trials had alarmingly high exclusion rate, sometimes reaching 93 percent. Importantly, similar biases can be found in laboratory studies; for example, mostly in an effort to avoid variations related to the oestrous cycle, animal (and even cellular) studies are massively biased toward males (the NIH has recently changed its policies to address this imbalance—see Clayton and Collins 2014).

5   In relation to this point, it is interesting to note that the 1998 report of the Committee on New and Emerging Models in Biomedical and Behavioral Research draws the line not between humans and other species but between primates and other species: "Studying human health involves two general experimental approaches: examining human or primate cells, tissues, and organs that constitute relatively direct models of human disease; and using a variety of model systems that offer special features and advantages that are not available for study in human beings or primates but can be applied to human health issues." (ILAR 1998, p. 6)

6   Pragmatic trials are designed to evaluate the effectiveness of interventions in real-life routine practice conditions, whereas explanatory trials aim to test whether an intervention works under optimal situations. Of note, there are ethical issues preventing a full representation of the population; to give an obvious example, children and developing fetuses cannot give or withhold consent for trial participation.

7   Frequentist thresholds of statistical significance are based on the frequency with which the threshold will lead us to commit type I errors: a p-value < 0.05 threshold will tend to wrongly reject the null hypothesis 5 percent of the time. When testing a large number of hypotheses (such as testing all genes in the genome for differential expression or testing many subgroups of a sample based on a combination of several parameters), this means that we would expect many "significant" positive results (rejection of the null hypothesis) to arise by chance. Several statistical methods have been proposed to address this issue (see Benjamini and Hochberg 1995 for the most popular method).

8   E.g., AIDS pathogenesis is decomposable in a consistently recognizable series of stages, namely transmission, acute infection, latent stage, relapse, complications associated with secondary infections, and death. Extensive knowledge about HIV infection and AIDS in humans was generated by studying SIV infection in rhesus macaques and HIV infection in humanized rodents; these models are also used to test treatments, such as vaccines and antiviral drugs. Each model is validated in respect to the aspects and degree of similarity with HIV infection/AIDS in humans. Macaques are phylogenetically, genetically, and anatomically closer to humans, thus making them better models for studying the mechanisms of disease transmission, but SIV is different from HIV (e.g., genes specific to HIV; simian AIDS develops within a year of SIV infection, while human AIDS develops 5–10 years after HIV infection). Furthermore, due to their long developmental cycle and maintenance costs, macaques are not ideal for genetic experiments; hence, researchers must cope with the potential confounding effects of genetic diversity and the absence of transgenic/knockout organisms. Humanized rodent models overcome these shortcomings. Rodent models are genetically more uniform, their T-cells can be infected by HIV, antiretroviral drugs affect HIV replication in the same way as in human clinical trials, and the same patterns of drug resistance develop over time; hence, these models are especially useful for predicting clinical antiviral efficacy in humans. At the same time, rodent models are less suitable for studying the mechanisms of viral transmission and AIDS progression due to genetic, anatomical, and life-span differences between rodents and humans.

9   The extrapolation would look like this: Lkb1 mutation (X1) provides resistance to the (mouse version of the) combined treatment (Y1) in mouse models (P1); therefore LKB1 mutation (X2) probably provides resistance to the combined treatment (Y2) in human patients (P2).

10 LKB1 mutation (X1) correlates with resistance to the combined treatment (Y1) in the clinical trial population (P1); therefore LKB1 mutation (X2) would probably provide resistance to the combined treatment (Y2) in the general population (P2).
11 See for instance Champions Oncology (http://championsoncology.com/). As with most of personalized medicine, the business model of Champions Oncology is that it profiles your tumor and records the success of different drugs on it, thereby accumulating a database of drug response patterns that it can then sell to pharmaceutical companies.

References

Ankeny, R. (2001) "Model organisms as models: understanding the 'Lingua Franca' of the human genome project," *Philosophy of Science*, 68-3, S251–S261

Ankeny, R.A. and S. Leonelli (2011) "What's so special about model organisms?" *Studies in History and Philosophy of Science Part A*, 42-2, pp. 313–323. doi:10.1016/j.shpsa.2010.11.039

Assmann, S.F., S.J. Pocock, L.E. Enos, and L.E. Kasten (2000) "Subgroup analysis and other (mis)uses of baseline data in clinical trials," *The Lancet*, 355(9209), pp. 1064–9. doi:10.1016/S0140-6736(00)02039-0

Baetu, T.M. (2014) "Models and the mosaic of scientific knowledge. The case of immunology." *Studies in History and Philosophy of Biological and Biomedical Sciences*, 45 (2014): 49–56. doi:10.1016/j.shpsc.2013.11.003

Baetu, T.M. (2015) "The 'Big Picture': The problem of extrapolation in basic research," *The British Journal for the Philosophy of Science*, advanced access. doi:10.1093/bjps/axv018

Barr, C. et al. (2015) "The PREEMPT study—evaluating smartphone-assisted n-of-1 trials in patients with chronic pain: Study protocol for a randomized controlled trial," *Trials*, 16-1 (2015), pp. 1–11. doi:10.1186/s13063-015-0590-8

Benjamini, Y. and Y. Hochberg (1995). "Controlling the false discovery rate: A practical and powerful approach to multiple testing," *Journal of the Royal Statistical Society. Series B (Methodological)*, 57-1 (1995), pp. 289–300.

Bolker, J.A. (2009) "Exemplary and surrogate models: Two modes of representation in biology," *Perspectives in Biology and Medicine*, 52-4, pp. 485–99. doi:10.1353/pbm.0.0125

Boyd, R. (1999) "Homeostasis, species, and higher taxa." In *Species: New Interdisciplinary Essays*, edited by R. A. Wilson, Cambridge: MIT Press.

Burian, R.M. (1993) "How the choice of experimental organism matters: Epistemological reflections on an aspect of biological practice," *Journal of the History of Biology*, 26-2, pp. 351–67.

Cardiff, R.D. et al. (2004) "Validation: The new challenge for pathology," *Toxicologic Pathology*, 32, pp. 31–9.

Cartwright, N. (1989) *Nature's Capacities and Their Measurement*. Oxford: Clarendon Press. doi:10.1093/0198235070.001.0001.

Cartwright, N. and E. Munro (2010) "The limitations of randomized controlled trials in predicting effectiveness," *Journal of Evaluation in Clinical Practice*, 16-2, pp. 260–266.

Chen, R.R., G.I. Mias, et al. (2012) "Personal omics profiling reveals dynamic molecular and medical phenotypes," *Cell*, 148-6, pp. 1293–307. doi:10.1016/j.cell.2012.02.009

Chen, Z., K. Cheng, et al. (2012) "A murine lung cancer co-clinical trial identifies genetic modifiers of therapeutic response," *Nature,* 483, pp. 1–5. doi:10.1038/nature10937

Clayton, J.A. and F.S. Collins (2014) "NIH to balance sex in cell and animal studies," *Nature*, 509, pp. 282–283.

Collins H.M. (1981) "'Son of seven sexes', the social destruction of a physical phenomenon," *Social Studies of Science*, 11-1, pp. 33–62.

Craver, C.F. (2007) *Explaining the Brain: Mechanisms and the Mosaic Unity of Neuroscience*. Oxford: Clarendon Press.

Creighton, C. et al. (2003) "Profiling of pathway-specific changes in gene expression following growth of human cancer cell lines transplanted into mice," *Genome Biology*, 4-7, R46. doi:10.1186/gb-2003-4-7-r46

Dekkers, O.M., E. von Elm, A. Algra, J.A. Romijn, and J.P. Vandenbroucke (2010) "How to assess the external validity of therapeutic trials: A conceptual approach," *International Journal of Epidemiology*, 39-1, pp. 89–94. doi:10.1093/ije/dyp174

DeRose, Y.S. et al. (2011) "Tumor grafts derived from women with breast cancer authentically reflect tumor pathology, growth, metastasis and disease outcomes," *Nature Medicine*, 17-11, pp. 1514–20. doi:10.1038/nm.2454

Evidence-Based Medicine Working Group (1992) "Evidence-based medicine: A new approach to teaching the practice of medicine," *JAMA*, 268-17, pp. 2420–2425.

Fisher, J.A. and C.A. Kalbaugh (2011) "Challenging assumptions about minority participation in US clinical research," *American Journal of Public Health*, 101-12, pp. 2217–2222. doi:10.2105/AJPH.2011.300279

Fletcher, R.H. and S.W. Fletcher (1996) *Clinical Epidemiology: The Essentials.* 3rd ed. Baltimore: Lippincott Williams and Wilkins.

Fuller, J. and L. Flores (2015) "The risk GP model: The standard model of prediction in medicine." *Studies in History and Philosophy of Biological and Biomedical Sciences*, advanced online publication. doi:10.1016/j.shpsc.2015.06.006.

Germain, P.-L. (2014a) "Living instruments and theoretical terms: Xenografts as measurements in cancer research." In *New Directions in the Philosophy of Science,* edited by M.C. Galavotti, D. Dieks, W.J. Gonzalez, et al., pp. 141–155. Cham: Springer International Publishing. doi:10.1007/978-3-319-04382-1

Germain, P.-L. (2014b) "Humans, animals, and Petri dishes: Biomedical modeling between experimentation and representation." Ph.D. thesis, Università degli Studi di Milano. http://hdl.handle.net/2434/234133

Godwin, M. et al. (2003) "Pragmatic controlled clinical trials in primary care: The struggle between external and internal validity," *BMC Medical Research Methodology*, 3-28. doi:10.1186/1471-2288-3-28

Hidalgo, M. et al. (2011) "A pilot clinical study of treatment guided by personalized tumorgrafts in patients with advanced cancer," *Molecular Cancer Therapeutics*, 0507, pp. 1311–1317. doi:10.1158/1535-7163.MCT-11-0233

Horton, R. (2000) "From star signs to trial guidelines," *The Lancet*, 355-9209, pp. 1033–1034. doi:10.1016/S0140-6736(00)02031-6

Howick, J. (2011) *The Philosophy of Evidence-Based Medicine.* Wiley-Blackwell.

Howick, J., P. Glasziou, and J.K. Aronson (2013) "Problems with using mechanisms to solve the problem of extrapolation," *Theoretical Medicine and Bioethics,* 34, pp. 275–291. doi:10.1007/s11017-013-9266-0

ILAR Committee on New and Emerging Models in Biomedical and Behavioral Research, NRC (1998) *Biomedical Models and Resources: Current Needs and Future Opportunities.* National Academies Press.

ISIS-2 (Second International Study of Infarct Survival) Collaborative Group (1988) "Randomised trial of intravenous streptokinase, oral aspirin, both, or neither among 17,187 cases of suspected acute myocardial infarction: ISIS-2". *The Lancet*, 2, pp. 349–360.

Khorsan, R. and C. Crawford (2014) "External validity and model validity: A conceptual approach for systematic review methodology," *Evidence-Based Complementary and Alternative Medicine*, 694804.

La Caze, A. (2008) "Evidence-based medicine can't be…", *Social Epistemology*, 22-4, pp. 353–370.

La Caze, A. (2009) "Evidence-based medicine must be…", *Journal of Medicine and Philosophy*, 34-5, pp. 509–527.

La Caze, A. (2011) "The role of basic science in evidence-based medicine," *Biology & Philosophy*, 26-1, pp. 81–98.

LaFollette, H. and N. Shanks (1996) *Brute Science. Dilemmas of Animal Experimentation.* London: Routledge.

Lagakos, S.W. (2006) "The challenge of subgroup analyses," *The New England Journal of Medicine*, 254-16, pp. 1667–1669. doi:10.1056/NEJMc061335

Langreth, R. and M. Waldholz (1999) "New era of personalized medicine targeting drugs for each unique genetic profile," *The Oncologist*, 4-5, pp. 426–427.

Leonelli, S. (2007) "What is in a model?" In M. Laubichler and G. Muller (Eds), *Modeling Biology: Structures, Behaviours, Evolution.* Cambridge, MA: MIT Press, pp. 15–36.

Levins, R. (1998) "The internal and external in explanatory theories," *Science as Culture*, 7-4, pp. 557–582.

Lillie, E. et al. (2011) "The n-of-1 Clinical Trial: The ultimate strategy for individualizing medicine?" *Perspectives in Medicine*, 8-2, pp. 161–73.

Mant, D. (1999) "Evidence and primary care: Can randomised trials inform clinical decisions about individual patients?" *The Lancet*, 353, pp. 743–746.

Maugeri, P. and A. Blasimme (2011) "Humanised models of cancer in molecular medicine: The experimental control of disanalogy," *History and Philosophy of the Life Sciences*, 33, pp. 603–622.

Meloni, M. and G. Testa (2014) "Scrutinizing the epigenetics revolution," *BioSocieties*, 9-4, pp. 1–26. doi:10.1057/biosoc.2014.22

Piotrowska, M. (2012) "From humanized mice to human disease: Guiding extrapolation from model to target," *Biology & Philosophy*, 28-3, pp. 439–455. doi:10.1007/s10539-012-9323-5

Rothwell, P. (2005) "External validity of randomised controlled trials: 'To whom do the results of this trial apply?'" *The Lancet*, 365.

Schaffner, K. (2001) "Extrapolation from animal models: Social life, sex, and super models." In *Theory and Method in the Neurosciences*, edited by P. Machamer, R. Grush, and P. McLaughlin. pp. 231–49. Pittsburgh: University of Pittsburgh Press.

Schwartz, D. and J. Lellouch (1967) "Explanatory and pragmatic attitudes in therapeutical trials," *Journal of Chronic Diseases*, 20-8, pp. 637–648.

Shanks, N., R. Greek, and J. Greek (2009) "Are animal models predictive for humans?" *Philosophy, Ethics, and Humanities in Medicine*, 4-2. doi:10.1186/1747-5341-4-2

Steel, D. (2007) *Across the Boundaries: Extrapolation in Biology and Social Science.* Oxford: Oxford University Press.

Steel, D. (2009) *Across The Boundaries: Extrapolation in Biology and Social Science.* New York: Cambridge University Press.

Steel, D. (2013) "Mechanisms and extrapolation in the abortion-crime controversy." In *Mechanism and Causality in Biology and Economics*, edited by C. Hsiang-Ke, C. Szu-Ting, and R. Millstein. Dordrecht: Springer.

Sweeney, K.G., D.P. Gray, R. Steele, and P. Evans (1995) "Use of warfarin in non-rheumatic atrial fibrillation: A commentary from general practice," *British Journal of General Practice*, 45-392, pp. 153–158.

Ware, J.H. and M.B. Hamel (2011) "Pragmatic Trials—Guides to Better Patient Care?" *The New England Journal of Medicine*, 364(18), pp. 1685–1687. doi:10.1056/NEJMp 1415160

Weber, M. (2005) *The Philosophy of Experimental Biology*. Cambridge University Press.

Willner, P. (1991) "Methods for assessing the validity of animal models of human psychopathology," *Animal Models in Psychiatry*, I-18, pp. 1–23.

Wimsatt, W.C. (1976) "Reductive explanation: A functional account," *Boston Studies in the Philosophy of Science*, 30, pp. 671–710.

Woodward, J. (2002) "What is a Mechanism? A counterfactual account," *Philosophy of Science*, 69, S366–S77.

Worrall, J. (2007) "Why there's no cause to randomize," *British Journal for the Philosophy of Science*, 58-3, pp. 451–488. doi:10.1093/bjps/axm024

11 Testing Oncological Treatments in the Era of Personalized Medicine

David Teira

Abstract

Should conventional randomized clinical trials provide the standard of safety and efficacy when testing targeted treatments for cancer? Should we make amendments to our current regulatory standard, stick to it, or dispense with it? I am going to maintain that, under certain circumstances, smaller phase II trials provide good enough grounds to grant regulatory approval for targeted therapies. My argument will hinge on the size of trial population, showing how this size is important not only for scientific considerations, but also for ethical and political reasons. The current system was designed to provide massive consumer protection at a point when our understanding of the biology of cancer was still relatively poor and statistical tests gave the only solid evidence about treatment effects. With targeted therapies, risks are hedged in a way that allows patients (if well informed) to make decisions for themselves, instead of deferring on pharmaceutical regulators.

1 Introduction

For the last five decades, medical treatments have been tested by pharmaceutical regulators with randomized clinical trials (RCTs). These are large comparative experiments in which an experimental therapy is compared with the standard alternative (or a placebo) according to a predefined statistical design. Regulatory agencies, such as the American Food and Drugs Administration (FDA) or the European Medicines Agency (EMA), require two positive (*phase III*) trials as proof of the safety and efficacy of a treatment before patients are granted access to it.[1] Phase III trials are large, often involving thousands of patients. But, as this volume illustrates, molecular medicine is changing our very concepts of disease and cure and, as we are going to defend here, it forces us to rethink the sort of regulatory standard that we expect treatments to meet in order to consider them safe and effective. Let us illustrate it with a recent episode from current research on cancer.

A molecular diagnostic of the genetic aberrations in each individual tumor opens the door for *targeted treatments*: drugs that selectively inhibit the products of these altered genes (Schilsky 2014). There are about a dozen such drugs available (Tursz and Bernards 2015), and many more should come. As Tsimberidou, Ringborg et al. (2013) contend, these molecular diagnostics pave the way to truly personalized treatments: for example, 26 of 32 (81 percent) melanoma patients bearing the V600E BRAF mutation had responded to a treatment based on the BRAF inhibitor vemurafenib; 47 of 82 (57 percent) patients with ALK-rearranged non-small-cell lung cancer responded to the ALK inhibitor crizotinib. However, as the figures in parentheses show, we are often speaking of evidence that comes from very few patients as compared to phase III trials. Oncological treatments have been so far tested, like any other drug, in RCTs in which patients are usually not selected according to their genotypes.

In addition, these phase III trials are not just large, but also *long*; they involve comparing a treatment with a standard alternative, following patients to a pre-designated endpoint after the administration. This is usually a point in time measured from the start of the trial; ideally, it should be *overall survival*—for how long patients who receive the treatment are still alive. In cancer, this end-point is often five years, and reaching it for a large number of patients takes time. Hence, RCTs for conventional cancer treatments are necessarily slow. But cancer patients cannot wait.

Testing targeted therapies thus poses an epistemic dilemma for pharmaceutical regulators: should they stick to large and long trials, when there are so few patients to test these treatments? Or should they decide on the basis of quicker tests? By way of motivation, we can illustrate this dilemma with two stories of success and failure. For some new targeted therapies, the effects observed in early studies led the FDA to grant accelerated regulatory approval. For example, in 2000, gemtuzumab ozogamicin was approved for the treatment of CD33 positive acute myeloid leukemia in first relapse. The evidential basis for this decision was provided by small tests: three trials on 142 subjects with the required mutation (*phase II* studies). However, it was apparently correct (Tsimberidou et al. 2009); more than a decade later, it is still on the market as a treatment for the same condition it was originally approved for.

In contrast, consider gefitinib, a drug initially approved by the FDA for the treatment of advanced non-small-cell lung cancer (after failure of standard chemotherapy). Gefitinib was granted accelerated approval in May 2003 on the basis of the tumor response rate, a surrogate endpoint for clinical efficacy. This endpoint should allow us to predict patients' survival. But in 2005, the FDA withdrew its approval for use in new patients after the completion of a large RCT comparing gefitinib against a placebo on unselected patients with overall survival as its endpoint. It showed no evidence that the former extended patients' life. A subsequent analysis of this randomized study demonstrated that patients with EGFR mutations had higher response rates than patients without EGFR mutations (37.5 percent vs. 2.6 percent). Yet only 26 of the former received gefitinib in the trial (Tsimberidou et al. 2009).

These two stories motivate the questions that I am going to address in this chapter. Should conventional phase III trials provide the standard of safety and efficacy for targeted therapies? Should we make amendments to this standard, stick to it, or dispense with it? I am going to maintain that, under certain circumstances, smaller phase II trials provide good enough grounds to grant regulatory approval for targeted therapies. My argument will hinge on the size of trial population, showing how this size is important not only for scientific considerations but also for ethical and political reasons. The current system was designed to provide massive consumer protection at a point when our understanding of the biology of cancer was still relatively poor and statistical tests gave the only solid evidence about treatment effects. Nowadays, what has shifted is not only medical practice, due to the molecular turn, but also the way in which we deal with the uncertainty of medical treatments. With targeted therapies, risks are hedged in a way that allows patients (if well informed) to make decisions for themselves instead of deferring to pharmaceutical regulators.

In the following section, I will briefly examine how the size of the population contributed to ground our current regulatory consensus. In Section 3, I will discuss how the trial of targeted cancer therapies challenges this consensus, requiring smaller trials. In Section 4, I will present three approaches to this challenge: a "reformist," a "revolutionary," and a "critical" one. In the fifth and final section, I will argue that dealing with smaller well-defined populations provides a good normative ground for impartial regulatory decisions in which patients decide their tolerance to the risks involved in targeted therapies.

2 The Size of the Trial Population: Why Does It Matter?

Clinical trials study the effects of a treatment on a population of patients that may potentially benefit from it (see also Germain and Baetu, this volume). The characteristics of this population of patients are defined in the eligibility criteria that grant admission in the trial. We test the treatment on a random sample drawn from this population under the assumption that if the treatment works on this sample, it will do the same for every other member of the population. This assumption is grounded on the statistical design of the trial from a frequentist standpoint (Teira 2011): we need a sample size that guarantees that there is only a very small chance of observing a statistical fluke, a treatment effect due to the particular characteristics of the patient sample that will not reappear if the trial is further replicated. In the assessment of the methodological quality of a trial, a proper calculation of its *statistical power* to reliably detect a true treatment effect is considered a plus.

For most methodologists, the statistical debate on the trial population ends here.[2] The size of the trial population has only received attention in the context of rare diseases (Tudur Smith, Williamson, and Beresford 2014); when a "life-threatening or chronically debilitating" condition affects about 1 in 2,000 people, we may not find enough patients for a properly powered trial. In these cases, the European Medicines Agency may relax its constraints on the levels of evidence required to grant regulatory approval according to given guidelines. Different randomized designs exist for gathering causal data with a limited number of patients,[3] but from a regulatory standpoint, such designs provide just a second best, the exception to an approval rule based on statistical power.

In sum, our regulatory system is built on the assumption that most treatments will target big enough populations of patients, on which large trials are possible. But how big is "big enough"? In my view, we need to go beyond statistical methodology in order to adequately answer this question.[4] Regulatory standards of proof are not exclusively built upon epistemic principles. In order to deserve regulatory protection, populations should be also big according to financial and political standards, as I am going to show.

On the one hand, the population of potential patients should be big enough to secure the financial viability of the drug development plan. Even if the actual cost of bringing a drug to the market is disputed, the figures are big enough to require substantial sales to make up for the investment; a recent rough informal estimate (by a journalist covering the industry[5]) put at $350 million the cost of launching a single drug over the last decade (2003–2013). With more drug approvals (between eight and thirteen), the cost may reach $5.5 billion. Yet, it is not size alone that affects costs: Treatments for ultrarare diseases may be occasionally lucrative if someone is wealthy enough to pay for them as much as $200,000 per patient per year. Lots of neglected diseases lack a cure because patients do not have the resources to fund their treatments. The philosophical debate on alternative ways of funding biomedical research in order to address neglected diseases shows that, for drug development purposes, a population should properly combine size and purchasing power to be well served by the pharmaceutical industry—see Reiss and Kitcher (2010) for a survey.

As to the political standards that populations of patients should meet, we should recall that pharmaceutical agencies like the EMA or the FDA are regulatory bodies established for one major goal: consumer protection. Such agencies require political support, and the number of required votes depends not only on the number of consumers protected but also on the sympathy they may elicit in their fellow citizens. For instance, Carpenter (2010) has shown that the FDA could only strengthen its power and demand stronger tests when Congress was persuaded that the victims of the 1930s sulfanilamide and the 1960s thalidomide tragedies were *influential* enough to deserve increased protection. Sulfanilamide was an antibacterial compound used to treat streptococcal infection that, in the late 1930s, was marketed in a toxic solution that caused more than 100 deaths in the U.S.. The supporters of granting stronger powers to the FDA framed the scandal in terms of the group of most likeable victims: White, virginal kids, avoiding any mention of the Black, male, and possibly sexually licentious consumers of sulfanilamide. The thalidomide tragedy affected mostly pregnant women, whose babies suffered phocomelia as a result of the ingestion of the sedative. Were it not for such influential groups of patients, the FDA may have not reached the level of regulatory powers it now enjoys.

In sum, when, in the 1960s, the FDA adopted randomized clinical trials as regulatory yardsticks to judge the safety and efficacy of medical treatments, there was an implicit twofold assumption about the populations targeted by such treatments: their members were numerous and politically significant enough as to deserve administrative protection. Hence, size, qualified by political influence, matters in pharmaceutical regulation.

The connection between these methodological and regulatory constraints is deeper than it may initially seem (Teira 2014). Scientists want trials to be *unbiased*, because they seek the truth about a treatment effect, uncontaminated by systematic interferences arising from the preferences of the participants in the experiment. The regulator wants to control for these preferences as well, because she needs the trial to be impartial regarding the interests in conflict about the tested treatments. The producer, its competitors, the medical community, patients, and health-care providers usually do not share their expectations about the outcome of a trial. Yet, they should all agree on the outcome, and the impartiality of the experiment (regarding their conflicting interests) is a prerequisite for their agreement.

For instance, nobody would agree on the outcome of a trial in which the sponsor had fiddled with the trial population in order to achieve a given outcome. If we are looking for a statistically significant difference between two treatments, one way to enlarge this difference is to delete the data of patients on which the treatment did not have a big enough effect. Such a manipulation can be justified with a revision of the eligibility criteria implemented, declaring *ex post* those low-effect patients not eligible for inclusion (Gøtzsche 2013). In order to control for this sort of manipulation, the eligibility criteria should be fully specified *ex ante* and the raw data registered in a publicly accessible database, so that everybody can verify that no patient has been lost for illegitimate reasons. A proper definition of the trial populations is a prerequisite for an unbiased and impartial trial, both in the (public) interest of the experimenter and the regulator.

Summing up, from a methodological standpoint, frequentist clinical trials are scientific experiments that presuppose big populations. If every disease had been rare, that is, a phenomenon of minorities instead of majorities, RCTs might have not provided the gold standard for testing treatments. But precisely because diseases generate big clusters of patients, they are a commercially interesting target, at least in rich developed countries. And these big populations are equally necessary to explain why the commercialization of treatments has been so strongly regulated in democratic countries. The point I am making here is that our current social consensus on RCTs as regulatory devices hinges on the confluence of methodological, commercial, and political approaches to the phenomenon of curing sickness for big numbers of people. With the possibility of individuating disease at the scale of the individual patient, these patients may cluster into significantly smaller groups, for which our consensus on RCTs as regulatory standards may not hold any further.

3 Redefining Cancer Populations

As of today, the size of a trial population ultimately depends on how we define the condition targeted by a treatment, that is, the eligibility criteria to enter a trial, and these depend in turn on our clinical and scientific understanding of such condition. Take, for instance, the current controversy on statins and stroke prevention; depending on how we define the "healthy" levels of cholesterol, the number of patients who may benefit from the treatment may significantly increase—by the millions (González-Moreno, Saborido, and Teira 2015). At the time phase III trials became a regulatory yardstick in the 1960s, our grasp of the biological mechanisms by which a drug cured a disease like cancer was often poor. Hence, the population targeted could be defined according to conflicting criteria, at least in the short run. If the aim was to test a pathobiological hypothesis under strict experimental conditions, for pure research purposes, the eligibility criteria would be more restrictive; if the trial was to test the effectiveness of a given treatment in regular clinical practice, the population would be more heterogeneous. In the former case, we are conducting, for example, basic research on a set of malignant cells without direct clinical implications, since cancer in real patients is a more complex pathology. When we try to find out how to treat the latter, we need to take into account eligibility criteria that capture such complexity. Nonetheless, according to some historians (Keating and Cambrosio 2012), from the 1970s onward the organization of large multicenter cancer trials allowed the gathering of big sample sizes, bringing the two approaches closer. In other words, the trial protocols created "criteria that attempted to generate homogeneous patient populations with regard to a constantly growing number of significant variables concerning response to therapy and the evolution of the disease under study" (Keating and Cambrosio 2012). Different treatment regimens could be thus tested with enough statistical power to detect significant effects. Bigger trials brought about a better understanding of the biology of cancer and a more precise definition of the target populations of its different treatments. From the 1970s onward, the size of the populations targeted in cancer trials met the three requirements stated above: the eligibility criteria were scientifically sound enough; the number of patients targeted was big enough to deserve commercial and regulatory attention.

However, for the last two decades, the genomic revolution in biomedical research has challenged the equilibrium of methodological, commercial, and regulatory considerations about population sizes. The possibility of targeting therapies according to genetic biomarkers involves, first, a change in the very definition of a trial population; we can now clearly define who is a potential participant with a perfect match of clinical and biological criteria. According to the Biomarkers Definitions Working Group (2001), a biomarker is "a characteristic that is objectively measured and evaluated as an indicator of normal biological processes, pathogenic processes, or pharmacologic responses to a therapeutic intervention" (see Nathan, this volume). Following Buyse et al. (2011), we should further distinguish between *prognostic* and *predictive* biomarkers. Whereas the former predict the likely course of disease in a defined clinical population, irrespective of treatment, the latter forecast instead the likely response to treatment. Assuming that a biomarker is properly *validated* from a biological and statistical standpoint, we can use it in a trial to allocate targeted treatments to the patients who may benefit most from them.[6] In a targeted trial design, only biomarker-positive patients are randomized. As Buyse et al. (2011) point out, "such trials have the capacity to confirm the usefulness of the marker in identifying a population in which there is a treatment benefit" (although they provide no information regarding the lack of benefit among marker-negative patients).[7] Granting access to a trial according to the presence or absence of a validated biomarker is as rigorous, if not more so, than any of our current eligibility criteria. The patient population in cancer trials has been usually defined in terms of the organ of origin, the extent of disease, and the previous treatment history. These latter two variables are in principle more open to interpretation, and therefore bias, than a biomarker assay. However, reliability comes at a price for the patient, and this will impact on the trial recruitment process (de Gramont et al. 2015). The amount of tissue needed might make the biopsy more or less difficult to bear to some of them. And it might happen, of course, that the screening reveals that the patient presents a target for which there is no effective therapy. Finally, even a reliable test does not guarantee that even the patients who present a given molecular alteration at the beginning of the treatment will all equally react to the therapy; the heterogeneity of tumors might generate various degrees of resistance in each individual—see Boniolo, this volume.

Leaving aside the ethical issues regarding the sort of consent patients should grant, we should notice how crucial this consent is in order to reach an appropriate sample size. If patient accrual was already a problem in regular cancer trials (fewer than 5 percent of adult patients with cancer participate), the situation gets worse in targeted trials, since only a subsample of the tested patients will turn out to be eligible for the study. Rodon et al. (2012) cite three identification strategies that might alleviate the problem: first, performing a retrospective screening for certain biomarkers once a regular trial is concluded; second, prescreening patients' tissue before they are considered for inclusion in the trial; and finally, screening patients who are receiving standard treatment. In the first strategy, a high number of patients may be put at risk of exposure to a study drug despite not presenting the target of interest. In the other two, their tolerance to a biopsy will be crucial.

In sum, populations become *de facto* smaller, since we are no longer targeting the undifferentiated cancer patient of previous eras. From a methodological standpoint, these smaller populations challenge the possibility of conducting properly powered trials. These smaller trials have though a clear antecedent in current trial design. As noted above, a clinical trial is a research process conducted in four stages (see footnote 1). Targeted therapies only admit a phase II trial, since they run short of patients for a properly powered phase III experiment.

Hence, we face the challenge of assessing the safety and efficacy of targeted therapies on the basis of an experiment that might not conclusively capture the true outcome of the treatment for lack of a large enough sample.

We will set aside the discussion of the commercial implications of the redefinition of cancer brought about by the genomic revolution, since producing targeted treatments according to genetic profiles requires an entirely different financial outlook for the pharmaceutical industry. Suffice it to say that they are willing to invest in it. We will focus instead, for the rest of the paper, on the challenge faced by the regulator: under which conditions are smaller phase II targeted trials acceptable as a proof of efficacy and safety? Do we need them at all? Let us spell out these methodological and normative challenges in more detail.

4 Three Approaches to the Regulation of Targeted Therapies

We are going to distinguish three approaches to the regulatory use of small phase II trials for targeted therapies: reformists, revolutionaries, and critics. *Reformists* accept the current regulatory system but argue that under certain circumstances pharmaceutical agencies can make exceptions and grant market access without phase III trials. *Revolutionaries* advocate for a radical reform of drug regulation in order to exploit the full potential of biomarkers. *Critics* question whether biomarkers have provided so far any evidential grounds for a reform, whether moderate or radical. We will present these three positions, providing our own argument for a moderate reform.

Starting with the reformists, Sharma and Schilsky (2012) argue, for instance, that there is evidence that we can make good regulatory decisions without top-quality evidence. The FDA has approved 31 oncology drugs between 1973 and 2006 without properly randomized trials and with a median number of patients per drug approval of 79. Had this decision been mistaken, the FDA would have later withdrawn the drugs from the market for lack of safety and efficacy. But 29 of these 31 treatments are still on the market.

Sharma and Schilsky (2012) argue that, in a targeted therapy tested on a small but well-validated sample, it is worth foregoing phase III trials if "the response rate and average response duration should indicate a clinically meaningful improvement over that which would be expected based on historical data for the existing standard of care in the same subset of selected patients," provided that these two outcome measures are interpreted in the context of the disease setting and there is no life-threatening safety concern about the therapy. The assumption in this argument (let us call it *assumption a*) is that this meaningful improvement, in the context of a cancer targeted therapy, is more likely to be caused by the action of this therapy on the cellular signaling pathways altered in malignant cells than by mere chance. With conventional therapies, if a small trial detected a large effect, we could not conclusively tell whether it was actually caused by the treatment or by a random coincidence, for example, the particular sample of patients it was tested on. With targeted therapies, our causal understanding of the biology of the tumor allows us to explain why such a large effect has risen in such a small group of patients. In other words, our biological background knowledge becomes as good as the statistical evidence provided by a phase III trial for establishing the safety and efficacy of a treatment.

However, Sharma and Schilsky do not question the normative inspiration of our current regulatory system: the protection of future pharmaceutical consumers is worth the costs of delaying the introduction of new treatments until their safety and efficacy is shown by a standard phase III trial. These costs are mainly the treatment opportunities that current patients lose for not having early access to untested treatments, provided they were willing to take the risks. In this regard, our current pharmaceutical regulation is clearly *paternalistic*; agencies such as the FDA interfere with the liberty and\or autonomy of individual patients, without their consent, for the sake of the patients' health. The social drive behind this normative position seems to be our fear of toxicity scandals, as discussed here; the actual victims of the 1930s sulfanilamide and the 1960s thalidomide tragedies are politically more relevant (in terms of pushing forward a paternalistic regulation) than the patients who suffer because of lack of access to untested treatments that may potentially benefit them (Wardell and Lasagna 1975). Could we arrange our regulatory system in a nonpaternalistic manner that gives these latter patients a chance?

This is an old debate that has been revived in our current controversy on the trial of targeted therapies, where antipaternalist critics are making a comeback, advocating for a revolution in our regulatory system (Stewart, Whitney, and Kurzrock 2010; Stewart and Kurzrock 2013). On the one hand, simulations allow us to estimate how many agents discarded in a standard phase III trial on unselected populations could have been shown effective in small targeted subpopulations. Hence, the old "lack of access" argument is now supported on a solid counterfactual: patients with a given genotype are losing access to treatments they could have benefited from only because these latter are tested on the wrong populations. Phase III trials would be now picking the most common target as winner, not the best drug for each subpopulation; they would more likely detect a very small advance affecting a high proportion of patients than to detect a very large advance affecting a small proportion of them. According to this argument, we would not only be losing effective treatments but we would also be spending more than we should both in running standard phase III trials and in delivering effective care according to the patients' real needs.

This is the position of our revolutionaries: for example, Stewart, Whitney, and Kurzrock argue for a different regulatory system for lethal diseases such as cancer, in which higher levels of risk than for benign and nonlethal diseases are accepted. Unlike in the previous case, the argument is not only methodological (about the evidential grounds for properly testing treatments, drawing on the assumption *a*), but also normative: instead of protecting consumers as if they were an undifferentiated population, we should personalize the protection according to their condition, genetic profile, and risk aversion. This position combines libertarian and individualist intuitions. As to the latter, the protection offered by regulatory agencies should not be judged in principle, but rather by its actual output; we should assess the safety and efficacy of treatments not on average populations but on genetic profiles as close as possible to the actual patient. As to the former, since we are now dealing with individuals, they should have their say on the degree of risk they are willing to tolerate; as I mentioned at the beginning of this section, with their consent we may have smaller and quicker targeted phase I/II trials that might make phase III trials dispensable.

Therefore, the reformist and the revolutionary concur in promoting a different evidential standard for the assessment of at least certain treatments. Where they differ is in the normative goals of regulation; whereas the reformist defends exceptions to a general paternalistic approach, considered on a case-by-case basis, the revolutionary defends a different regulatory approach to a whole class of treatments in which the patients should be protected in an individualized manner.

Before we discuss these two positions, it is useful to consider a critical stance regarding them both. Reformists and revolutionaries accept what I have called *assumption* α. In standard RCTs, we can remain agnostic as to the causal mechanisms behind the tested treatments; we just rely on the statistical power of the test to detect the true difference between the effects of both therapies. The power of a trial to detect what a treatment really does depends crucially on the size of the sample. If the number of patients is not big enough, we may be unable to differentiate a true treatment effect from a random spike that may disappear once the sample size grows. Reformists and revolutionaries assume α instead; our superior understanding of the biological underpinnings of a targeted therapy would allow us to distinguish true effects from random spikes even with small samples. John Ioannidis and various coauthors have been challenging this assumption in a series of papers. In a relevant sample of targeted trials (those reported in highly cited papers), the effect estimates for postulated associations are larger in the trial outcome than in subsequent meta-analyses evaluating the same associations. In other words, the effect disappears as the sample size grows.

According to Ioannidis and his coauthors, this is not a problem just with targeted therapies but with all sorts of very large treatment effects of medical interventions (Pereira, Horwitz, and Ioannidis 2012). They highlight two major points in their analysis. First, these very large effects typically become smaller or lose their statistical significance once additional evidence is obtained, since they usually arise in small trials with few events. According to Ioannidis (2008), biomedical researchers tend to claim discoveries based exclusively on p-values, disconnecting significance from statistical power. But those statistically significant outcomes are difficult to reproduce without the backup of a proper sample size.

Moreover, following still Pereira, Horwitz, and Ioannidis (2012), statistically significant outcomes may not have a clear clinical interpretation. As I mentioned in the introduction, the ideal endpoint of a cancer treatment should be survival, but it takes a lot of time to follow patients for years. This is why we consider alternative endpoints; according to (Pignatti et al. 2015), "there has been a tendency to recognize progression-free survival as a clinical benefit endpoint in itself, leading to standard approvals"—that is, the time during and after the treatment of the disease in which the patient lives with it without getting worse. This surrogate point, though, is always context dependent: what are the expected clinical benefits of delaying progression and how big should they be? And, again, in order to judge this size, we need to a proper sample size.

According to Ioannidis, such large effects (at least for mortality conditions) are "exceedingly rare." Targeted therapies will not be an exception; as Ioannidis and Khoury (2013) put it: "Most of the emerging genomic information that is meandering its way toward health applications is still either non-validated noise or true signals with validated small effects, which are not suitable for applying to clinical practice." Against reformists and revolutionaries, Ioannidis and coauthors argue first that assumption α has yet to receive statistical confirmation in properly powered studies. Without it, small targeted trials do not provide firm enough grounds for unbiased regulatory decisions. Hence, the normative case for a change in our regulatory system rests on purely theoretical conjectures and, as Ioannidis and Khoury argue, we might better improve it by investing in larger conventional RCTs with reliable measures and clinically relevant outcomes.

5 Where Do We Stand?

The three approaches presented here are just a snapshot of an ongoing debate in a field in very quick progress; the sort of biomarkers discussed above is just a first step in the personalization of oncology, clustering patients according to a particular genetic aberration depending on the organ affected (Boniolo, Boem, and Pavelka 2015). But there is not just interpatient variability but also intra-patient, intratumor heterogeneity, and the very pressure of the treatment on the affected cells, making them evolve, may question the value of a single sample to capture the complete genomic landscape of a patient's cancer (Dienstmann, Rodon, and Tabernero 2015). In the words of Donald Berry (2015), "soon every cancer patient will have an ultra-orphan disease." If personalized medicine keeps progressing, treatments will target smaller and smaller populations, forcing us to reconsider our experimental standards for safety and efficacy. Citing again Berry (2015), "large clinical trials in narrowly defined diseases are impossible." For the time being, sample size is being reached through alternative trial designs for which the terminology is still evolving—see, for example, Ocana et al. (2013).[8]

However, even if these designs allow us to increase our samples, they may do it slowly enough to put regulatory agencies in a difficult position regarding the approval of targeted therapies for cancer. When and on what basis should they make their decision? I think that a *reformist approach* to the regulatory use of targeted trials is defensible. On the one hand, I think that the critics are correct from a purely *methodological* standpoint; we need large phase III trials in order to grasp conclusively the true effects of targeted therapies. On the other hand, the revolutionaries are right from a *normative* standpoint; with targeted therapies, there is no need for pharmaceutical paternalism. We can combine these two points in a reformist approach as follows: regulatory agencies should make exceptions and approve targeted therapies on the basis of small phase II trials, with two provisos. First, we need to conduct larger trials in order to validate the decision (*pace* revolutionaries). Second, those targeted therapies should only be administered to patients with the proper biomarkers who are informed about the uncertainty about the treatment (*pace* critics). Is this third way tenable?

Let us first take stock of the discussion so far. Our current consensus on the size of regulatory RCTs is grounded on a combination of scientific, commercial, and normative considerations. As to the former, we started testing cancer treatments at a point, 40 years ago, in which our causal understanding of the disease was often poor and there was a huge element of chance in finding treatment with large effects. Today, our understanding of cancer is solid enough to make assumption α compelling for a great number of cancer scholars. It is not statistically validated yet, since we have not observed a big enough number of large effects in smaller trials that did not vanish in larger ones. So, in this particular regard, it is not wise to rearrange part of our regulatory system on the basis of smaller trials alone. Here I stand with the critics; we still need large phase III trials in order to reach a final conclusion as to the safety and efficacy of treatments.

However, if we agree on the quality of the basic science and on its potential for real pharmaceutical innovation, we need to consider what incentives would make the industry invest in a business as risky as targeted therapies. Investing in drugs for smaller populations will only make sense if the effect is large enough to catch the attention of consumers (and third-party payers), but we are not seeing such effects yet. If we want the industry to invest in targeted therapies, we need indeed quicker trials that might separate winners and losers early on, and the price to pay is, of course, to see some false positives go through. Eventually, some unexpected adverse effects may harm patients.

This is primarily a normative question: are we willing to take the risks involved in the development of targeted therapies? Our current consensus on the undesirability of adverse effects dates from the 1960s, but, in my view, there are enough grounds to revise it, at least when it comes to the regulation of targeted treatments. On the one hand, we are no longer dealing with the protection of big size populations such as the victims of the Thalidomide scandal (potentially any pregnant woman requiring a sedative). As I have argued above, we are now focusing on patients with a given genetic profile identifiable with reasonable precision. And, by definition, these patients cluster in increasingly smaller groups. Targeted therapies will not pose massive public health threats; or, at least, we might identify how big is the mass, according to the DNA profile, and require bigger trials if necessary. From a political standpoint, the risk of a toxicity scandal (even if not always correctly estimated in targeted therapies) is smaller and manageable, at least if we stick to the following two principles.

First, we should only give access to targeted therapies to the patients who have the proper biomarkers to benefit from them. Second, they should provide their informed consent; they should know that our understanding of the disease pathways is good enough to expect them to benefit from the treatment even if we lack conclusive statistical evidence about it.

In this respect, *I think the revolutionaries are correct*: there is room for relaxing our paternalistic approach to pharmaceutical regulation and leave patients to consent either to take part in a trial or receive treatment tested in a small one according to their own degree of risk aversion. However, in exchange for this access, the liabilities that may arise from unanticipated adverse effects should be negotiated in advance. If an adverse effect occurs, patients may not be able to sue the physicians or manufacturers. It will all depend on the terms of the agreement they reached to start the treatment. There are limits, however, to the relaxation of pharmaceutical paternalism. As Dan Carpenter reminds us,[9] in the U.S., "access to medicines and technological advances is an important value, but it's neither a constitutional nor a legal right." As long as this is the case, we are just working within the current scheme of the FDA for accelerated approval; this is just a contract between the FDA and a pharmaceutical company: "In return for promises of further clinical studies, the company receives provisional approval and rapid market access" (Carpenter, Kesselheim, and Joffe 2011) The rights of the patients are created by this contract and, if further studies contradict the initial evidence, the regulator is entitled to withdraw the approval, depriving future patients of the therapy.

Hence, it is possible to strike a reformist compromise between critics and revolutionaries: we still need large trials for methodological reasons, but there are grounds to relax our current regulatory paternalism. However, there can be no compromise about the impartiality of our regulatory trials, be they small or large. We expect regulatory agencies to act in the public interest. That is, they should make decisions on impartial grounds, unprejudiced by the particular interests of the industry, patients, or any other stakeholder in a trial. If we are going to see more and more cases of accelerated approval, can we expect pharmaceutical regulators to preserve their impartiality?

Whereas phase III RCTs should ideally provide conclusive evidence as to the effects of a treatment, small phase II trials with surrogate outcomes provide much less conclusive grounds for a regulatory decision. Such uncertainty can be exploited in the interest of the sponsor, who might be more willing to take risks (for financial reasons) than any other stakeholder in the trial. My previous case applies here: on the one hand, if the patient is equally willing to take his risks, it is just a matter of negotiating liabilities under the supervision of the regulator; on the other hand, accelerated approval is only conditional and should be withdrawn if the initial decision is proven incorrect.

Nonetheless, as a general principle, if large effects in small trials are false positives, we want them to be random events, not fakes engineered by spurious trial design or data analysis. There are many different sources of bias, of course, but targeted trials control a major one: the definition of the population we are dealing with through precise biomarkers. As I mentioned above, a notorious strategy of pharmaceutical disease-mongering has been to fiddle with the trial populations in order to achieve positive outcomes. When the inclusion and exclusion criteria are open to interpretation, the *ex post* elimination of a few patients at the stage of data analysis might bring about a statistically significant result. If we reach a consensus on the proper validation of biomarkers, there will be less room for this particular bias in targeted trials than in conventional ones.

We should open a debate on which other biases might creep in smaller trials, but probably the most contentious issue is their endpoint. A potential compromise in order to make smaller trials more acceptable is perhaps to focus on the hardest clinical outcomes at a first stage until we obtain a better understanding of targeted therapies. These are more attractive in principle for patients and regulators to accept the risk.

To close, I summarize my case as follows. Small phase II trials provide enough grounds for regulatory agencies to grant advanced access to targeted treatments if we observe the following principles: (i) we need to make sure that these trials are impartial; (ii) we should restrict the access to the therapies to those patients who have the proper biomarkers, under informed consent agreements about the possible side effects; and (iii) we need to conduct larger trials to validate the advanced access. Revolutionaries and critics will surely find this compromise objectionable, but for the development of personalized medicine, we need a consensus that somehow brings together the best of both approaches.

Notes

1 Clinical trials are usually divided into four phases: in the first one there are experiments that seek the correct dosage, pharmacokinetics, and so on; in the second one, we find trials with a small number of patients designed as pilots for the third phase trials, on which regulatory agencies ground their approval decisions. In the fourth phase, there is pharmacovigilance of the actual use of the approved treatment in the market. For a quick overview, see Hackshaw (2009).

2 Except, perhaps, for some principled Bayesians who consider trial populations an abstract entity from which we are not actually sampling. The patients in the sample usually share more traits that those explicitly stated in the eligibility criteria: their geographical location, socioeconomic status, and so on. See Urbach (1993) for a quick discussion.

3 For an updated discussion, check out the website of the European Union funded research project "Integrated design and analysis of small population group trials": www.ideal.rwth-aachen.de/ (accessed on September 9th, 2015)

4 See Edwards et al. (1997) for another take on this same problem. We owe this reference and a fruitful discussion of the topic to Cecilia Nardini.

5 Matthew Herper (2013) "How much does pharmaceutical innovation cost? A look at 100 companies," Forbes/Pharma & Healthcare, www.forbes.com/sites/matthewherper/2013/08/11/the-cost-of-inventing-a-new-drug-98-companies-ranked/ (accessed on September 9th, 2015).

6   How to carry out this validation is a controversial topic in itself. We will assume, for the sake of the argument, that we can have properly validated biomarkers. Without them, the *reformist* and *revolutionary* positions we will examine in the next section become, in my view, untenable.

7   Again, we are going to focus on pretreatment biomarkers, measured prior to initiation of therapy, that allow us to estimate the drug efficacy for a particular class of patients. There is an increasing advocacy for posttreatment predictive biomarkers, but we won't discuss it here; see, for example, Stone and Schmitt (2014).

8   A major contender to overcome the dilemma of size is the Bayesian approach to clinical trials; see Berry (2012) for a review and Teira (2011) for a discussion.

9   E. Silverman, "Avastin & FDA were both on trial: Dan explains," *Pharmalot*, June 30th, 2011, available at http://people.hmdc.harvard.edu/~dcarpent/fdaproject/avastin-fda-were-both-on-trial-carpenter-explains.pdf (accessed on September 9th, 2015).

References

Berry, D.A. (2012) "Adaptive clinical trials in oncology," *Nat Rev Clin Oncol* 9 (4), pp. 199–207.

Berry, D.A. (2015) "The Brave New World of clinical cancer research: Adaptive biomarker-driven trials integrating clinical practice with clinical research," *Mol Oncol* 9 (5), pp. 951–9.

Biomarkers Definition Working Group (2001) "Biomarkers and surrogate endpoints: preferred definitions and conceptual framework," *Clin Pharmacol Ther* 69 (3), pp. 89–95.

Boniolo, G., F. Boem, and Z. Pavelka (2015) "Stratification and biomedicine. How philosophy stems from medicine and biotechnology." In *The Future of Scientific Practice: Bio-Techno-Logos*, edited by M. Bertolaso. London: Pickering & Chatto.

Buyse, M., S. Michiels, D.J. Sargent, A. Grothey, A. Matheson, and A. de Gramont (2011) "Integrating biomarkers in clinical trials," *Expert Rev Mol Diagn* 11 (2), pp. 171–82.

Carpenter, D., A.S. Kesselheim, and S. Joffe (2011) "Reputation and precedent in the bevacizumab decision," *N Engl J Med* 365 (2), e3.

Carpenter, D.P. (2010) *Reputation and Power: Organizational Image and Pharmaceutical Regulation at the FDA (Princeton Studies in American Politics)* Princeton: Princeton University Press.

de Gramont, A., S. Watson, L.M. Ellis, J. Rodon, J. Tabernero, and S.R. Hamilton. (2015) "Pragmatic issues in biomarker evaluation for targeted therapies in cancer," *Nat Rev Clin Oncol* 12 (4), pp. 197–212.

Dienstmann, R., J. Rodon, and J. Tabernero (2015) "Optimal design of trials to demonstrate the utility of genomically-guided therapy: Putting precision cancer medicine to the test," *Mol Oncol* 9 (5), pp. 940–50.

Edwards, S.J., R.J. Lilford, D. Braunholtz, and J. Jackson (1997) "Why 'underpowered' trials are not necessarily unethical," *The Lancet* 350 (9080), pp. 804–7.

González-Moreno, M., C. Saborido, and D. Teira (2015) "Disease-mongering through clinical trials," *Studies in History and Philosophy of Biological and Biomedical Sciences* 51, pp. 11–18.

Gøtzsche, P.C. (2013) *Deadly Medicines and Organised Crime: How Big Pharma Has Corrupted Healthcare*. London: Radcliffe Health.

Hackshaw, A.K. (2009) *A Concise Guide to Clinical Trials*. Chichester, UK; Hoboken, NJ: Wiley-Blackwell.

Ioannidis, J.P. (2008) "Why most discovered true associations are inflated," *Epidemiology* 19 (5), pp. 640–8.

Ioannidis, J., and M. Khoury (2013) "Are randomized trials obsolete or more important than ever in the genomic era?" *Genome Medicine* 5 (4), p. 32.

Keating, P., and A. Cambrosio (2012) *Cancer on Trial: Oncology as a New Style of Practice*. Chicago: The University of Chicago Press.

Ocana, A., E. Amir, F. Vera-Badillo, B. Seruga, and I.F. Tannock (2013) "Phase III trials of targeted anticancer therapies: Redesigning the concept," *Clin Cancer Res* 19 (18), pp. 4931–40.

Pereira, T.V., R.I. Horwitz, and J.A. Ioannidis (2012) "Empirical evaluation of very large treatment effects of medical interventions," *JAMA* 308 (16), pp. 1676–1684.

Pignatti, F., B. Jonsson, G. Blumenthal, and R. Justice (2015) "Assessment of benefits and risks in development of targeted therapies for cancer—The view of regulatory authorities," *Mol Oncol* 9 (5), pp. 1034–41.

Reiss, J., and P. Kitcher (2010) "Biomedical research, neglected diseases, and well-ordered science," *THEORIA. An International Journal for Theory, History and Foundations of Science* 24 (3), pp. 263–282.

Rodon, J., C. Saura, R. Dienstmann, A. Vivancos, S. Ramon y Cajal, J. Baselga, and J. Tabernero (2012) "Molecular prescreening to select patient population in early clinical trials," *Nat Rev Clin Oncol* 9 (6), pp. 359–366.

Schilsky, R.L. (2014) "Implementing personalized cancer care," *Nat Rev Clin Oncol* 11 (7), pp. 432–8.

Sharma, M.R., and R.L. Schilsky (2012) "Role of randomized phase III trials in an era of effective targeted therapies," *Nat Rev Clin Oncol* 9 (4), pp. 208–214.

Stewart, D.J., S.N. Whitney, and R. Kurzrock (2010) "Equipoise lost: Ethics, costs, and the regulation of cancer clinical research," *Journal of Clinical Oncology* 28 (17), pp. 2925–35.

Stewart, D., and R. Kurzrock (2013) "Fool's gold, lost treasures, and the randomized clinical trial," *BMC Cancer* 13 (1), p.193.

Stone, A., and N. Schmitt (2014) "Can a treatment be licenced on the basis of post-treatment predictive biomarkers?" *Pharm Stat* 13 (4), pp. 214–21.

Teira, D. (2011) "Frequentist versus Bayesian clinical trials." In F. Gifford (ed), *Philosophy of Medicine*. Amsterdam: Elsevier.

Teira, D. (2014) "On the impartiality of British clinical trials," *Studies in History and Philosophy of Biological and Biomedical Sciences* 44 (3), pp. 412–418.

Tsimberidou, A.M., F. Braiteh, D.J. Stewart, and R. Kurzrock (2009) "Ultimate fate of oncology drugs approved by the US food and drug administration without a randomized trial," *J Clin Oncol* 27 (36), pp. 6243–50.

Tsimberidou A.M, U. Ringborg, and R.L. Schilsky (2013) "Strategies to overcome clinical, regulatory, and financial challenges in the implementation of personalized medicine," *American Society for Clinical Oncology Educational Book*. Asco, pp. 118–125.

Tudur Smith, C., P.R. Williamson, and M.W. Beresford (2014) "Methodology of clinical trials for rare diseases," *Best Pract Res Clin Rheumatol* 28 (2), pp. 247–62.

Tursz, T., and R. Bernards (2015) "Hurdles on the road to personalized medicine," *Mol Oncol* 9 (5), pp. 935–9.

Urbach, P. (1993) "The value of randomization and control in clinical trials," *Statistical science* 12, pp. 1421–31.

Wardell, W.M., and L. Lasagna (1975) *Regulation and Drug Development, Evaluative Studies 21*. Washington: American Enterprise Institute for Public Policy Research.

12 Opportunities and Challenges of Molecular Epidemiology

Federica Russo and Paolo Vineis

Abstract

Epidemiology studies the distribution and variation in exposure and disease in populations. Molecular epidemiology does so by measuring exposure and disease at the deepest biological level. Such move required important changes at the methodological level as well as the conceptual level—notably, by developing the "meeting-in-the-middle" methodology and the concept of "exposome." In this chapter, we discuss how molecular methodology offers an opportunity to reflect upon traditional problems, such as the use of statistical analyses and the interpretation of data, and the role of technology in the scientific process. These, in turn, raise new conceptual and practical challenges, for instance the need to reconceptualize productive causality, and to design public health policies in the light of the results of molecular epidemiology.

1 Introduction: The Rise of Molecular Epidemiology

Epidemiology studies the distribution and variation in exposure (risk factors) and disease in populations. It does so by collecting data via observational studies of different types, most typically questionnaires that gather information about characteristics of individuals and of the environment.

Over the years, epidemiology specialized in several subfields such as environmental, social, or clinical epidemiology. Yet, methods have not changed too much, relying on similar basic designs and statistical models that analyze the relations between "macro variables." Most often, these are "coarse-grained" categorizations of diseases and of types of exposures, for instance air and water pollution, dietary habits and obesity (or rather, body-mass index), behavioral variables, types of infection, and so on.

In particular, "environmental" epidemiology has long established links between environmental factors or hazards and numerous diseases. Rappaport and Smith (2010), however, argued that "traditional" epidemiology has reached a limit and needs to change approach in order to generate new hypotheses and to study different exposure categories *together*, rather than separately. To do so, argue Rappaport and Smith, we have to change our concept of "environment." *Molecular* epidemiology is the way forward, as it contributes to studying the relation between exposure and disease by changing the scale of measurement and, in so doing, to enriching its toolbox with a new concept—the *exposome*—and with a new methodology—the *meeting-in-the-middle* (henceforth, MITM). Another reason for moving on to the molecular scale is that epidemiology needs to improve on the issue of misclassification of exposure and disease variables, and the hope is that we can investigate, in this way, more homogeneous and meaningful categories (Schulte 1993).

It is important to make clear the respects in which molecular epidemiology is different from, or similar to, traditional epidemiology. Wild et al. (2008) trace the history of molecular epidemiology back to Perera and Weinstein (1982), who defined it in the following terms:

> Advanced laboratory methods are used in combination with analytic epidemiology to identify at the biochemical or molecular level specific exogenous and/or host factors that play a role in human cancer causation.[1]

Although the coinage of the term *molecular epidemiology* is relatively recent, Schulte (1993) points out how it is rooted in the long tradition of biologic measurements, which can be traced back to the early days of immunology, bacteriology, pathology, clinical chemistry, and so on.

What changes with molecular epidemiology is the *scale* of measurement: measurement for both exposure and disease (at various stages) is done at the molecular level, not at the level of classes or categories (e.g., types of diseases, types of exposures, types of socioeconomic factors; see also Boniolo, this volume). Schulte emphasizes that molecular epidemiology doesn't represent a radical change in the *principles* of traditional epidemiology. But by changing the scale of measurement we can improve on a number of aspects: group comparison, clarification of disease mechanisms, and more specialized assessment of individual risks. This, continues Schulte, should defuse another worry, namely that molecular epidemiology is trapped in an intrinsic, twofold contradiction. On the one hand, epidemiology is about groups, while "molecular" and "cellular" indicate component levels of individuals; thus, if we study exposure and disease at the molecular level, we miss the specificity of epidemiology, namely its being concerned with *populations*. On the other hand, epidemiology relies on observation and the inference of associations between variables, while molecular biology is largely experimental in character (however, see Section 3.1 for a discussion of how the results of experiments are analyzed with the tools of statistics). According to Schulte, we should see molecular epidemiology as an "evolutionary step" of epidemiology, one that enables us to integrate different approaches and methods and consequently does not aim to replace the established methods of epidemiology but rather to complement them.

Another important addition to the traditional tools of epidemiology is the use of biomarkers in order to identify exposure and development of disease *within the body* (see also Nathan, this volume). Biomarkers are key to understanding the molecular basis of exposure and of disease development because they are the "traces" that exposure to hazards such as air or drinking contaminant leaves in our bodies; yet, as we later discuss in Section 3.2, their meaning and ontological status remains controversial. It is important to note that the idea of measuring the mechanisms of exposure and of disease development at the molecular level using biomarkers can only be implemented by using the appropriate *instruments* for the analysis of biological specimens and of hazards and for the analysis of data thus generated. Consequently, to appreciate the novelties that molecular epidemiology introduces, we must include a thorough examination of the several technological devices—from the "omics technologies"[2] to softwares for data analysis—that make it possible. By changing the scale of measurement, molecular epidemiology improves on traditional epidemiology also in another important respect: it goes beyond associations and is in principle able to shed light on the *mechanisms* of disease causation, rather than just hypothesize them, because it can make appropriate tests (Schulte 1993).

This brief reconstruction of the motivation and methods of molecular epidemiology is complemented, in Section 2, with the presentation of some recent projects in molecular epidemiology. We highlight their conceptual basis—the exposome—and their methodology of MITM. This should give the reader sufficient ground in order to follow the discussion of the opportunities and challenges in Section 3. We explain how molecular epidemiology offers an opportunity to revive debates on traditional topics such as the use of statistics and of the interpretation of data (Section 3.1) and the role of technology (Section 3.2). However, as we hope it will become clear throughout the discussion, molecular epidemiology also offers *novel* elements to these discussions, which in turn give rise to new challenges. We discuss two of them in Section 3.3 and Section 3.4, respectively: the conceptualization of productive causality, and the design of public health policies in response to the results of molecular epidemiology. In the conclusion section we offer some general reflections on the status of molecular epidemiology and on the prospects of the lines of research identified in Section 3.

2 Exposomics Research: The Molecular Roots of Exposure to Environmental Factors

In this section, we briefly describe some recent projects in the field of exposome research. Exposome research has the potential to uncover and explain how diseases develop as a consequence of exposure to some hazards. The prospects of this new area of research have been widely recognized within the scientific community and by international funding bodies. As a result, several projects have already been funded and some are currently in progress. We mention here two of them.

EnviroGenomarkers (2009–2013) was a network funded within the 7th Framework Programme of the European Commission, with eleven partners from six European countries,[3] and it investigated the effects of environmental exposure on various diseases by using omic technologies and biomarkers. The project focused on the role of environmental agents in breast cancer and non-Hodgkin's lymphoma, and in childhood diseases including allergy, neurological and immune diseases, and thyroid disruption. The underlying idea was to measure the effects of environmental agents through the evolution of biomarkers that predict the increased risk of the aforementioned diseases.

EXPOsOMICS—Enhanced Exposure Assessment and Omic Profiling for High Priority Environmental Exposures in Europe—involves twelve partners from seven countries.[4] The project is currently funded within the FP7 and runs between 2012 and 2016. The objective is to study the "internal and external exposome" for exposures such as air and drinking water contaminants in order to predict risks of disease (notably, asthma, cardiovascular disease, cancer, and neurodevelopmental changes) at the individual level. Exposure and risk are assessed at different critical stages in life, which includes *in utero* exposure.

There is a sense in which the emphasis on exposure is a natural follow up after the limited results achieved in genomics research, which aimed at explaining health and disease via inherited genetic features (see Liu, Love and Travisano, this volume). Despite the fact that genomics research managed to produce and collect massive quantities of data, relatively little has been explained about disease causation, using genes *alone*. This suggests that mapping the genome, or using genome-wide association studies (GWASs), is not enough. Although scientists found many gene–disease relationships, comparatively less of them proved to be useful to explain disease. One reason why genomics alone cannot tell the whole story about how diseases develop is that we are missing important information about *exposure*, which is where many diseases start.

Exposome research is based on the concept of *exposome* as developed by Wild (2005, 2009, 2011) and by Rappaport and Smith (2010). This is the idea that exposure does not simply reduce to individuals being in contact with "external" factors such as pollution—this had long been established by environmental epidemiology. Instead, once exposed to certain hazards, our body also becomes an environment. So to study exposure, we need to study *both* the external and the internal exposure. The external component of the exposome is assessed by measuring the levels of chemicals, for instance, of disinfectant products in the water of swimming pools. These chemicals are in contact with our body, and we then can also measure the exposure *inside* the body, which, as just mentioned, is also an environment. This means that biochemical and molecular processes do happen inside the body as a consequence of environmental exposure. The internal component of the exposome may be also assessed, and this is done by repeated measurements in biosamples before and after the exposure, at critical stages in life, and in exposed versus nonexposed groups, looking for biomarkers of changes due to exposure.

These measurements have to be carried out at different "microlevels" in the body. More specifically: metabolomics studies chemical processes involving metabolites (small molecules); adductomics studies DNA or protein adducts, that is, compounds that bind to macromolecules (in the case of binding to DNA causing damages and mutations in the cell); epigenomics studies epigenetic changes in the genetic material of a cell; transcriptomics studies mRNA expression profiling; proteomics studies proteins, especially their structure and functions. Exposure to environmental factors can leave traces at any of these levels, which we try to capture using *biomarkers*.

Identifying biomarkers, however, is not enough. What we need to identify is how signals of exposure evolve, or develop, over time, in a *continuum* from exposure to early clinical stages and then to disease development. Tracing this continuum will allow us to better understand the underlying mechanisms of some diseases and to make better predictions based on known exposures.

The development of biomarkers from exposure to disease requires a methodology that has been called *meeting-in-the-middle* (MITM). This was originally put forward by Vineis and Perera (2007) and then further developed by Chadeau-Hyam et al. (2011) and Vineis et al. (2013). The MITM is based on a multilayer causal framework. Its implementation aims at addressing the challenge of identifying causal relationships that link exposures and diseases. This approach is usually based on a combination, within a population study, of a prospective search for intermediate biomarkers—which are changed in subjects who eventually develop disease—and a retrospective search for links between such biomarkers and past environmental exposures. The reasoning underlying this approach comes in three steps. The first step consists in the investigation of the association between exposure and disease. The next step consists in the study of the relationship between (biomarkers of) exposure and intermediate omics biomarkers of early effects; third, the relation between the disease and intermediate biomarkers is assessed. The MITM stipulates that the causal nature of an association is reinforced if it is found in all three steps.

In a recent study, Fasanelli et al. (2015) present the results of an epigenome-wide study of DNA hypomethylation[5] and its association with tobacco exposure. DNA methylation has been shown to be an important marker of current and past smoking habits. While the association between tobacco and numerous diseases has been extensively demonstrated, what is not yet known is whether methylation levels eventually result in an increased risk of smoking-related cancers. The study compares analyses from two sets: a discovery set with data coming from the Norwegian Women and Cancer (NOWAC) longitudinal cohort, and a validation set with data coming from two prospective cohorts (the MCCS and NSHDS). The comparison of the results of the analyses in these two sets allowed scientists to conclude that the smoking-induced hypomethylation in specific genes is indeed associated with an increased risk of lung cancer, thus suggesting that methylation alterations mediate the carcinogenic effect of tobacco. More specifically, in a series of epigenome-wide studies of DNA from prediagnostic blood samples, the authors observed that the most robust associations with lung cancer risk were for cg05575921 in AHRR and cg03636183 in F2RL3,[6] previously shown to be strongly hypomethylated in smokers. These associations remained significant after adjustment for smoking and were confirmed in statistical mediation analyses, suggesting that residual confounding is unlikely to explain the observed associations and that hypomethylation of these CpG sites may mediate the effect of tobacco on lung cancer risk, that is, be a genuine MITM marker.

Studies like this, besides identifying important biomarkers in the continuum from exposure to disease, contribute to further illuminating the molecular roots of disease mechanisms.

In these projects, massive data sets are produced. But, once data is generated, it has to be analyzed and interpreted. This is far from being an obvious task, as it doesn't solely rely on expertise in statistics or the computational capacity of statistic software. This is research that lies right at the border of what molecular biology knows about disease mechanisms and what it still does *not* know. We witness therefore a fascinating *va et vient* between discovery (or exploration) and confirmation. Studies are simultaneously confirmatory and exploratory, walking that thin line where we need solid knowledge to set up further research *and* we need to grope around in order to find what we are looking for. So the stake is high as there is a lot that we will understand from this research, but it might take quite some time. The possible impact on public health policy is also potentially significant but no less problematic. Projects like EXPOsOMICS or EnviroGenomarkers involve several institutions and researchers having very heterogeneous expertise, from molecular biology to statistics to computer science. The European Union (EU) is also pushing toward involving small and medium enterprises (SMEs) in fundamental research. This poses important questions about research and innovation or about the relation between science and technology. Some of these questions are political in character, in that they concern the vision of research that funding bodies endorse and promote; and some others are more epistemological as they concern important concepts, such as reality, observability, or knowledge. Some of these issues will be further discussed in Section 3, while for others we point to new lines of research in the conclusion section.

3 Opportunities and Challenges

So far we have presented molecular epidemiology as a relatively new field within the multifaceted realm of epidemiology and biomedicine more broadly, and we have highlighted how it introduces novelties both in the methods and in the concepts used to study the relation between exposure and disease. No wonder, therefore, that these methodological and conceptual changes also call for philosophical reflection. In the following, we discuss only a selection of issues. We begin by discussing the use of statistics and data interpretation (Section 3.1) and the role of technology (Section 3.2). These issues are not new *per se*, but we aim to show how molecular epidemiology offers an opportunity to reconsider them. We also hope to show that once we rethink these traditional issues, other challenges emerge, namely the conceptualization of productive causality and the design of public health policies. This selection is, admittedly, largely influenced by the background and research interests of the authors. In particular, it is largely influenced by the belief that a timely philosophy (in the sense of Floridi 2011) arises from a deep engagement with the scientific practice. Needless to say, this selection of topics does not intend to lessen other possible issues, as we discuss in the concluding section.

3.1 Statistical Analyses and Data Interpretation in Omics Research

Molecular epidemiologists set the search for causal links between exposure and disease—not just correlations—as an explicit objective. Statistical analyses belong to the traditional methodological toolkit of epidemiology, but a large part of omics research consists of the data mining of big data sets (see Boem, Ratti, this volume). Finding correlations is essential to the exploratory *and* confirmatory moments of omics research. However, *pace* Anderson (2008), correlations will not supersede causation, and we are far from being at the "end of theory." The exploratory and confirmatory aspects of omics research are worth explaining further. In what follows, we aim to make clear the sense in which molecular epidemiology is not merely agnostic. Instead, the formulation of prior hypotheses and background knowledge still play an important role in the analysis and interpretation of data.

There are two steps in the study of associations: one is agnostic, meaning that associations are searched without prior hypotheses; the other, instead, is guided by prior hypotheses, for instance biologically informed (e.g., through prior experimental knowledge) pathways affected by a certain exposure and/or involved in the disease development. For example, oxidative damage to DNA is a typically investigated pathway. It is worth noting that there has been quite a radical change driven by technology (see also later in Section 3.2). Omics research can be thus characterized as "technology-driven" besides "data-driven." Data-driven research started already with GWASs, where researchers looked at hundreds of thousands/millions of genetic variants (single nucleotide polymorphisms). In studies like these, there is literally no prior hypothesis, in the sense that all the genetic variants are investigated, and prior knowledge is usually not used to filter out some *a priori* (ir)relevant variants, so everything is *a posteriori*. However, the *interpretation* of data is clearly done on the basis of existing knowledge; for instance, if one observes that the variant of a certain gene influences the risk of heart arrhythmia, one looks at the function of the gene and whether the association has biological plausibility (considering that the confounding problem in the case of genetic associations is less problematic than with environmental exposures).

There has been a shift because epidemiology has always insisted on the formulation of prior hypotheses, and on the importance of the study design, which should be guided by prior hypotheses—examples in epidemiology textbooks testify to this. This hypothesis-driven philosophy has been largely abandoned with omics epidemiology, which places research in an exploratory context, seeking for the identification of novel, unreported findings from highly complex and large data sets; omics research becomes significantly technology-driven (see also later in Section 3.2) and can be viewed as a "fishing" exercise. Much of current research is done, for example, with metabolomics and epigenetics: the researcher generates a large amount of data and then *a posteriori* looks at their meaning. While these investigations do not necessarily call upon prior knowledge, it is interesting to note that the understanding and, ultimately, the validation of these novel results cannot be achieved without integrating prior knowledge to establish their biological relevance. Prior knowledge is also important in order to reduce computational burden and to ensure both the feasibility and interpretability of statistical analyses, and Bayesian approaches are typically used for that purpose.

Different kinds of data require different statistical approaches, depending, for example, on the type, size, and complexity of the data. There is also a hierarchy of approaches for a given type of data. In an exploratory approach, one usually begins with the univariate analysis and then checks whether in the data there is evidence for potential combined/cluster effects; sets of omic signals may be more strongly associated with exposures or disease risk than the sum of the single signals. In the latter case (potentially identified through the existence of correlations), one uses multivariate models, which include dimension reduction techniques, for instance, principal components analysis (PCA), which builds upon the correlation across omic signals to summarize the data without losing information; such models also make data more tractable and define homogeneous pools of informative signals.

The typical example of the search for links between signals and relevant outcomes is metabolomics, where blood samples are chemically analyzed through mass spectrometry or nuclear magnetic resonance, and thousands of signals are derived. The first step is a sort of data curation and is called preprocessing. It means that one looks at out-of-range data points, which are clearly outliers; then, in the case of mass spectrometry, one removes the peaks, which are due to those molecules that we are not interested in but that are nonetheless abundant; then one looks at those variables that influence the quality of data, called nuisance parameters. To be sure, there is a whole series of preprocessing steps, not just one. For instance, metabolomics, or other techniques, may be influenced by the temporal period in which the measurement has been carried out, or slight differences in the reagents used (the batch effects). This is quite evident in the case of epigenetics analyses. Imagine we have 500 samples at our disposal, but we can analyze only 100 of them at any given time (due to technical constraints). We therefore divide them into five batches but, nearly always, we will find some degree of discrepancy between one batch and another: if we use the same sample for five different batches, we might obtain five slightly different results.

Preprocessing is a way to curate the data in order to adjust, for example, for the batch effect, one of the nuisance variables. There are some solutions to calibrate the data across batches, including the use of quality control samples that are run in each batch. When preprocessing the data, measurements are slightly transformed to ensure that the same values from the quality control sample (which is the same sample across batches) are consistently found across batches. Further steps are related to the interpretation of the findings. For example, in metabolomics, we use databases that are key to annotation. This consists in interpreting the spectrometry peaks, that is, identifying which signal corresponds to which molecule. For instance, suppose we find ten peaks that are statistically associated with colon cancer in our observations; those ten are usually the ones that have withstood all the statistical tests (including correction for multiple comparisons) and the pre-processing procedures. At this point, we try to understand what those ten peaks mean in chemical terms (mass spectrometry only tells us which are the peaks, not which molecules they are signals of). In order to discover what these molecules are, we can look at databases; this phase is called *annotation*. If we are lucky enough, some of those peaks can be found in the databases (based on their mass and other chemicophysical properties like retention time); this is usually not the case, so annotation requires other work at the chemical level in order to identify specific molecules.

There has been a huge increase in what is called "throughput": this is the possibility of quickly analyzing thousands of samples. For instance, with GWASs, thousands of samples for millions DNA variants can be analyzed in weeks. So there has been a significant increase in the analytical power, but while tremendous improvements in the computational power have been achieved in the past decade, in-depth statistical analyses and, more importantly, biological interpretations, together with full exploitation of the results, still remain suboptimal.

One of the reasons why apologetic discussions of big data theories *à la* Anderson (mentioned earlier) do not work is because by looking at correlations *only*, we cannot disentangle the interfering causal chains. Recently, this problem has attracted a lot of attention in epidemiology through the theory of "colliders" (Porta et al. 2015). The theory has led to deconstructing paradoxes, for example involving body weight, such as the putative benefits of obesity in diabetics. Although obesity is a well-established risk factor for type 2 diabetes, among those people who have already developed diabetes, the obese ones are often observed to live longer than those having normal weight. However, the observation lacks causal significance because of "conditioning on a collider"; in fact, focusing only on individuals who developed diabetes creates a spurious association between obesity and survival. A collider is a variable directly affected by two or more other variables in a causal chain; in our example, diabetes is a collider since it is affected by obesity and, independently, by an unknown variable (e.g., some environmental chemicals or genetic variants). Controlling (conditioning) on diabetes will artificially create a noncausal association—referred to as *collider bias*—between the shared causes of the collider. The theory is supported by a new and rapidly developing branch of causality based on graphic representations, the directed acyclic graphs (DAGs) (Porta et al. 2015). This simplified example shows that a simple computational (i.e., correlational) approach to making sense of omic data is insufficient and that prior knowledge is required to disentangle plausible causal chains.

3.2 The Role of Technology

In exposome research, technology is present at all stages, from data generation to data analysis. Its prominence should license the appellative *technoscience* rather than just "science" (of exposure), a concept widely used in social studies of science, much less in philosophy of science, and hardly at all in science itself. This is clearly not unique to molecular epidemiology, as this is the case in much of particle physics and, one might argue, in the social sciences too. Even if molecular epidemiology is not a special case in this respect, it is nonetheless an excellent occasion to revive old discussions about the relations between science and technology and to point toward new directions for the debate.

Setting linguistic and terminological elucidations aside, it is important to clarify what "technoscience" implies in practice. In technoscientific practices, the border between science and technology—if such a thing has ever existed (see, e.g., Mayr 1976)—is getting more and more blurred until it disappears. This is not the place to engage in a detailed discussion on nature or status of technoscience. We refer the interested reader to some relevant contributions, with respect to its historical roots in the work of French epistemologist Bachelard (on which, see Rheinberger 2005), to its hybrid status between pure and applied sciences (Boon 2011), to its role in scientific experiments and theory (Hacking 1983; Radder 2003) or in its relations to science policy (Bensaude-Vincent 2009; Guchet 2011).

We will instead be interested in some of the conceptual consequences of using "technoscience" rather than just "science." In order to fully appreciate the deep conceptual change that this brings about, we must start with identifying the various technologies used in exposomics research and spell out their use and role in the whole scientific process. Exposome research takes advantage mainly, but not exclusively, of the following technologies: omic technologies, sensor technologies in smartphones, imaging technologies, electronic diaries, and statistics softwares.

Omic technologies stand for a whole range of technological devices used to analyze human biological specimens such as blood or urine at the micro level. These technologies are used in order to characterize the internal exposome, namely what happens inside the body in response to exposure to environmental factors and any other (internal) process that such exposure may trigger. These analyses allow scientists to collect quantitative data on global biological response to environmental agents, as explained earlier in Section 3.1. Measurements are performed at different levels, for instance at the cellular level, multiple external chemicals (metabolites) or profiles of normal biomolecules as they are configured under the influence of external environmental factors.

The goal of omics research is to identify biomarkers of the effects of environmental exposure at the different micro levels in the body (which we briefly explained in Section 2). Omic technologies, however, are not just a type of technology but are themselves a family of technological tools including (but not limited to) high-resolution analytical platforms such as liquid chromatography coupled with mass spectrometry and/or nuclear magnetic resonance spectroscopy. Simply put, liquid chromatography is a chemistry technique to detect the presence of chemicals in other chemicals. Mass spectrometry allows scientists to measure the mass-to-charge ratio in charged particles, often exploiting processes such as the ionization of energy in molecules. Nuclear magnetic resonance spectroscopy also allows the detection of physical and chemical properties of atoms and molecules, exploiting the magnetic properties of atomic nuclei.

A peculiarity of the latest projects in the "European exposome initiative" is the use of "personal devices" to generate data about environmental exposure and also about individual life styles and habits. Devices that can record, say, air pollution have been developed. Such devices are connected to smartphones, so that transfer of the records to the databases of scientists is afterwards made possible. GPSs and motion sensors record location of users, the type of physical activity, and accurate estimates of rate of inhalation. Other applications are currently under study. These technologies are already available, and participation in the FP7/H2020 projects is also aimed at enhancing business possibilities of SMEs.

Omics, sensors, and smartphones produce very large data sets containing potentially useful information about the external and internal exposome. But such information has to be "extracted" from the data sets. Statistics and, more specifically, statistics software, are an obvious choice to analyze large data sets. But standard statistical tools, such as those used by "traditional" epidemiology, are insufficient for this task. This is because high-throughput techniques (i.e., the omics and sensors described above) generate high-dimensional data, for which new algorithms for analysis have to be developed. Such sophisticated softwares allow calibrating measurements, i.e. "polish the data," from possible measurement errors (see also earlier in Section 3.1). They also seek the best combination of biomarkers that predict exposure by combining data from retrospective and prospective studies (see the MITM methodology presented in Section 2). These softwares can handle hundreds of thousands of predictors. They check potential correlations within the data that would therefore hinder any (possible) causal conclusion. Statistical analyses of the untargeted biomarkers are used to discover and validate biomarkers that are then investigated in more targeted approaches. Cross-omics analyses are also performed in order to investigate common patterns. It is worth noting that all this requires excellent skills in statistics or informatics, as deep understanding of the biology of these processes is likewise required both to design the algorithms *and* to interpret the results. A particularly delicate problem is that, usually, the number of variables (recording signals of, e.g., many thousands of metabolites in metabolomics or a million probes in methylation epigenetics) is much larger than the numbers of subjects investigated. This goes against not only common sense but also against the principles regimenting the tools developed in classical statistics, calibrated for investigations in which many subjects were used to study a relatively small number of variables. Also, omics data usually have a very complex structure with substantial internal correlations.

This brief overview should make clear that technology is essential to exposome research. However, this should not be read as an argument for the priority of technology over "pure science," whatever that means. We contend that establishing whether technology or "pure science" comes first is not the interesting issue at stake. In fact, while one might argue that without technology there would be no exposome research, one might rebut that any of the technological devices just mentioned could not be thought of, designed, built, or even used without the great advancements made in chemistry, biology, particle physics, or informatics—a point that has been convincingly made, using different case studies, by Hacking (1983) or Radder (2003), for example.

Instead, the fact that technology is so deeply embedded in science *and vice versa* should lead us to rethink some traditional concepts, from "reality" to "observable" to "knowledge." It goes far beyond the scope of the present contribution to discuss all these concepts in detail. We will content ourselves with just pointing to some of the reasons why "reality," "observable," or "knowledge" may need reconceptualization. The debate on scientific realism is quite useful to illustrate the point.

In philosophy of science, the debate on realism has been by and large polarized around the question of whether we can legitimately infer the objective existence of nonobservable and *human-independent* entities (say, electrons), given the current status of our best experiments and theories, mainly in physics (for an introduction to the topic, see Chakravartty 2015). Transposed onto the domains of molecular biology, a relevant question would concern the objective existence of genes, various molecules, and, of course, biomarkers. The role of technology has been—in these debates—largely neglected, with the exception of authors such as Hacking (1983) or Ihde (1991), who problematized the *instrumental* role of technology. Technological devices, for instance microscopes, have been used to support a form of *instrumental* realism, that is, to provide reasons to believe in the objective existence of entities that we cannot see by the naked eye, even when measurements and observations are *mediated* through them. Technology backs up a realist position (i.e., that unobservable entities are real) in a rather sophisticated way. Hacking, in particular, also emphasized the kind of *expertise* needed in order to see through a microscope, as this requires a fair amount of knowledge at the theoretical level (say, in biology or chemistry or other) as well as of a practice. So if we "see" small things such as viruses and bacteria, it is not simply because the instrument makes our sensory apparatus more powerful, but also because we have the right knowledge, expertise, and skills to operate it. This is also known as the Sneed–Stegmüller thesis of "entrance knowledge" (see Sneed 1979 and Stegmüller 1976): for example, the interpretation of an x-ray by a radiologist is made possible by a long training that includes notions of anatomy, physiology, radiation physics, and so on; in other words, it is a "thick" description that requires entrance knowledge (once again, *pace* positions à la Anderson; some of these issues are also addressed by Boem, Ratti, this volume).

The role of technology in issues related to scientific realism has been discussed from a different angle in the work of Radder (2003, 2012). In particular, Radder (2012) develops quite a sophisticated view on scientific realism, which he calls *referential realism*. Two issues, in particular, are relevant to our discussion. On the one hand, Radder expresses skepticism about the possibility of establishing "independence claims" (i.e., claims about the objective and human-independent existence of unobservable entities). This is because most of the phenomena on which such claims are based do not occur "naturally" but are artificially created in the lab using machines and are in need of much theoretical interpretation. On the other hand, the kind of scientific realism of "mainstream" philosophy of science aimed to "locate" or identify unobservable entities by "carving nature at its joints." Yet, as Radder notices, the location (and the nature) of the joints also depends on the carving *tools*. Considerations like these lend support to the idea that thorough reconceptualizations of reality, or knowledge, should be put (back) on our research agenda.

Let us see how these considerations apply to the case of molecular epidemiology. To begin with, omics technologies do much more than just allow us to see things that are too small or too big to be seen with the naked eye. To be sure, the biomarkers that we try to identify using these machines are not simply out there as minuscule entities in our bodies. Differently put, even a cursory analysis of the role and use of technology in molecular epidemiology should instill the doubt that any naïve form of realism becomes quickly untenable. So what is a biomarker? Is it a real, tiny, human-independent entity to be found in our biological specimen? Witness, for instance, Schulte (1993, pp. 14–15):

> Biologic markers (or biomarkers) generally include biochemical, molecular, genetic, immunologic, or physiologic signals of events in biologic systems. [...] When considering how biologic markers can be used in epidemiologic research, it is useful to reflect on the nature of the relationship between the marker and the event it marks [...] Is a marker different from a test or an assay? Often, biologic markers and tests or assays for a marker are considered the same because, without the assay, the marker cannot be demonstrated. Strictly speaking, they are different and care should be taken not to gloss over the differences. Analytically, a test or assay is said to be valid if it performs "truthfully" in the presence or absence of a marker. The attributes of a test are not necessarily those of the marker. The marker's attributes may pertain to its nature and natural history. A marker may exist although no assay is sensitive enough to detect it. The measurements of a marker involves differentiation of a signal above background noise.

Biomarkers, rather than being objective out there, are largely *constructed* by cross-checking data that are generated by some machines and subsequently analyzed using other machines. So what kind of ontological status should we give to biomarkers? Schulte (1993, pp. 14–15) describes biomarkers in terms of "events" in the continuum from exposure to disease and then warns the reader that, however, we might still ask different questions: whether a marker represents an event, whether it is an event itself that is correlated with the event, or whether it is a predictor of the event. Schulte (1993, p. 15) then concludes:

> The answers to these questions may affect who is sampled, how and when they are sampled, and what confounders or effect modifiers are considered. [...] Thus, a biologic marker often refers to the use made of a piece of biologic information rather than to a specific type of information.

The question about the ontological status of biomarkers gets even more complex if we consider that molecular epidemiology is not interested in finding biomarkers *per se* but in understanding the continuum of disease development from early exposures *via* biomarkers. Here, there seems to be an emphasis on *processes* rather than entities, which again may shift the meaning of "reality" from something essentially made of "things" to something that is instead made of "processes" (for a discussion, see Floridi 2008).

So, if the interesting role of technology is not to see "the small" or "the big," what can it be? Technologies make it possible to generate data, then to curate them, and finally to analyze them. And once analyzed, we interpret the results and organize them in a way to express our understanding of the exposome. Technology has therefore a *poietic* character, namely it enables the *production* of data first and of knowledge later. How much these technological devices are themselves imbued with "pure science" is beyond the point and in fact misses the interesting thing at stake. Any of these technoscientific objects allows us to do something we couldn't do otherwise, namely to generate data and knowledge. But once we grant technoscience such a poietic character, we have to be simultaneously ready to revise not only the notions of observations and of reality—as just mentioned —but also most likely many others, for instance "process," "entity," or "experiment," as suggested in the chapter by Boem and Ratti. In the next section, we will investigate the prospects of revising the notion of productive causality.

3.3 The Conceptualization of Productive Causality

As argued earlier in Section 3.1, all research—including omics "agnostic" research—is in the end hypothesis-driven, at least in the phase of data interpretation. The projects mentioned previously in Section 2, while being largely data-driven, are not totally conceptually agnostic. Their "gnostic" component is twofold. On the one hand, molecular epidemiology needs support from available biological theories in the search for and validation of biomarkers at given molecular levels. On the other hand, as Russo and Williamson (2012) point out, projects like these are simultaneously exploratory and confirmatory. Both these tasks—exploration and confirmation—concern *causal links* between exposure and disease, and consequently molecular epidemiology needs an appropriate conceptualization of causation.

The question of how molecular epidemiology establishes causal claims about exposure and disease can also be reframed in terms of the *evidence* needed in order to support said claims. Russo and Williamson (2012) point out that in carrying out exploratory and confirmatory tasks, molecular epidemiologists look at several sources of evidence: evidence *that* exposure leads to disease development and evidence about *how* this happens. Statistical associations, as described earlier in Section 3.1, provide evidence *that* exposure (possibly) leads to disease. Scientists need to establish chains of *difference-making*: whether biomarkers are good predictors of disease, at different stages of disease development, or at different stages of life, and so on. At the same time, we also look for evidence about *how* exposure leads to developing disease. Typically, "how" exposure leads to disease has been understood in terms of the mechanisms that produce disease. Clearly, available biological theories about molecular functions (or dysfunctions) loom large here.[7] Such position has been called in the philosophical literature *evidential pluralism* to emphasize the need for multifold evidence in order to establish causal claims. The thesis has been discussed and supported for the medical sciences in general (see, e.g., Russo and Williamson 2007; Clarke et al. 2013) and for current research in molecular epidemiology in particular (see, e.g., Russo and Williamson 2012).

The difference-making component of evidential pluralisms is, in a sense, less controversial than the productive component, as even theorizers of agnostic data-driven approaches will readily agree that the search for robust statistical associations lies at the very heart of data-intensive science. What remains controversial is what biomarkers are marks *of*. In fact, as mentioned at the end of the previous section, biomarkers are *indicators* of difference makers that we try to get a grip on by means of these extensive omics analyses. In turn, these associations need to be supported by a story about *how* exposure leads to disease or, in other words, how exposure *produces* disease. Productive causality, we suggest, is in need of further discussion and, possibly, *re*conceptualization.

In the recent debate on evidential pluralism, mechanisms have been most often proposed as appropriate candidates to cash out the notion of causal production. In a sense, this is the case for molecular epidemiology too. Vineis and Perera, explaining the use of biomarkers, say:

> When combined with the best of the earlier validated biomarkers of dose, effect, and susceptibility, such new markers have the potential to add considerably to knowledge about the *mechanistic pathways* that relate pathogenic

exposures to disease onset and also to serve as informative early markers of disease risk.

(Vineis and Perera, 2007, p. 1955, emphasis ours)

Of course, biomarkers are not a panacea, nor do they allow for an immediate and direct identification of causal links. Vineis and Perera further explain:

> One of the main challenges with intermediate biomarkers is to understand whether they belong to the causal pathway between exposure and disease, whether they are simply a side effect of exposure or disease, or whether their measurement is confounded by some other exposure. For example, it is likely that certain mutations are genuine intermediate markers in the causal pathway, whereas others are a consequence of the disease, such as genomic instability that arises in cancer cells.
>
> (Vineis and Perera, 2007, p. 1961)

But what are these *mechanistic pathways*? And especially, are these the same as the "continuum from exposure to disease" that molecular epidemiology seeks establish? Often, in the scientific literature, "mechanistic pathways" are cashed out in terms of "processes" that lead from exposure to disease and that we can track using the signals of the biomarkers. This strongly resembles, at least *prima facie*, the Salmon–Dowe process—tracing approach to causality, while evidential pluralism famously emphasized the explanatory role of *mechanisms* rather than processes. So, how strict is evidential pluralism with respect to this requirement of mechanisms?

Illari and Russo (2014a; 2014b, Ch. 13; forthcoming) make the point that what is key for evidential pluralism is to provide evidence of *production*, that is, how the cause brings about or *produces* the effect. In turn, production needs not be cashed out solely or exclusively in terms of mechanisms, although admittedly these occupy a large part of the recent philosophical literature (for an overview of the philosophy of mechanisms and their relation to causality, see Illari and Williamson, 2012; Illari and Russo, 2014b, Ch. 12).

Productive causality, in the most general terms, is about how causes and effects are *linked*. But the sought type of linking may be different depending on the scientific context or the purpose of the causal question. Let us confine the discussion to molecular epidemiology. What gives us this linking here? Mechanisms? Salmon–Dowe processes? Illari and Russo conjecture that although "linking" is the most general way in which we should understand production, we still need to explicate the concept, or in other words, to say what linking amounts to. There are a number of candidates in the philosophical literature, most notably processes and complex-systems mechanisms. Illari and Russo explain why processes and complex-system mechanisms, while having many epistemic virtues, are nonetheless unable to provide a *general* concept of linking, one that would be also applicable to the context of molecular epidemiology. Let us examine them.

In the "combined" Salmon–Dowe view (Salmon 1984, 1997; Dowe 1992, 2000), processes are conceptualized as *world lines of objects*. An airplane flying in the sky is a world line, but so is its shadow on the ground. An important question concerns how to discriminate between world lines, or processes, that are causal and those that are not. In the Salmon–Dowe approach, causal processes are capable of transmitting conserved quantities, such as mass-energy, linear momentum, or charge. So two airplanes that move in the sky and that eventually collide are causal processes, as either process is modified after the interaction. This is, however, not true of the corresponding shadows on the ground, which are not modified when they intersect.[8] But in this way, Salmon–Doweprocesses do not illuminate the way in which molecular epidemiologists talk about processes and, mostly, the way they describe processes that take place at the molecular level. Yet, as we shall see later, molecular epidemiologists *do* conceptualize linking in terms of the notion of "process," which suggests that we need to broaden this notion and make it more widely applicable across scientific domains.

The second candidate for productive causality is the concept of "mechanism." In the last two decades, philosophers of science revived debates around mechanisms and their relation to causation. Illari and Williamson (2012, p. 120) propose the following characterization that can be considered the current consensus[9]:

A mechanism for a phenomenon consists of entities and activities organized in such a way that they are responsible for the phenomenon.

This characterization has the advantage of being general enough, as it doesn't constrain *a priori* the kind of entities and activities to consider. Instead, relevant entities and activities can be identified once we specify the scientific contexts and the level of abstraction at which we pitch the explanation of the phenomenon. This characterization can thus accommodate understanding mechanisms at the biochemical level (say, protein synthesis) as well as the socioeconomic level (say, the economic bubble causing the financial crisis). But what counts for a mechanistic *explanation* is the identification of how these entities and activities are *organized* to bring about, or produce, the phenomenon.

Mechanisms are an appealing candidate to cash out the concept of productive causality because we know a lot about the entities (or parts) that partake in disease causation. Molecular biology, in particular, has made stunning progress in uncovering the mechanisms at the molecular level, which includes the regulation of gene expression, the role of miRNA, and so on. Undoubtedly, this is important in exposome science too, as Vineis and Perera (2007, p. 1955) explain:

When combined with the best of the earlier validated biomarkers of dose, effect, and susceptibility, such new markers have the potential to add considerably to knowledge about the mechanistic pathways that relate pathogenic exposures to disease onset and also to serve as informative early markers of disease risk.

While mechanisms are clearly important in molecular epidemiology—for instance, to set up specific research hypotheses or to direct specific "omic" analyses—a mechanistic account of productive causality puts emphasis on the organization of different entities and activities rather than the *continuum* linking exposure to disease. Thus, for instance, a mechanistic explanation sheds light on the way a gene normally is methylated and how it is hypomethylated when exposed to tobacco smoking. We can shed light on these mechanistic aspects by identifying the relevant entities and activities involved and, mainly, their *organization*. But it does not help, *per se*, in conceptually reconstructing the *continuum* from being exposed to tobacco smoking to the development of lung cancer.

Illari and Russo (2014a) further suggest that we need instead to explore the prospects of the notion of information. The hint comes from the language used in biomarkers research, where the idea of "picking up signal" recurs, for instance:

> While classical statistical models to analyzing -omics data serve the purpose of identifying signals and separating them from noise, little has been done in chronic diseases to model time into the exposure- biomarker-disease continuum.
>
> (Vineis and Chadeau-Hyam 2011, sec. 4)

> From these two parallel analyses [statistical analyses], we obtained lists of putative markers of (i) the disease outcome, and (ii) exposure. These were compared in a second step in order to identify possible intersecting signals, therefore defining potential intermediate biomarkers.
>
> (Chadeau-Hyam et al. 2011, p. 85)

What is the *signal* that we have to pick up? In what sense will this give us the sought production relation between exposure and disease? Illari and Russo suggest conceptualizing the detecting and tracing of signals in terms of *information transmission*. Simply put, the transmission of information will give us—conceptually—the continuum from exposure to disease. It is worth noting that this is not in sharp contrast with the earlier conceptualizations in terms of processes. It is instead a natural development of Salmon's mark transmission theory. But information works better as it is more general than the "physics" marks that are transmitted in Salmon's theory.

The key difference with Salmon–Dowe processes consists in the marking aspect. In Salmon's view, we need to introduce a mark in a process and see whether the mark persists at a later stage. A stock example is the dented car, and the dent is transmitted along with the movement of the car. However, even if we dent the car, its *shadow* will not further transmit the mark. But the whole approach rests on the *introduction* of the mark and on what *would* happen without marking the process (we counterfactually evaluate what would be the case, had we not marked it). In molecular epidemiology, instead, *in the process* from exposure to disease, we look for marks (without introducing marks ourselves) and for their transmission along the process.[10] Once we reframe the question about productive causality as a question about information transmission, rather than mark transmission, *we don't have to introduce marks*. Information is already present in any process—it is the signal that we have to pick up and separate from noise. This is precisely the hard task, because marks—which, in exposome research, are the biomarkers—may be marks of something *else*. To be sure, the issue at stake here is the ontological nature of biomarkers—see also later in Section 3.2. For instance, if we conduct a methylome study, we can find that obesity leaves a mark in DNA methylation (e.g., in CpG probes of gene ABCG1).[11] The latter, however, can be due to an external factor such as a drug to treat diabetes, since diabetes is associated with obesity. As a result, that mark may not be due to obesity itself but to the "collision" of two causal chains (see earlier about colliders).

In the example of the dented car, the car transmits information about the modification of the dent on the car, and this is causal in character. The shadow of the car transmits information too, but in this case the modification of the dent is not transmitted, and from this we infer the noncausal character of the process. In exposome research, we seek to track the signal, or information, that certain biomarkers carry with them. Using the MITM methodology, we attempt to reconstruct the whole transmission from the early exposure events to disease development. We are thus taking full advantage of a conceptualization in terms of processes but without being tied to physics quantities, say energy or momentum, being transmitted. We are also taking full advantage of a conceptualization in terms of mechanisms, because knowledge of relevant molecular or biochemical mechanisms will indicate where to look for signals, for instance, choosing appropriate omics levels for the analyses of biological specimens. It is in this sense that Illari and Russo (2014a) say that mechanisms are *information channels*.

In sum, information transmission avoids strong metaphysical assumptions about causal production. On the one hand, it doesn't presuppose that reality eventually reduces to physics. On the other hand, although we can recast physical, biological, or communication processes in terms of information, this remains at the level of their description and conceptualization rather than constitution. In this sense, information transmission and information channels have primarily an *epistemic* function and connotation.

One advantage of information transmission is that it is potentially widely applicable and capable of explaining how inhomogeneous factors such as micro and macro— biological and social—are linked; this is arguably a pressing issue in the light of results of epigenetic studies and also for the design of public health policies, which we discuss in Section 3.4. It goes without saying that a full-blown account of productive causality in terms of information has not yet been developed. While the idea is promising in several respects, more work is needed. In particular, whether information is ultimately to be understood as semantic information or other (see, e.g., Floridi 2011) isn't settled as yet. Likewise, while molecular epidemiology provides a tough sounding board for productive causality, other areas are no less challenging, for instance, productive causality in social contexts. If, as Illari and Russo (2014a) suggest, information can live up to the challenge of generality, it should be able to be applicable in the case of molecular epidemiology too.

3.4 The Design of Public Health Policies

Funding bodies at national and international level put increasing pressure on researchers in order to obtain results that can be exploitable for technological innovation and for policy making. Specifically, results coming from research in molecular epidemiology should be maximally usable for designing public health policies. While the promise is easy to make in research proposals, the objective of translating results into decisions and actions that aim to reduce the burden of disease at the population level is difficult to meet.

We will not discuss in this contribution the different sorts of political and social pressure that make policy making a difficult enterprise. We will instead confine the discussion to some conceptual reasons at the basis of said difficulty. One issue concerns passing from the individual, molecular level to the population level. Another issue concerns the conceptualization of chronic diseases and relevant interventions. These two issues, while distinct, are clearly related to each other. Moreover, they are also related to questions about the regulatory standards (see Teira, this volume).

As it should be clear from the presentation of exposome research in Section 2, these projects will be successful to the extent that they identify the "right" biomarkers of exposure, of early clinical changes, and of diseases. These biomarkers will allow scientists to understand "how much" and "how long for" we need to be exposed to certain chemicals in order to trigger the mechanisms of the targeted mechanisms. However, once we know all this, for instance, about the chemicals used to disinfect swimming pool water or about the chemicals we breathe in a busy town center, how do we design an effective public health intervention? Do we target the group of individuals who are more likely to be exposed and/or are more susceptible? If so, how do we identify them? Exposome research is collecting data about the *molecular* characteristics of individuals and population. This implies that we would need information about sociodemographic or economic characteristics, or about lifestyles, or about individual preferences; that is, once again, we need to go from the molecular to the population level. In case we don't have this data from the very beginning, do we need to plan additional epidemiologic and sociologic analyses? And if we do that, how to ensure that these "macro characteristics" map onto the "micro characteristics" previously identified?

Let us consider now the types of disease that exposome research is targeting in the first instance: mainly chronic diseases, rather than infectious diseases. No doubt it is absolutely important to understand their biological, molecular basis. However, understanding the biology behind a disease may not help with the decision about how to intervene to reduce the burden of disease. This is because the mechanisms of disease do not necessarily coincide with the mechanisms of intervention or prevention (Kelly and Russo, under review). The belief that reduction of exposure to a risk factor implies a reduction in the burden of disease may be read as a consequence of the discovery of pathogenic agents causing infectious diseases since the second half of the nineteenth century. The discovery and isolation of, say, *Vibrio cholerae* has certainly been a cornerstone in biology and in medicine. At the same time, progress in understanding the biological components of (infectious) diseases also led to *biologizing* the phenomena of health and disease, which instead *also* involve socio–demo–psycho–economic factors. Together, the biological and nonbiological components of disease belong to what sociologists have called the "life world" of individuals. Starting from these considerations, Kelly et al. (2014) argue for the need to *re*conceptualize health and disease, taking into account their mixed—biological and social—components. This line of argument is a promising path of research, especially because the mechanisms of disease do not necessarily coincide with the mechanisms of intervention or prevention. In other words, even if we manage to elucidate the molecular mechanisms of a given disease, we might still need to design an intervention that targets a different level, say behavior, and to change it.

While sociology of health and social epidemiology have long pleaded for a more explicit integration of social factors in the explanation of health and disease, this need is gradually emerging from within the biomedical side too. The increasing importance of epigenetics and human social genomics testifies to this fact.

An example of the integration of social factors in biological research is the study of epigenetic modifications induced by the experience of psychosocial adversity in initiating physiological dysregulations. More specifically, human and animal studies have shown that socioeconomic status (SES) influences DNA methylation and gene expression, particularly across genomic regions regulating immune function (Stringhini et al. 2015).[12] A pivotal study in macaques detected altered levels of expression and methylation in inflammatory genes (in particular *NFATC1*, *IL8RB* CXCR2 *in humans* and *PTGS2*) in relation to changes in hierarchical status (dominance rank, a proxy for social status). To date, the few studies that have addressed this issue in humans reported associations between SES in early and adult life and DNA methylation of genes regulating the immune function (Stringhini et al. 2015). However, it remains to be established (a) to what extent SES has an impact on DNA methylation independently of unhealthy lifestyles, thus directly altering gene regulation; and (b) which are the features of low SES that are directly influencing cellular activity and its regulation.

We are gradually getting a better understanding of how the epigenome reflects both acquired and inherited genetic modifications as a result of pressures coming from environmental factors, and we are gathering increasing evidence that the resulting epigenetic patterns may strongly impact on phenotype (Anderson, Sant, and Dolinoy 2012; Glier, Green, and Devlin 2014). An interesting case to exemplify this is the follow-up of children and grandchildren of those individuals affected by the "Dutch famine" since November 1944. This famine was caused by a German embargo that severely affected the transport of food to the Western Netherlands. In studies of the descendants of the people affected by the famine, scientists noticed an increase in infant mortality and the development of chronic diseases such as diabetes, obesity, coronary heart disease, and neurological conditions. The incidence of these conditions was then reported to be associated with differential methylation in several genes, including those involved in growth and metabolic control (Heijmans et al. 2008). Following up these individuals nearly 60 years later provides an incredibly valuable basis for studying the epigenetic patterns and supports the possible influence of early exposures and experiences *in utero* and in early life and of maternal status.

The concept of *human social genomics* has been developed to encompass the overall impact of social adversities on biology (Slavich and Cole 2013). A core element is the modulation of transcription factors by environmental and social circumstances, through the so-called *conserved transcription response to adversity (CTRA)*. The latter is represented by the pattern of responses (largely inflammatory and immunological) that have developed during evolution to cope with stressful environments, and that implicate networks of transcription factors in their mechanisms. Although such patterns were originally developed to respond to microbial and traumatic sources of stress, they have evolved into generalized responses to environmental pressure, including psychological stress or sleep loss (Slavich and Cole 2013).

In sum, while it is undoubtedly important to understand the biological basis of disease, down to the molecular aspects of exposure and of disease development, health and disease cannot be reduced to their biological components. Time is high to reintroduce socio–psycho–behavioral factors into the picture, and the development of epigenetics and human social genomics is a clear step in this direction, which will hopefully help with translating the results of molecular epidemiology into effective public health policies.

4 Conclusion

Epidemiology has a longstanding tradition in studying the distribution of health, disease, and exposure in populations. Since the very beginning, the goal has not just been to map, in a descriptive way, the associations between hazards and diseases. Instead, epidemiological studies have always shown their importance and relevance for interventions, namely for *reducing* the burden of disease. Arguably, said interventions have to be based on *causal* knowledge rather than mere correlations. This, it goes without saying, does not imply that epidemiology is always in a position to establish causal relations. Many epidemiological claims remain, still today, conjectural in character, and interventions have to be designed on the basis of robust correlations, considerations about plausible or conceivable disease mechanisms, and precautionary considerations.

The methods of epidemiology coincide, by and large, with a conceptual framework of "traditional" study designs and with statistical methods for the analysis of data, which have changed greatly with the increased sophistication of statistics itself. But epidemiological methods underwent increased sophistication also in other respects, notably about *measurement* and scientific instrumentation. We might argue, in fact, that the possibility (and willingness— see Kronfeldner 2014) of measuring exposure and disease at the *molecular* level points to important changes in epidemiology, not so much in its goals and objectives but at the conceptual level.

A first change concerns the concept of *exposure*. The exposome denotes exposure to factors that are "outside," in the environment, as well as factors that are "inside," in our body, and that interact with the external ones. The claim of molecular epidemiology is that we can fully understand the impact of external, environmental factors when we *also* understand what happens inside us, once we are externally exposed. This has licensed talking about the *science of exposure*. To be sure, understanding of internal changes is not necessary to take action; the carcinogenicity of many chemicals and mixture (tobacco, asbestos) has been demonstrated without any molecular knowledge. But once we work with a different concept of exposure, we must likewise modify our methods— this is done with the MITM methodology, which tries and combines the identification of biomarkers of exposure, of early clinical changes, and of disease from prospective and retrospective studies. In the mind of the researchers, this methodology will allow us to trace the continuum of disease from early exposure events to the development of disease. Thus we will be in a position to better understand disease mechanisms at the molecular level *and* to predict disease events in individuals with more accuracy.

These changes about the level of measurement, the concept of exposure, and the methods for data analysis create the conditions for *other* important opportunities and conceptual challenges. In this chapter, we highlighted four of them. The first opportunity is to rethink the use of background knowledge in setting up statistical analyses on omics data. Here the issue at stake is the alleged agnostic character of data-intensive science. However, we have suggested that while part of the research is "agnostic"—in the sense that there isn't always a specified prior hypothesis in searching for, for example, genetic variants—this does not imply that research is entirely agnostic, thus marking "the end of theory."

The second opportunity concerns reviving the debate on the role of technology in science, as motivated by exposome research *inter alia*. While it is uncontroversial to say that much of contemporary science is in fact *techno*science, we still owe an explanation of what such turn may imply. On the one hand, a large part of philosophy of science debates has neglected the technological element of science. On the other hand, a large part of philosophy of technology debates examined the technological element to either argue for a primacy of technology over "pure science" or to offer arguments for scientific realism, but one that takes into account the role of *instruments* in establishing the existence of nonobservable entities. But there seem to be *other* pressing questions to ask, for instance about how the very concepts of "observable" or "reality" change given the way technologies produce data and give us access to certain mechanisms. Is the ontological status of a biomarker the same as of electrons or Higgs bosons? Or perhaps we must change altogether our conception of reality, from entity-based to process-based? The stake is of course high but, again, the opportunity is there for us to drive change.

One such change that we might drive concerns the conceptualization of productive causality. If we buy the idea that molecular epidemiology is after the *process* that leads from early exposure to disease, then we must find an alternative to traditional conceptualizations in terms of necessary and sufficient conditions. But even the most sophisticated accounts of productive causality, such as processes *à la* Salmon–Dowe or complex systems mechanisms *à la* Illari and Williamson, won't do. This is for several reasons, most importantly because the type of *linking* that exposome research tries to establish cannot be captured by concepts that are tailor made for physics, nor by coarse-grained interactions between entities as described in mechanistic explanations. The type of linking that exposomics is after resembles more the transmission of information *via* certain biochemical or molecular processes. It is therefore no accident that "picking up signal from noise" is a most recurrent way of explaining what the MITM methodology is supposed to do. We face here at once a challenge—reconceptualizing productive causality—and an opportunity—to see whether, and to what extent, the concepts of information and of information transmission can illuminate productive causality.

The last challenge we discuss in this contribution concerns the design of public health policies in the light of the results coming from exposome research. This is far from being an easy task because public health interventions typically target the behavior of individuals, while exposome research identifies hazards at the molecular level. So how do we translate information about stuff that happens at molecular level into something exploitable in public health? This is where, in our view, lies the biggest promise of molecular epidemiology: reintegrating the social dimension of health and disease into the most updated understanding of their biological dimension.

After all, epidemiology and public health delivered their better results when they studied and intervened at the level of what sociologists call the "life world," namely the whole sphere of social relations, economic factors, and behavioral mechanisms of individuals. This is true of nineteenth-century public health as well as present-day interventions in developing countries. One might say that interventions aimed to improve sanitation and hygiene or to provide basic access to health infrastructures worked (and still work nowadays) when populations are underdeveloped, or developing, according to Western–capitalist economic parameters.

But what if the modern epidemics of Western, capitalist, developed countries, such as cancer, cardiovascular diseases, or obesity, had too a social component that is as yet underexplored (see Christakis and Fowler 2007)? Agreed, the life world that is relevant here concerns less basic hygiene rules or access to sanitary infrastructure and much more changes in lifestyle, including dietary habits, new occupational hazards, and modified social relations. But it is precisely here that molecular epidemiology has an opportunity: to counteract the trend of (over)biologizing disease that has been ongoing since we have been able to isolate and understand the mechanisms of pathogens such as *Vibrio cholerae*, and instead explain how the biological and social dimension of health and disease are but two aspects of a same phenomenon. This calls for an *integrated explanation* of health and disease, one that simultaneously invokes micromolecular factors and macrosocial factors.

This was noticed and emphasized as early as 1993 by Schulte:

> Finally, from an organizational viewpoint, epidemiologists may potentially be polarized into two worlds: one of molecular epidemiologists who emphasize the molecular and genetic causes of disease and the other of social epidemiologists who stress the role of social, psychological, and economic factors in health. Neither approach alone will satisfactorily address the health issues of the current era. History has shown that complete reliance on reductionist approaches is antithetical to public health, yet failure to use the powerful tools available also will not safeguard public health. A synthesis of the two approaches is needed that can address the entire scope of health issues.

There are numerous interesting topics that we haven't addressed in this contribution. For instance, molecular epidemiology is an excellent case to reflect upon the nature of and the practical aspects of interdisciplinary research (Andersen and Wagenknecht 2013), from the challenge of integrating methods and approaches proper of different fields such as biology and epidemiology; to the one of building a common vocabulary where concepts such as "validity," "sensitivity," "bias," and so on are intersubjectively understood; to the one of building procedures that allow departments, institutes, or laboratories to collaborate while being located in different places (issues related to the organization of research, for instance). Another pressing issue concerns data storage and data transfer. This is not simply addressing considerations about the logistics behind data but also about deep questions on the *nature* of data, their portability or transferability (on this topic, see Leonelli 2012, 2014, 2015).

In sum, molecular epidemiology is a vibrant research field. Thanks to its new concepts and methods, it is opening up new and unwalked paths for scientific, philosophical, and sociological investigations of health and disease.

Acknowledgments

We wish to thank the editors for the opportunity of writing jointly on these themes, which we have been discussing together for a few years. Their comments on an earlier draft were very useful and stimulating. Precious comments and suggestions have been also provided by Phyllis Illari and Jan Stam and by the participants in the philosophy of science seminar at the Vrije Universiteit (Amsterdam), in particular Hans Radder and Dingmar van Eck. Any remaining mistakes or inaccuracies remain, of course, our own.

Notes

1 Admittedly, this definition of molecular epidemiology focuses on cancer. For a more general discussion and review of various available definitions, see also Foxman and Riley (2001).

2 The term *omics* is being routinely used to indicate fields of study within biology at different molecular levels, for instance gen*omics*, prote*omics*, or transcript*omics*.

3 National Hellenic Research Foundation, Greece; University of Maastricht, Netherlands; Imperial College London, United Kingdom; Umeå University, Sweden; Istituto per lo Studio e la Prevenzione Oncologica, Italy; University of Crete, Greece; University of Utrecht, Netherlands; Istituto Superiore di Sanità, Italy; National Public Health Institute (KTL), Finland H.; University of Leeds, United Kingdom; Lund University, Sweden.

4 Imperial College, King's College, University of Bristol, U.K.; Universiteit Utrecht and Universiteit Maastricht, Netherlands; IARC, France; Fundacio Centre de Recerca en Epidemiologia Ambiental and CRIC, Spain; Ethniko Idryma Erevnon, Greece; Swiss TPH and Genedata, Switzerland; University of California, U.S.

5 Methylation is one of the key epigenetic changes. The addition of a methyl group to DNA cytosines at selected CpG sites usually leads to suppression of transcription and therefore to a reduced expression of the corresponding gene, especially if these sites are located at CpG islands at promoter regions.

6 These are CpG sites (particular chemical subunits in DNA) in two genes involved in key cellular functions and carcinogenesis.

7 In the philosophy of causality, the terminology of difference-making (or dependence) versus production is a development of the account originally proposed by Hall (2004). Illari and Russo (2014b) explain the fruitfulness of this idea (that there are two concepts of cause), once reframed in terms of the evidence needed to establish causal claims in given scientific contexts.

8 In a more recent approach, developed by Boniolo et al. (2011), the quantities that causal processes are able to transmit are extensive rather than conserved. Thus, the authors argue, the account can also make sense of stationary cases, which the Salmon–Dowe could not do. Let us explain this by means of a simple example. In Boniolo et al.'s approach, we can causally explain the interaction between a stove and

the surrounding environment, which gets hotter when the stove is heated up. The stove exchanges with the environment "extensive" quantities rather than conserved quantities, and so they wouldn't count as legitimate causal processes in Salmon–Dowe. Instead, they do in Boniolo et al.'s approach. In either version, however, processes are tailored to physics, as the concepts to explicate causal processes and causal transmission come from physics.

9   The characterization of Illari and Williamson (2012) has a lot in common with those provided by Machamer, Darden, and Craver; Bechtel and Richardson; and Glennan, all of whom are referenced and thoroughly discussed therein. Illari and Williamson explain at length why their formulation is a consensus, capturing what is essential and what is not. For instance, organization is essential to all mechanisms, but regularity is not.

10   It should be noted that this is not an argument against the use of counterfactuals *tout court*. Counterfactual reasoning widely used in science, for instance, to generate scientific hypothesis (see for instance Illari and Russo 2014b). In molecular medicine, Nathan (this volume) explains how counterfactuals are used in diagnosis and prognosis. But here, the focus is on biomarker discovery and validation.

11   The methylome is the totality of methylation changes (see footnote 3) in the DNA of an individual or a population. ABCG1 is a gene involved in cholesterol metabolism whose methylation has been found to be associated with obesity.

12   This study used the EPIC cohort (see http://epic.iarc.fr), which contains information about several sociodemographic characteristics of individuals, lifestyles, and so on; for the same individuals, anthropometric data and blood samples were also available.

References

Andersen, H. and S. Wagenknecht (2012) "Epistemic dependence in interdisciplinary groups," *Synthese*, Volume 190 (11), pp. 1881–1898.

Anderson, C. (2008) "The end of theory: The data deluge makes the scientific method obsolete," *WIRED Magazine*.

Anderson, O.S., K.E. Sant, and D.C. Dolinoy (2012) "Nutrition and epigenetics: An interplay of dietary methyl donors, one-carbon metabolism and DNA methylation," *J Nutr Biochem*, 23 (8), pp. 853–59.

Bensaude-Vincent, B. (2009) "Technoscience and convergence: A transmutation of values?" Summerschool on Ethics of Converging Technologies, Dormotel Vogelsberg, Omrod/Alsfeld, Germany, September 2008.

Boniolo, G., R. Faraldo, and A. Saggion (2011) "Explicating the notion of 'causation': The role of extensive quantities." In P. Illari, F. Russo, and J. Williamson (eds) *Causality in the Sciences* (Chapter 24). Oxford: Oxford University Press.

Boon, M. (2011) "In defence of engineering sciences: On the epistemological relations between science and technology," *Techné* 15(1), pp. 49–71.

Chadeau-Hyam M., T.J. Athersuch, H.C. Keun, M.D. Iorio, T.M. Ebbels, M. Jenab, C. Sacerdote, S.J. Bruce, E. Holmes, and P. Vineis (2011) "Meeting-in-the-middle using metabolic profiling—a strategy for the identification of intermediate biomarkers in cohort studies," *Biomarkers* 16(1), pp. 83–88.

Chakravartty A. (2015) "Scientific realism." In E.N. Zalta (ed.), *The Stanford Encyclopedia of Philosophy* (Fall 2015 Edition), http://plato.stanford.edu/archives/fall2015/entries/scientific-realism/>.

Christakis, N.A. and J.H. Fowler (2007) "The spread of obesity in a large social network over 32 years," *N Engl J Med*; 357, pp. 370–379.

Clarke B., D. Gillies, P. Illari, F. Russo, and J. Williamson (2013) "Mechanisms and the evidence hierarchy," *Topoi*. 33, pp. 339–360. doi:10.1007/ s11245-013-9220-9.

Dowe P. (1992) "Wesley Salmon's process theory of causality and the conserved quantity theory," *Philos Sci* 59(2), pp. 195–216.

Dowe P. (2000) "Causality and explanation: Review of Salmon," *Br J Philos Sci* 51, pp. 165–174.

Fasanelli, F., L. Baglietto, E. Ponzi, F. Guida, G. Campanella, M. Johansson, K. Grankvist, M. Johansson, M.B. Assumma, A. Naccarati, M. Chadeau-Hyam, U. Ala, C. Faltus, R. Kaaks, A. Risch, B. De Stavola, A. Hodge, G.G. Giles, M.C. Southey, C.L. Relton, P.C. Haycock, E. Lund, S. Polidoro, T.M. Sandanger, G. Severi, and P. Vineis (2015) "Hypomethylation of smoking-related genes is associated with future lung cancer in four prospective cohorts," *Nat Commun*. Dec 15(6), p. 10192. doi: 10.1038/ncomms10192.

Floridi, L. (2008) "A defence of informational structural realism," *Synthese*, 161.2, pp. 219–253.

Floridi, L. (2011) *The Philosophy of Information*. Oxford: Oxford University Press.

Foxman, B. and L. Riley. (2001) "Molecular epidemiology: Focus on infection," *Am. J. Epidemiol*. (2001) 153 (12), pp. 1135–1141.

Glier, M.B., T.J. Green, and A.M. Devlin (2014) Methyl nutrients, DNA methylation, and cardiovascular disease," *Mol Nutr Food Res* 58 (1), pp. 172–82. doi: 10.1002/mnfr.201200636.

Guchet, X. (2011) "Les technosciences: Essai de definition," *Philonsorbonne* (5).

Hacking, I. (1983) *Representing and Intervening*. Cambridge University Press.

Hall, N. (2004) "Two concepts of cause." In L. Paul, N. Hall, and J. Collins (eds), *Causation and Counterfactuals*, pp. 225–76. Cambridge, Massachussets: MIT Press.

Heijmans, B.T., et al. (2008) "Persistent epigenetic differences associated with prenatal exposure to famine in humans," *Proc Natl Acad Sci USA* 105 (44), pp. 17046–49.

Ihde, D. (1991) *Instrumental Realism. The Interface between Philosophy of Science and Philosophy of Technology*. Indiana University Press.

Illari, P. and F. Russo (2014a) "Information channels and biomarkers of disease," *Topoi*. doi: 10.1007/s11245-013-9228-1.

Illari, P. and F. Russo (2014b) *Causality: Philosophical Theory Meets Scientific Practice*. Oxford University Press.

Illari, P. and F. Russo (forthcoming) "Causality and information." In L. Floridi, (ed), *The Routledge Handbook of Philosophy of Information*.

Illari, P.M., and J. Williamson (2012) "What is a mechanism? Thinking about mechanisms across the sciences," *Eur J Philos Sci* 2 (1), pp. 119–135.

Kelly, M.P., R.S. Kelly, and F. Russo (2014) "The integration of social, behavioural, and biological mechanisms in models of pathogenesis," *Perspectives in Biology and Medicine*, 57(3), pp. 308–328.

Kelly, M.P. and F. Russo (under review) "The Whig interpretation of the history of public health and its impact on the prevention of non-communicable diseases: The case of England."

Kronfeldner M. (2014) "Commentary: How norms make causes," *International Journal of Epidemiology*, 43(6), pp. 1707–1713.

Leonelli, S. (2012) "Making sense of data-driven research in the biological and the biomedical sciences," *Studies in the History and Philosophy of the Biological and Biomedical Sciences* 43(1), pp. 1–3.

Leonelli, S. (2014) "What difference does quantity make? On the epistemology of big data in biology," *Big Data & Society*.

Leonelli, S. (2015) "What counts as scientific data? A relational framework," *Philosophy of Science* 82, pp. 1–12.

Mayr, O. (1976) "The Science-Technology relationship as a historiographic problem," *Technology and Culture* 17(4).

Perera, F.P., and I.B. Weinstein. (1982) "Molecular epidemiology and carcinogen–DNA adduct detection: New approaches to studies of human cancer causation," *J Chron Dis* 35, pp. 581–600.

Porta M., P. Vineis, and F. Bolúmar (2015) "The current deconstruction of paradoxes: One sign of the ongoing methodological 'revolution'," *Eur J Epidemiol.* 30(10), pp. 1079–1087.

Radder, H. (2003) "Technology and theory in experimental science." In H. Radder (ed.) *The Philosophy of Scientific Experimentation* (Ch 8). University of Pittsburgh Press.

Radder, H. (2012) *The Material Realization of Science*. Boston Studies in Philosophy of Science. Springer.

Rappaport, S.M. and M.T. Smith (2010) "Environment and disease risks," *Science* 330, pp. 460–461.

Rheinberger, H.-J. (2005) "Gaston Bachelard and the notion of 'Phenomenotechnique'," *Perspectives on Science* 13(3), pp. 313–328.

Russo, F. and J. Williamson (2007) "Interpreting causality in the health sciences," *International Stud Philos Sci* 21(2), pp. 157–170.

Russo, F. and J. Williamson (2012) "Envirogenmarkers. The interplay between difference-making and mechanisms," *Med Stud* 3, pp. 249–262.

Salmon, W.C. (1984) *Scientific Explanation and the Causal Structure of the World*. Princeton, NJ: Princeton University Press.

Salmon, W.C. (1997) "Causality and explanation: A reply to two critiques," *Philos Sci* 64(3), pp. 461–477.

Schulte, P.A. (1993) "A conceptual and historical framework for molecular epidemiology." In P.A. Schulte and F. Perera (eds) *Molecular Epidemiology. Principles and Methods* (Ch. 1). Academic Press.

Slavich G.M., and S.W. Cole (2013) "The emerging field of human social genomics," *Clinical Psychological Science* 1(3), pp. 331–348.

Sneed, J. (1979) *The Logical Structure of Mathematical Physics*. Dordrecht: D. Reidel, 2nd edition.

Stegmüller, W. (1976) *The Structure and Dynamics of Theories*. New York: Springer.

Stringhini, S., S. Polidoro, C. Sacerdote, R.S. Kelly, K. van Veldhoven, C. Agnoli, S. Grioni, R. Tumino, M.C. Giurdanella, S. Panico, A. Mattiello, D. Palli, G. Masala, V. Gallo, R. Castagné, F. Paccaud, G. Campanella, M. Chadeau-Hyam, and P. Vineis (2015) "Life-course socioeconomic status and DNA methylation of genes regulating inflammation," *Int J Epidemiol.* 44(4), pp. 1320–30.

Vineis, P. and M. Chadeau-Hyam (2011) "Integrating biomarkers into molecular epidemiological studies," *Curr Opin Oncol* 23(1), pp. 100–105.

Vineis, P. and F. Perera (2007) "Molecular epidemiology and biomarkers in etiologic cancer research: The new in light of the old," *Cancer Epidemiol Biomarkers Prev* 16(10), pp. 1954–1965.

Vineis, P., K. van Veldhoven, M. Chadeau-Hyam, and T.J. Athersuch (2013) "Advancing the application of omics-based biomarkers in environmental epidemiology," *Environ Mol Mutagen* 54(7), pp. 461–467. doi: 10.1002/em.21764. Epub 2013 Mar 21.

Wild, C.P. (2005) "Complementing the genome with an 'exposome': The outstanding challenge of environmental exposure measurement in molecular epidemiology," *Cancer Epidemiol Biomarkers Prev* 14, pp. 1847–1850.

Wild, C.P. (2009) "Environmental exposure measurement in cancer epidemiology," *Mutagenesis* 24(2), pp. 117–125.

Wild, C.P. (2011) "Future research perspectives on environment and health: The requirement for a more expansive concept of translational cancer research," *Environ Health* 10 (Suppl 1).

Wild, C., S. Garte, and P. Vineis (2008) "Introduction: Why molecular epidemiology?" In C. Wild, S. Garte, and P. Vineis (eds) *Molecular Epidemiology of Chronic Diseases*, Ch. 1. John Wiley & Sons.

Index